Emulsion Science and Technology

Edited by
Tharwat F. Tadros

Related Titles

T.F. Tadros (Ed.)

Colloids and Interface Science Series

Volume 1: The Role of Surface Forces – Part 1

2007
ISBN: 978-3-527-31462-1

Volume 2: The Role of Surface Forces – Part 2

2007
ISBN: 978-3-527-31503-1

Volume 3: Colloids Stabilty and Application in Pharmacy

2007
ISBN: 978-3-527-31463-8

Volume 4: Colloids in Cosmetics and Personal Care

2008
ISBN: 978-3-527-31464-5

Volume 5: Colloids in Agrochemicals

2009
ISBN: 978-3-527-31465-2

D. Platikanov, D. Exerowa (Eds.)

Highlights in Colloid Science

2009
ISBN: 978-3-527-32037-0

K.J. Wilkinson, J.R. Lead (Eds.)

Environmental Colloids and Particles

Behaviour, Separation and Characterisation

2007
ISBN: 978-0-470-02432-4

A. Aserin

Multiple Emulsion

Technology and Applications

2007
ISBN: 978-0-470-17093-9

L.L. Schramm

Emulsions, Foams, and Suspensions

Fundamentals and Applications

2005
ISBN: 978-3-527-30743-2

T.F. Tadros

Applied Surfactants

Principles and Applications

2005
ISBN: 978-3-527-30629-9

Emulsion Science and Technology

Edited by
Tharwat F. Tadros

WILEY-
VCH

WILEY-VCH Verlag GmbH & Co. KGaA

The Editor

Prof. Dr. Tharwat F. Tadros
89 Nash Grove Lane
Wokingham, Berkshire, RG40 4HE
United Kingdom

All books published by Wiley-VCH are carefully produced. Nevertheless, authors, editors, and publisher do not warrant the information contained in these books, including this book, to be free of errors. Readers are advised to keep in mind that statements, data, illustrations, procedural details or other items may inadvertently be inaccurate.

Library of Congress Card No.: applied for

British Library Cataloguing-in-Publication Data
A catalogue record for this book is available from the British Library.

Bibliographic information published by the Deutsche Nationalbibliothek
The Deutsche Nationalbibliothek lists this publication in the Deutsche Nationalbibliografie; detailed bibliographic data are available on the Internet at http://dnb.d-nb.de.

© 2009 WILEY-VCH Verlag GmbH & Co. KGaA, Weinheim

All rights reserved (including those of translation into other languages). No part of this book may be reproduced in any form – by photoprinting, microfilm, or any other means – nor transmitted or translated into a machine language without written permission from the publishers. Registered names, trademarks, etc. used in this book, even when not specifically marked as such, are not to be considered unprotected by law.

Printed in the Federal Republic of Germany
Printed on acid-free paper

Cover Design Adam Design, Weinheim
Typesetting Thomson Digital, Noida, India
Printing   Strauss GmbH, Mörlenbach
Bookbinding Litges & Dopf Buchbinderei GmbH, Heppenheim

ISBN: 978-3-527-32525-2

Contents

Preface *XIII*

List of Contributors *XV*

1	**Emulsion Science and Technology: A General Introduction** *1*	
	Tharwat F. Tadros	
1.1	Introduction *1*	
1.2	Industrial Applications of Emulsions *3*	
1.3	The Physical Chemistry of Emulsion Systems *4*	
1.3.1	The Interface (Gibbs Dividing Line) *4*	
1.4	The Thermodynamics of Emulsion Formation and Breakdown *5*	
1.5	Interaction Energies (Forces) Between Emulsion Droplets and Their Combinations *7*	
1.5.1	Van der Waals Attraction *7*	
1.5.2	Electrostatic Repulsion *9*	
1.5.3	Steric Repulsion *11*	
1.6	Adsorption of Surfactants at the Liquid/Liquid Interface *12*	
1.6.1	The Gibbs Adsorption Isotherm *13*	
1.6.2	Mechanism of Emulsification *16*	
1.6.3	Methods of Emulsification *18*	
1.6.4	Role of Surfactants in Emulsion Formation *19*	
1.6.5	Role of Surfactants in Droplet Deformation *21*	
1.7	Selection of Emulsifiers *25*	
1.7.1	The Hydrophilic-Lipophilic Balance (HLB) Concept *25*	
1.7.2	The Phase Inversion Temperature (PIT) Concept *27*	
1.7.3	The Cohesive Energy Ratio (CER) Concept *29*	
1.7.4	The Critical Packing Parameter for Emulsion Selection *31*	
1.8	Creaming or Sedimentation of Emulsions *32*	
1.8.1	Creaming or Sedimentation Rates *33*	
1.8.2	Prevention of Creaming or Sedimentation *35*	
1.9	Flocculation of Emulsions *37*	
1.9.1	Mechanism of Emulsion Flocculation *38*

1.9.1.1	Flocculation of Electrostatically Stabilized Emulsions	38
1.9.1.2	Flocculation of Sterically Stabilized Emulsions	40
1.9.2	General Rules for Reducing (Eliminating) Flocculation	41
1.10	Ostwald Ripening	41
1.11	Emulsion Coalescence	43
1.11.1	Rate of Coalescence	44
1.11.2	Phase Inversion	45
1.12	Rheology of Emulsions	46
1.12.1	Interfacial Rheology	46
1.12.2	Measurement of Interfacial Viscosity	47
1.12.3	Interfacial Dilational Elasticity	47
1.12.4	Interfacial Dilational Viscosity	48
1.12.5	Non-Newtonian Effects	49
1.12.6	Correlation of Interfacial Rheology with Emulsion Stability	49
1.12.6.1	Mixed Surfactant Films	49
1.12.6.2	Protein Films	49
1.12.7	Bulk Rheology of Emulsions	50
1.12.8	Rheology of Concentrated Emulsions	51
1.12.9	Influence of Droplet Deformability on Emulsion Rheology	53
1.12.10	Viscoelastic Properties of Concentrated Emulsions	53
	References	55
2	**Stabilization of Emulsions, Nanoemulsions and Multiple Emulsions Using Hydrophobically Modified Inulin (Polyfructose)**	**57**
	Tharwat F. Tadros, Elise Vandekerckhove, Martine Lemmens, Bart Levecke, and Karl Booten	
2.1	Introduction	57
2.2	Experimental	58
2.2.1	Materials	58
2.2.2	Methods	58
2.2.2.1	Preparation of Emulsions, Nanoemulsions and Multiple Emulsions	58
2.2.2.2	Investigation of Emulsion Stability	59
2.3	Results and Discussion	59
2.3.1	Emulsion Stability Using INUTEC®SP1	59
2.3.2	Nanoemulsion Stability Using INUTEC®SP1	60
2.3.3	Multiple Emulsion Stability Using INUTEC® SP1	64
2.4	Conclusions	65
	References	65
3	**Interaction Forces in Emulsion Films Stabilized with Hydrophobically Modified Inulin (Polyfructose) and Correlation with Emulsion Stability**	**67**
	Tharwat Tadros, Dotchi Exerowa, Georgi Gotchev, Todor Kolarov, Bart Levecke, and Karl Booten	
3.1	Introduction	67

3.2	Materials and Methods	68
3.3	Results and Discussion	69
3.4	Conclusions	73
	References	73

4 Enhancement of Stabilization and Performance of Personal Care Formulations Using Polymeric Surfactants *75*
Tharwat F. Tadros, Martine Lemmens, Bart Levecke, and Karl Booten
4.1 Introduction *75*
4.2 Experimental *76*
4.3 Results and Discussion *76*
4.3.1 Massage Lotion *76*
4.3.2 Hydrating Shower Gel *79*
4.3.3 Soft Conditioner *80*
4.3.4 Sun Spray SPF19 *81*
4.4 Conclusions *81*
References *81*

5 Effect of an External Force Field on Self-Ordering of Three-Phase Cellular Fluids in Two Dimensions *83*
Waldemar Nowicki and Grażyna Nowicka
5.1 Introduction *83*
5.2 The Model *84*
5.3 Results and Discussion *85*
5.3.1 Energies of Cluster Insertion and Transformation *85*
5.3.2 Evolution of the System in a Gravitational Field *90*
5.4 Conclusions *93*
References *94*

6 The Physical Chemistry and Sensory Properties of Cosmetic Emulsions: Application to Face Make-Up Foundations *97*
Frédéric Auguste and Florence Levy
6.1 Introduction *97*
6.2 Materials and Methods *98*
6.2.1 Selection of the Foundations to be Studied *98*
6.2.2 Characterization Methods *98*
6.3 Experimental Results and Discussion *99*
6.3.1 Drying of the Foundation Bulk and Drift in Composition During Drying *99*
6.3.2 Evolution of Viscosity During Drying *100*
6.3.3 Play-Time and Disposition of Foundation on the Skin *102*
6.4 Conclusions *104*
References *104*

7	**Nanoparticle Preparation by Miniemulsion Polymerization** *107*
	Man Wu, Elise Rotureau, Emmanuelle Marie, Edith Dellacherie, and Alain Durand
7.1	Introduction *107*
7.2	Experimental *108*
7.2.1	Materials *108*
7.2.2	Emulsion Preparation *108*
7.2.3	Polymerization *108*
7.2.4	Size Measurement of the Emulsion Droplets *108*
7.2.5	Particle Characterization *109*
7.3	Results and Discussion *109*
7.3.1	Synthesis of Hydrophobically Modified Dextrans *109*
7.3.2	Preparation of O/W Miniemulsions *111*
7.3.2.1	Control of Initial Droplet Size by Process Variables *111*
7.3.2.2	Influence of Polymer Structure on Initial Droplet Size *112*
7.3.3	Stability of Miniemulsions within Polymerization Duration *114*
7.3.3.1	Mechanism and Kinetics of Miniemulsion Polymerization *114*
7.3.3.2	Mechanism and Rate of Emulsion Aging *116*
7.3.3.3	Variation of the Rate of Emulsion Aging with Polymerization Conditions *118*
7.3.4	Preparation of Defined Nanoparticles with Various Monomers *123*
7.3.4.1	Poly(styrene) Nanoparticles Covered by Dextran *123*
7.3.4.2	Poly(butylcyanoacrylate) Nanoparticles *126*
7.3.5	Colloidal Properties of the Obtained Suspensions *128*
7.4	Conclusions *129*
	References *130*
8	**Recent Developments in Producing Monodisperse Emulsions Using Straight-Through Microchannel Array Devices** *133*
	Isao Kobayashi, Kunihiko Uemura, and Mitsutoshi Nakajima
8.1	Introduction *133*
8.2	Principles of Microchannel Emulsification *135*
8.3	Straight-Through MC Array Device and Emulsification Set-Up *137*
8.4	Effect of Channel Shapes on Emulsification Using Symmetric Straight-Through MC Arrays *139*
8.4.1	Effect of Channel Cross-Sectional Shape *139*
8.4.2	Effect of the Aspect Ratio of Oblong Channels *139*
8.4.3	Computational Fluid Dynamics (CFD) Simulation and Analysis *141*
8.5	Effect of Process Factors on Emulsification Using Symmetric Straight-Through MC Arrays *144*
8.5.1	Effect of Surfactants and Emulsifiers *144*
8.5.2	Effect of To-Be-Dispersed Phase Viscosity *146*
8.5.3	Effect of To-Be-Dispersed Phase Flux *148*
8.6	Scaling-Up of Straight-Through MC Array Devices *149*

8.7	Emulsification Using an Asymmetric Straight-Through MC Array	*150*
8.8	Conclusions and Outlook	*152*
	References	*154*

9 Isotropic and Anisotropic Metal Nanoparticles Prepared by Inverse Microemulsion *157*
Ignác Capek

9.1	Introduction	*157*
9.1.1	Properties of Nanoscale Particles	*157*
9.1.2	Production of Nanoparticles and Microemulsions	*158*
9.2	General Aspects of Microemulsions	*159*
9.2.1	Droplet Dimensions	*160*
9.2.2	The Use of W/O Microemulsions	*161*
9.2.3	Nanoparticle Preparation	*162*
9.2.4	Surfactant-Based Methods	*164*
9.2.5	Coprecipitation	*166*
9.3	Isotropic Nanoparticles	*166*
9.4	Anisotropic Nanoparticles	*167*
9.5	Conclusions and Outlook	*179*
	References	*185*

10 Preparation of Nanoemulsions by Spontaneous Emulsification and Stabilization with Poly(caprolactone)-*b*-poly(ethylene oxide) Block Copolymers *191*
Emmanuel Landreau, Youssef Aguni, Thierry Hamaide, and Yves Chevalier

10.1	Introduction	*191*
10.1.1	Block Copolymers	*192*
10.1.1.1	Spontaneous Emulsification	*193*
10.1.1.2	Biodegradability and Biocompatibility	*194*
10.1.2	Block Copolymer Micelles	*194*
10.1.3	Diblock Copolymers	*195*
10.2	Materials and Methods	*195*
10.2.1	Materials	*195*
10.2.2	Synthesis of Block Copolymers (PCL-*b*-PEO)	*196*
10.2.3	Methods	*196*
10.2.4	Emulsification of Oils or PCL	*197*
10.3	Results and Discussion	*197*
10.3.1	Emulsions of PCL by Spontaneous Emulsification	*198*
10.3.1.1	Fabrication of the Emulsions	*198*
10.3.1.2	Particle Sizes	*200*
10.3.1.3	Stability of the Emulsions	*201*
10.3.2	Emulsions of Various Oils by Spontaneous Emulsification	*203*
10.4	Conclusions	*205*
	References	*206*

11		**Routes Towards the Synthesis of Waterborne Acrylic/Clay Nanocomposites** 209
		Gabriela Diaconu, Maria Paulis, and Jose R. Leiza
	11.1	Introduction 209
	11.2	Experimental 211
	11.2.1	Materials 211
	11.2.2	Synthesis of Waterborne (MMA-BA)/MMT Nanocomposites by Emulsion Polymerization 213
	11.2.3	Synthesis of Waterborne (MMA-BA)/MMT Nanocomposites by Miniemulsion Polymerization 214
	11.2.4	Characterization and Measurements 215
	11.3	Results and Discussion 217
	11.3.1	Waterborne Nanocomposites by Emulsion Polymerization 217
	11.3.2	Waterborne Nanocomposites by Miniemulsion Polymerization 219
	11.4	Conclusions 226
		References 226
12		**Preparation Characteristics of Giant Vesicles with Controlled Size and High Entrapment Efficiency Using Monodisperse Water-in-Oil Emulsions** 229
		Takashi Kuroiwa, Mitsutoshi Nakajima, Kunihiko Uemura, Seigo Sato, Sukekuni Mukataka, and Sosaku Ichikawa
	12.1	Introduction 229
	12.2	Materials and Methods 230
	12.2.1	Materials 230
	12.2.2	Preparation of W/O Emulsions Using MC Emulsification 230
	12.2.3	Formation of GVs 231
	12.2.4	Measurement of Droplet and Vesicle Diameters 232
	12.2.5	Determination of Entrapment Yield 232
	12.3	Results and Discussion 233
	12.3.1	Preparation of GVs Using Monodisperse W/O Emulsions 233
	12.3.2	Size Control of GVs and Entrapment of a Hydrophilic Molecule into GVs 234
	12.3.3	Formation Characteristics of GVs 237
	12.4	Conclusions 240
		References 241
13		**On the Preparation of Polymer Latexes (Co)Stabilized by Clays** 243
		Ignác Capek
	13.1	Introduction 243
	13.2	Cloisite Clays and Organoclays 247
	13.3	Radical Polymerization 260
	13.3.1	Solution/Bulk Polymerization 260
	13.3.2	Radical Polymerization in Micellar Systems 263
	13.4	Collective Properties of Polymer/MMT Nanocomposites 281

13.4.1	Kinetic and Molecular Weight Parameters	*281*
13.4.2	X-Ray Diffraction Studies	*284*
13.4.2.1	Homopolymers	*284*
13.4.2.2	Copolymers	*288*
13.4.3	Thermal and Mechanical Properties	*290*
13.4.3.1	Polystyrene and Poly(methyl methacrylate) Nanocomposites	*290*
13.4.3.2	Poly(ethylene oxide) Nanocomposites	*293*
13.5	Polymer–Inorganic Nanocomposites	*296*
13.6	General	*301*
13.7	Conclusions and Outlook	*302*
	References	*310*

Index *317*

Preface

Today, emulsions are applied in a wide variety of industrial products, including food, cosmetics, pharmaceuticals, agrochemicals, and paints. With this in mind, a series of World congresses has recently been held – the first in Paris in 1993, the second in Bordeaux in 1997, the third in Lyon in 2002, and the most recent again in Lyon, in 2006. Following each meeting, a number of topics were selected, the details of which were subsequently published in the journals *Colloids and Surfaces* and *Advances in Colloid and Interface Science*.

This book contains selected topics from the Fourth World Congress, the title of which – "Emulsion Science and Technology" – reflects the importance of applying scientific principles to the preparation and stabilization of emulsion systems.

As a "introduction" to the subject, Chapter 1 provides a general description of the physical chemistry of emulsion systems, with particular attention being paid to the interaction forces that occur between emulsion droplets. The adsorption of surfactants at liquid/liquid interfaces is analyzed, and the methods and mechanism of emulsification and role of surfactants described. Those methods applicable to emulsifier selection are also detailed, as are the various emulsion breakdown processes such as creaming or sedimentation, flocculation, Ostwald ripening, coalescence and phase inversion. Methods used to prevent such breakdown processes are also detailed. Chapter 2 relates to the special application of a polymeric surfactant (a hydrophobically modified inulin) for the stabilization of emulsions, nanoemulsions, and multiple emulsions, while Chapter 3 provides the details of a fundamental study of the interaction forces in emulsion films stabilized with hydrophobically modified inulin and the correlation with emulsion stability. In Chapter 4, the application of polymeric surfactants for enhancing the stabilization and performance of personal care formulations – such as massage lotions, hydrating shower gel, soft conditioners, and sun sprays – is described, while Chapter 5 provides the details of a more fundamental study of the effect of external force fields on the self-ordering of three-phase cellular fluids in two dimensions. Here, attention is focused on the energies of cluster insertion and transformation, and the evolution of the system in a gravitational field. Chapter 6 relates to the application of the physical chemistry and sensory properties of cosmetic formulations, with the example of facial make-up being used to illustrate the principles involved in both drying and the evolution of viscosity. In Chapter 7, a detailed account is provided of nanoparticle preparation using miniemulsion (nanoemulsion) polymerization, and for which a variety of monomers (e.g., styrene and butylcyanoacrylate) are used to illustrate the principles. In Chapter 8, the details of some recent developments in the production of monodisperse emulsions using straight-through microchannel array devices are provided, while Chapter 9 outlines not only the preparation of isotropic and anisotropic nanoparticles (using inverse microemulsions) but also the properties of the nanoparticulate product. The preparation of nanoemulsions by spontaneous emulsification and stabilization of the resulting nanodroplets by block copolymers, namely poly(caprolactone-b-poly(ethylene oxide), are described in Chapter 10, while the routes for the synthesis of waterborne acrylic/clay nanocomposites (prepared by miniemulsion polymerization) are outlined in Chapter 11. The preparation of giant vesicles with a controlled size and a high entrapment efficiency, by using monodisperse water-in-oil emulsions, is detailed in Chapter 12, while the final chapter describes the preparation of polymer latexes stabilized with clay particles, and the possible preparation of nanocomposites, using the same approach.

Based on the above descriptions and details, it is clear that this book covers a wide range of topics, both fundamental and applied, and also highlights the importance of emulsion science in many modern-day industrial applications. It is hoped that the book will be of great help to emulsion research scientists in both academia and industry.

Finally, I would like to thank the organizers of the Fourth World Congress – and in particular Dr Alain Le Coroller and Dr Jean-Erik Poirier – for inviting me to edit this book.

January 2009 *Tharwat Tadros*

List of Contributors

Youssef Aguni
UMR 5007 CNRS – Université de Lyon
Laboratoire d'Automatique et de Génie
des Procédés – LAGEP
Bât 308, 43 Bd du 11 Novembre
69622 Villeurbanne Cedex
France

Frédéric Auguste
L'Oréal – Centre de Chevilly-Larue
188 rue Paul Hochart
94150 Chevilly Larue
France

Karl Booten
ORAFTI Bio Based Chemicals
Aandorenstraat 1
3300 Tienen
Belgium

Ignác Capek
Slovak Academy of Sciences
Polymer Institute
Institute of Measurement Science
Dúbravská cesta 9
842 36 Bratislava
Slovakia
and
Trenčín University
Faculty of Industrial Technologies
Ul. I. Krasku 30
020 01 Púchov
Slovakia

Yves Chevalier
UMR 5007 CNRS – Université de Lyon
Laboratoire d'Automatique et de Génie
des Procédés – LAGEP
Bât 308, 43 Bd du 11 Novembre
69622 Villeurbanne Cedex
France

Edith Dellacherie
CNRS-Nancy-University ENSIC
Laboratoire de Chimie Physique
Macromoléculaire
1 rue Grandville
54001 Nancy Cedex
France

Gabriela Diaconu
University of the Basque Country
Facultad de Ciencias Químicas
POLYMAT, Joxe Mari Korta zentroa
Tolosa Etorbidea 72
20018 Donostia-San Sebastián
Spain

Alain Durand
CNRS-Nancy-University ENSIC
Laboratoire de Chimie Physique
Macromoléculaire
1 rue Grandville
54001 Nancy Cedex
France

Emulsion Science and Technology. Edited by Tharwat F. Tadros
Copyright © 2009 WILEY-VCH Verlag GmbH & Co. KGaA, Weinheim
ISBN: 978-3-527-32525-2

Dotchi Exerowa
Bulgarian Academy of Sciences
Institute of Physical Chemistry
Acad. G. Bonchev Str.
Sofia 1113
Bulgaria

Georgi Gotchev
Bulgarian Academy of Sciences
Institute of Physical Chemistry
Acad. G. Bonchev Str.
Sofia 1113
Bulgaria

Thierry Hamaide
Laboratoire de Chimie et Procédés
de Polymérisation LCPP
CPE Lyon
69622 Villeurbanne Cedex
France
Present address:
Université de Lyon
Ingénierie des Matériaux Polymères
LMPB, UMR 5223
15 Bd Latarjet
69622 Villeurbanne Cedex
France

Sosaku Ichikawa
University of Tsukuba
Graduate School of Life and
Environmental Sciences
Tennodai 1-1-1
Tsukuba
Ibaraki 305-8572
Japan

Takashi Kuroiwa
University of Tsukuba
Graduate School of Life and
Environmental Sciences
Tennodai 1-1-1
Tsukuba
Ibaraki 305-8572
Japan
and
National Food Research Institute
Food Engineering Division
Kannondai 2-1-12
Tsukuba
Ibaraki 305-8642
Japan

Isao Kobayashi
National Food Research Institute
Food Engineering Division
2-1-12 Kannondai
Tsukuba
Ibaraki 305-8642
Japan

Todor Kolarov
Bulgarian Academy of Sciences
Institute of Physical Chemistry
Acad. G. Bonchev Str.
Sofia 1113
Bulgaria

Emmanuel Landreau
UMR 5007 CNRS – Université de Lyon
Laboratoire d'Automatique et de Génie
des Procédés LAGEP
Bât 308, 43 Bd du 11 Novembre
69622 Villeurbanne Cedex
France

Jose R. Leiza
University of the Basque Country
Institute for Polymer Materials
POLYMAT, Joxe Mari Korta zentroa
Tolosa Etorbidea 72
20018 Donostia-San Sebastián
Spain

Martine Lemmens
ORAFTI Bio Based Chemicals
Aandorenstraat 1
3300 Tienen
Belgium

Bart Levecke
ORAFTI Bio Based Chemicals
Aandorenstraat 1
3300 Tienen
Belgium

Florence Levy
L'Oréal – Centre de Chevilly-Larue
188 rue Paul Hochart
94150 Chevilly Larue
France

Emmanuelle Marie
CNRS-Nancy-University ENSIC
Laboratoire de Chimie Physique
Macromoléculaire
1 rue Grandville
54001 Nancy cedex
France

Sukekuni Mukataka
University of Tsukuba
Graduate School of Life and
Environmental Sciences
Tennodai 1-1-1
Tsukuba
Ibaraki 305-8572
Japan

Mitsutoshi Nakajima
University of Tsukuba
Graduate School of Life and
Environmental Sciences
Tennodai 1-1-1
Tsukuba
Ibaraki 305-8572
Japan
and
National Food Research Institute
Food Engineering Division
Kannondai 2-1-12
Tsukuba
Ibaraki 305-8642
Japan

Waldemar Nowicki
A. Mickiewicz University
Faculty of Chemistry
Grundwadzka 6
60-780 Poznań
Poland

Grażyna Nowicka
A. Mickiewicz University
Faculty of Chemistry
Grundwadzka 6
60-780 Poznań
Poland

Maria Paulis
University of the Basque Country
Institute for Polymer Materials
POLYMAT, Joxe Mari Korta zentroa
Tolosa Etorbidea 72
20018 Donostia-San Sebastián
Spain

Elise Rotureau
CNRS-Nancy-University ENSIC
Laboratoire de Chimie Physique
Macromoléculaire
1 rue Grandville
54001 Nancy Cedex
France

Seigo Sato
University of Tsukuba
Graduate School of Life and
Environmental Sciences
Tennodai 1-1-1
Tsukuba
Ibaraki 305-8572
Japan

Tharwat F. Tadros
89 Nash Grove Lane
Wokingham, Berkshire RG40 4HE
UK

Kunihiko Uemura
National Food Research Institute
Food Engineering Division
2-1-12 Kannondai
Tsukuba
Ibaraki 305-8642
Japan

Elise Vandekerckhove
ORAFTI Bio Based Chemicals
Aandorenstraat 1
3300 Tienen
Belgium

Man Wu
CNRS-Nancy-University ENSIC
Laboratoire de Chimie Physique
Macromoléculaire
1 rue Grandville
54001 Nancy Cedex
France

1
Emulsion Science and Technology: A General Introduction
Tharwat F. Tadros

1.1
Introduction

Emulsions are a class of disperse systems consisting of two immiscible liquids [1–3]. The liquid droplets (the disperse phase) are dispersed in a liquid medium (the continuous phase). Several classes of emulsion may be distinguished, namely oil-in-water (O/W), water-in-oil (W/O) and oil-in-oil (O/O). The latter class may be exemplified by an emulsion consisting of a polar oil (e.g. propylene glycol) dispersed in a nonpolar oil (paraffinic oil), and *vice versa*. In order to disperse two immiscible liquids a third component is required, namely the *emulsifier*; the choice of emulsifier is crucial not only for the formation of the emulsion but also for its long-term stability [1–3].

Emulsions may be classified according to the nature of the emulsifier or the structure of the system (see Table 1.1).

Several processes relating to the breakdown of emulsions may occur on storage, depending on:

- the particle size distribution and the density difference between the droplets and the medium;
- the magnitude of the attractive versus repulsive forces, which determines flocculation;
- the solubility of the disperse droplets and the particle size distribution, which in turn determines Ostwald ripening;
- the stability of the liquid film between the droplets, which determines coalescence; and
- phase inversion.

The various breakdown processes are illustrated schematically in Figure 1.1.

The physical phenomena involved in each breakdown process is not simple, and requires an analysis to be made of the various surface forces involved. In addition, the above processes may take place simultaneously rather then consecutively, which in turn complicates the analysis. Model emulsions, with monodisperse droplets, cannot be easily produced and hence any theoretical treatment must take into account the effect of

Emulsion Science and Technology. Edited by Tharwat F. Tadros
Copyright © 2009 WILEY-VCH Verlag GmbH & Co. KGaA, Weinheim
ISBN: 978-3-527-32525-2

1 Emulsion Science and Technology: A General Introduction

Table 1.1 Classification of emulsion types.

Nature of emulsifier	Structure of the system
Simple molecules and ions	Nature of internal and external phase:
Nonionic surfactants	O/W, W/O
Surfactant mixtures	Micellar emulsions (microemulsions)
Ionic surfactants	Macroemulsions
Nonionic polymers	Bilayer droplets
Polyelectrolytes	Double and multiple emulsions
Mixed polymers and surfactants	Mixed emulsions
Liquid crystalline phases	
Solid particles	

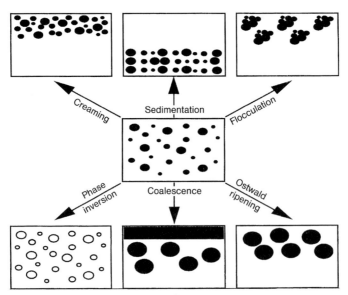

Figure 1.1 Schematic representation of the various breakdown processes in emulsions.

droplet size distribution. Theories that take into account the polydispersity of the system are complex, and in many cases only numerical solutions are possible. In addition, the measurement of surfactant and polymer adsorption in an emulsion is not simple, and such information must be extracted from measurements made at a planar interface.

A summary of each of the above breakdown processes is provided in the following sections, together with details of each process and methods for its prevention.

Creaming and Sedimentation This process results from external forces, usually gravitational or centrifugal. When such forces exceed the thermal motion of the droplets (Brownian motion), a concentration gradient builds up in the system such that the larger droplets move more rapidly either to the top (if their density is less than that of the medium) or to the bottom (if their density is greater than that of the medium) of the container. In the limiting cases, the droplets may form a close-packed (random or ordered) array at the top or bottom of the system, with the remainder of the volume occupied by the continuous liquid phase.

Flocculation This process refers to aggregation of the droplets (without any change in primary droplet size) into larger units. It is the result of the van der Waals attractions which are universal with all disperse systems. Flocculation occurs when there is not sufficient repulsion to keep the droplets apart at distances where the van der Waals attraction is weak. Flocculation may be either 'strong' or 'weak', depending on the magnitude of the attractive energy involved.

Ostwald Ripening (Disproportionation) This effect results from the finite solubility (etc.) of the liquid phases. Liquids which are referred to as being 'immiscible' often have mutual solubilities which are not negligible. With emulsions which are usually polydisperse, the smaller droplets will have a greater solubility when compared to larger droplets (due to curvature effects). With time, the smaller droplets disappear and their molecules diffuse to the bulk and become deposited on the larger droplets. With time, the droplet size distribution shifts to larger values.

Coalescence This refers to the process of thinning and disruption of the liquid film between the droplets, with the result that fusion of two or more droplets occurs to form larger droplets. The limiting case for coalescence is the complete separation of the emulsion into two distinct liquid phases. The driving force for coalescence is the surface or film fluctuations; this results in a close approach of the droplets whereby the van der Waals forces are strong and prevent their separation.

Phase Inversion This refers to the process whereby there will be an exchange between the disperse phase and the medium. For example, an O/W emulsion may with time or change of conditions invert to a W/O emulsion. In many cases, phase inversion passes through a transition state whereby multiple emulsions are produced.

1.2
Industrial Applications of Emulsions

Several industrial systems consist of emulsions of which the following are worthy of mention:

- *Food emulsions*, such as mayonnaise, salad creams, deserts and beverages.
- Personal care and cosmetic products, such as hand-creams, lotions, hair-sprays and sunscreens.
- *Agrochemicals* - self-emulsifiable oils which produce emulsions on dilution with water, emulsion concentrates (droplets dispersed in water; EWs) and crop oil sprays.

- *Pharmaceuticals*, such as anesthetics of O/W emulsions, lipid emulsions, double and multiple emulsions.
- *Paints*, such as emulsions of alkyd resins and latex emulsions.
- *Dry-cleaning formulations*; these may contain water droplets emulsified in the dry-cleaning oil, which is necessary to remove soils and clays.
- Bitumen emulsions are prepared stable in their containers but, when applied the road chippings, they must coalesce to form a uniform film of bitumen.
- *Emulsions in the oil industry* - many crude oils contain water droplets (e.g. North Sea oil); these must be removed by coalescence followed by separation.
- Oil slick dispersants - oil spilled from tankers must be emulsified and then separated. The emulsification of unwanted oil is a very important process in pollution control.

The above-described utilization of emulsions in industrial processes justifies the vast amount of basic research which is conducted aimed at understanding the origins of the instability of emulsions and developing methods to prevent their break down. Unfortunately, fundamental research into emulsions is not straightforward, as model systems (e.g. with monodisperse droplets) are difficult to produce. In fact, in many cases, the theoretical bases of emulsion stability are not exact and consequently semi-empirical approaches are used.

1.3
The Physical Chemistry of Emulsion Systems

1.3.1
The Interface (Gibbs Dividing Line)

An interface between two bulk phases, such as liquid and air (or liquid/vapor) or two immiscible liquids (oil/water), may be defined provided that a dividing line is introduced (Figure 1.2). The interfacial region is not a layer that is one molecule thick; rather, it has a thickness δ with properties that differ from those of the two bulk phases α and β.

By using the Gibbs model, it is possible to obtain a definition of the surface or interfacial tension γ.

The surface free energy dG^σ is composed of three components: an entropy term $S^\sigma dT$; an interfacial energy term $Ad\gamma$; and a composition term $\sum n_i d\mu_i$ (where n_i is the

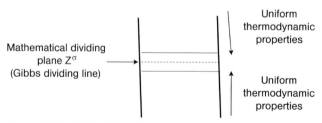

Figure 1.2 The Gibbs dividing line.

number of moles of component i with chemical potential μ_i). The Gibbs–Deuhem equation is therefore,

$$dG^\sigma = -S^\sigma dT + Ad\gamma + \sum n_i d\mu_i \qquad (1.1)$$

At constant temperature and composition,

$$dG^\sigma = Ad\gamma$$

$$\gamma = \left(\frac{\partial G^\sigma}{\partial A}\right)_{T,n_i} \qquad (1.2)$$

For a stable interface γ is positive – that is, if the interfacial area increases, then G^σ increases. Note that γ is energy per unit area (mJ m^{-2}), which is dimensionally equivalent to force per unit length (mN m^{-1}), the unit usually used to define surface or interfacial tension.

For a curved interface, one should consider the effect of the radius of curvature. Fortunately, γ for a curved interface is estimated to be very close to that of a planar surface, unless the droplets are very small (<10 nm). Curved interfaces produce some other important physical phenomena which affect emulsion properties, such as the Laplace pressure Δp which is determined by the radii of curvature of the droplets,

$$\Delta p = \gamma \left(\frac{1}{r_1} + \frac{1}{r_2}\right) \qquad (1.3)$$

where r_1 and r_2 are the two principal radii of curvature.

For a perfectly spherical droplet $r_1 = r_2 = r$ and

$$\Delta p = \frac{2\gamma}{r} \qquad (1.4)$$

For a hydrocarbon droplet with radius 100 nm, and $\gamma = 50$ mN m^{-1}, $\Delta p \sim 10^6$ Pa (10 atm).

1.4
The Thermodynamics of Emulsion Formation and Breakdown

Consider a system in which an oil is represented by a large drop 2 of area A_1 immersed in a liquid 2, which is now subdivided into a large number of smaller droplets with total area A_2 (such that $A_2 \gg A_1$), as shown in Figure 1.3. The interfacial tension γ_{12} is the same for the large and smaller droplets as the latter are generally in the region of 0.1 µm to few microns in size.

The change in free energy in going from state I to state II is made from two contributions: a surface energy term (that is positive) that is equal to $\Delta A \gamma_{12}$ (where

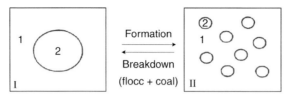

Figure 1.3 Schematic representation of emulsion formation and breakdown (see text for details).

$\Delta A = A_2 - A_1$). An entropy of dispersions term which is also positive (since the production of a large number of droplets is accompanied by an increase in configurational entropy) which is equal to $T\Delta S^{conf}$.

From the second law of thermodynamics,

$$\Delta G^{form} = \Delta A \gamma_{12} - T\Delta S^{conf} \tag{1.5}$$

In most cases, $\Delta A \gamma_{12} \gg -T\Delta S^{conf}$, which means that ΔG^{form} is positive – that is, the formation of emulsions is nonspontaneous and the system is thermodynamically unstable. In the absence of any stabilization mechanism, the emulsion will break by flocculation, coalescence, Ostwald ripening, or a combination of all these processes. This situation is illustrated graphically in Figure 1.4, where several paths for emulsion breakdown processes are represented.

In the presence of a stabilizer (surfactant and/or polymer), an energy barrier is created between the droplets and therefore the reversal from state II to state I becomes noncontinuous as a result of the presence of these energy barriers. This is illustrated graphically in Figure 1.5 where, in the presence of the above energy barriers, the system becomes kinetically stable.

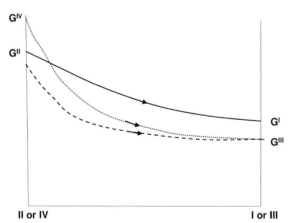

Figure 1.4 The free energy path in emulsion breakdown.
Solid line: flocculation + coalescence.
Broken line: flocculation + coalescence + sedimentation.
Dotted line: flocculation + coalescence + sedimentation + Ostwald ripening.

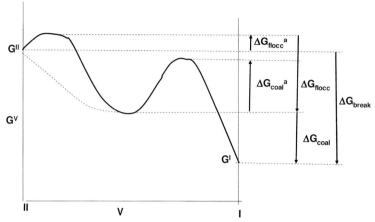

Figure 1.5 Schematic representation of the free energy path for the breakdown (flocculation and coalescence) of systems containing an energy barrier.

1.5
Interaction Energies (Forces) Between Emulsion Droplets and Their Combinations

Generally speaking, there are three main interaction energies (forces) between emulsion droplets, the details of which are discussed in the following sections.

1.5.1
Van der Waals Attraction

The van der Waals attraction between atoms or molecules are of three different types: (i) dipole–dipole (Keesom); (ii) dipole-induced dipole ((Debye-)interactions); and (iii) dispersion (London interactions). The Keesom and Debye attraction forces are vectors, and although the dipole–dipole or dipole-induced dipole attraction is large they tend to cancel due to the different orientations of the dipoles. Thus, the most important are the London dispersion interactions, which arise from charge fluctuations. With atoms or molecules consisting of a nucleus and electrons that are continuously rotating around the nucleus, a temporary dipole is created as a result of charge fluctuations. This temporary dipole induces another dipole in the adjacent atom or molecule. The interaction energy between two atoms or molecules G_a is short range and is inversely proportional to the sixth power of the separation distance r between the atoms or molecules,

$$G_a = -\frac{\beta}{r^6} \tag{1.6}$$

where β is the London dispersion constant that is determined by the polarizability of the atom or molecule.

Hamaker [4] suggested that the London dispersion interactions between atoms or molecules in macroscopic bodies (such as emulsion droplets) could be added, and this would result in a strong van der Waals attraction, particularly at close distances of separation between the droplets. For two droplets with equal radii R, at a separation distance h, the van der Waals attraction G_A is given by the following equation (due to Hamaker),

$$G_A = -\frac{AR}{12h} \tag{1.7}$$

where A is the effective Hamaker constant,

$$A = (A_{11}^{1/2} - A_{22}^{1/2})^2 \tag{1.8}$$

and where A_{11} and A_{22} are the Hamaker constants of droplets and dispersion medium, respectively.

The Hamaker constant of any material depends on the number of atoms or molecules per unit volume q and the London dispersion constant β,

$$A = \pi q^2 \beta \tag{1.9}$$

G_A is seen to increase very rapidly with a decrease of h (at close approach). This is illustrated in Figure 1.6 which shows the van der Waals energy–distance curve for two emulsion droplets with separation distance h.

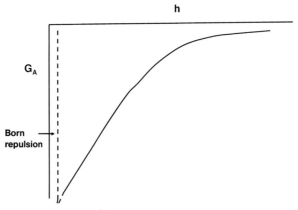

Figure 1.6 Variation of the Van der Waals attraction energy with separation distance.

In the absence of any repulsion, flocculation occurs very rapidly to produce large clusters. In order to counteract the van der Waals attraction, it is necessary to create a repulsive force. Two main types of repulsion can be distinguished depending on the nature of the emulsifier used: (i) *electrostatic*, which occurs due to the creation of double layers; and (ii) *steric*, which occurs due to the presence of adsorbed surfactant or polymer layers.

1.5.2
Electrostatic Repulsion

This can be produced by the adsorption of an ionic surfactant, as shown in Figure 1.7, which shows a schematic representation of the structure of the double layer according to Gouy-Chapman and Stern pictures [5]. The surface potential ψ_o decreases linearly to ψ_d (Stern or zeta potential) and then exponentially with the increase of distance x. The double-layer extension depends on electrolyte concentration and valency (the lower the electrolyte concentration and the lower the valency, the more extended is the double layer).

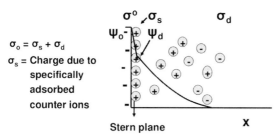

Figure 1.7 Schematic representation of double layers produced by the adsorption of an ionic surfactant.

When charged colloidal particles in a dispersion approach each other such that the double layer begins to overlap (i.e. the particle separation becomes less than twice the double-layer extension), then repulsion will occur. The individual double layers can no longer develop unrestrictedly, as the limited space does not allow complete potential decay [5, 6]. This is shown schematically in Figure 1.8 for two flat plates. This shows clearly that, when the separation distance h between the emulsion droplets become smaller than twice the double-layer extension, the potential at the mid plane between the surfaces is not equal to zero (which would be the case when h is more than twice the double-layer extension) plates.

The repulsive interaction G_{el} is given by the following expression:

$$G_{el} = 2\pi R \varepsilon_r \varepsilon_o \psi_o^2 \ln[1 + \exp(-\kappa h)] \tag{1.10}$$

where ε_r is the relative permittivity, ε_o is the permittivity of free space, κ is the Debye–Huckel parameter; $1/\kappa$ is the extension of the double layer (double-layer

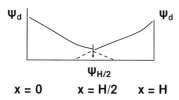

Figure 1.8 Schematic representation of a double-layer overlap.

thickness) that is given by the expression,

$$\left(\frac{1}{\kappa}\right) = \left(\frac{\varepsilon_r \varepsilon_0 k T}{2 n_0 Z_i^2 e^2}\right) \tag{1.11}$$

where k is the Boltzmann constant, T is the absolute temperature, n_o is the number of ions per unit volume of each type present in bulk solution, Z_i is the valency of the ions and e is the electronic charge.

Values of $(1/\kappa)$ at various 1 : 1 electrolyte concentrations (C) are as follows:

C (mol dm^{-3})	10^{-5}	10^{-4}	10^{-3}	10^{-2}	10^{-1}
$1/\kappa$ (nm)	100	33	10	3.3	1

The double layer extension decreases with increase of electrolyte concentration, which means that the repulsion decreases with increase of electrolyte concentration, as illustrated in Figure 1.9.

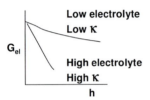

Figure 1.9 Variation of G_{el} with h at low and high electrolyte concentrations (κ).

A combination of van der Waals attraction and double-layer repulsion results in the well-known theory of colloid stability due to Deryaguin, Landau, Verwey and Overbeek (DLVO theory) [5, 6].

$$G_T = G_{el} + G_A \tag{1.12}$$

A schematic representation of the force (energy) distance curve according to the DLVO theory is given in Figure 1.10.

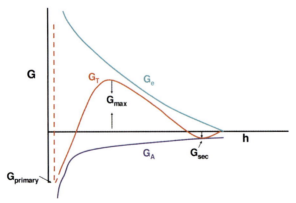

Figure 1.10 Total energy–distance curve according to the DLVO theory.

The above presentation is for a system at low electrolyte concentration. At large h, attraction prevails which results in a shallow minimum (G_{sec}) of the order of few kT units. At very short h, $V_A \gg G_{el}$, resulting in a deep primary minimum (several hundred kT units). At intermediate h, $G_{el} > G_A$, resulting in a maximum (energy barrier) the height of which depends on ψ_o (or ζ) and electrolyte concentration and valency. The energy maximum is usually kept > 25 kT units, and prevents not only a close approach of the droplets but also flocculation into the primary minimum. The higher the value of ψ_o and the lower the electrolyte concentration and valency, the higher the energy maximum. At intermediate electrolyte concentrations, weak flocculation into the secondary minimum may occur.

1.5.3
Steric Repulsion

This is produced by using nonionic surfactants or polymers, such as alcohol ethoxylates, or A–B–A block copolymers PEO–PPO–PEO (PEO = polyethylene oxide; PPO = polypropylene oxide), as illustrated in Figure 1.11.

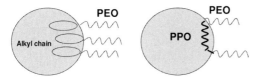

Figure 1.11 Schematic representation of the adsorbed layers.

The 'thick' hydrophilic chains (PEO in water) produce repulsion as a result of two main effects [7]:

- Unfavorable mixing of the PEO chains
  When this occurs in good solvent conditions (moderate electrolyte and low temperatures) it is referred to as the osmotic or mixing free energy of interaction that is given by the expression,

$$\frac{G_{mix}}{kT} = \left(\frac{4\pi}{V_1}\right)\phi_2^2 N_{av} \left(\frac{1}{2}-\chi\right)\left(\delta-\frac{h}{2}\right)^2 \left(3R+2\delta+\frac{h}{2}\right) \quad (1.13)$$

where V_1 is the molar volume of the solvent, ϕ_2 is the volume fraction of the polymer chain with a thickness δ and χ is the Flory–Huggins interaction parameter.
When $\chi < 0.5$, G_{mix} is positive and the interaction is repulsive; when $\chi > 0.5$, G_{mix} is negative and the interaction is attractive; when $\chi = 0.5$, $G_{mix} = 0$ and this is referred to as the θ-condition.

- Entropic, volume restriction or elastic interaction, G_{el}
  This results from the loss in configurational entropy of the chains on significant overlap. Entropy loss is unfavorable and, therefore, G_{el} is always positive.

A combination of G_{mix}, G_{el} with G_A gives the total energy of interaction G_T (theory of steric stabilization),

$$G_T = G_{mix} + G_{el} + G_A \tag{1.14}$$

A schematic representation of the variation of G_{mix}, G_{el} and G_A with h is shown in Figure 1.12.

In Figure 1.12 it is clear that there is only one minimum (G_{min}), the depth of which depends on R, δ and A. When $h_o < 2\delta$, a strong repulsion occurs and this increases very sharply with any further decrease in h_o. At a given droplet size and Hamaker constant, the larger the adsorbed layer thickness, the smaller the depth of the minimum. If G_{min} is made sufficiently small (large δ and small R), then thermodynamic stability may be approached, and this explains the case with nanoemulsions.

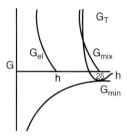

Figure 1.12 Schematic representation of the energy–distance curve for a sterically stabilized emulsion.

1.6
Adsorption of Surfactants at the Liquid/Liquid Interface

When surfactants accumulate at interfaces, the process is described as *adsorption*. The simplest interfaces are air/water (A/W) and oil/water (O/W). The surfactant molecule positions itself at the interface, with the hydrophobic portion oriented towards the hydrophobic phase (air or oil) and the hydrophilic portion oriented at the hydrophilic phase (water). This is shown schematically in Figure 1.13. As a result of adsorption, the surface tension of water is reduced from its value of $72\,\text{mN m}^{-1}$ before adsorption to $\sim 30\text{–}40\,\text{mN m}^{-1}$, while the interfacial tension for the O/W system decreases from a value of $50\,\text{mN m}^{-1}$ (for an alkane oil) before adsorption to a value of $1\text{–}10\,\text{mN m}^{-1}$, depending on the nature of the surfactant.

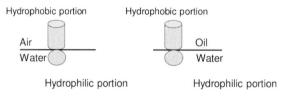

Figure 1.13 Schematic representation of the orientation of surfactant molecules.

1.6 Adsorption of Surfactants at the Liquid/Liquid Interface

Two approaches can be applied to treat surfactant adsorption at the A/L and L/L interfaces [5]:

- The 'Gibbs approach', which treats the process as an equilibrium phenomenon and in which case the second law of thermodynamics can be applied.
- The 'equation of state approach', whereby the surfactant film is treated as a 'two-dimensional' layer with a surface pressure π.

The Gibbs approach allows the surfactant adsorption to be obtain from surface tension measurements, whereas the equation of state approach allows the surfactant orientation to be studied the at the interface. In the following section details of only the Gibbs approach will be described.

1.6.1
The Gibbs Adsorption Isotherm

Gibbs derived a thermodynamic relationship between the variation of surface or interfacial tension with concentration and the amount of surfactant adsorbed Γ (moles per unit area), referred to as the 'surface excess'. At equilibrium, the Gibbs free energy $dG^\sigma = 0$ and the Gibbs–Duehem equation becomes,

$$dG^\sigma = -S^\sigma dT + Ad\gamma + \sum n_i^\sigma d\mu_i = 0 \tag{1.15}$$

At constant temperature,

$$Ad\gamma = -\sum n_i^\sigma d\mu_i \tag{1.16}$$

or

$$d\gamma = -\sum \frac{n_i^\sigma}{A} d\mu_i = -\sum \Gamma_i^\sigma d\mu_i \tag{1.17}$$

For a surfactant (component 2) adsorbed at the surface of a solvent (component 1),

$$-d\gamma = \Gamma_1^\sigma d\mu_1 + \Gamma_2^\sigma d\mu_2 \tag{1.18}$$

If the Gibbs dividing surface is used and the assumption $\Gamma_1^\sigma = 0$ is made, then

$$-d\gamma = \Gamma_{2,1}^\sigma d\mu_2 \tag{1.19}$$

The chemical potential of the surfactant μ_2 is given by the expression,

$$\mu_2 = \mu_2^0 + RT \ln a_2^l \tag{1.20}$$

where μ_2^0 is the standard chemical potential, a_2^l is the activity of surfactant that is equal to $C_2 f_2 \sim x_2 f_2$, where C_2 is the concentration (in moles dm^{-3}) and x_2 is the mole fraction that is equal to $C_2/(C_2 + 55.5)$ for a dilute solution and f_2 is the activity coefficient that is also ~ 1 in dilute solutions.

By differentiating Equation 1.20, one obtains,

$$d\mu_2 = RTd \ln a_2^l \tag{1.21}$$

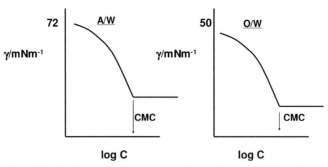

Figure 1.14 Surface or interfacial tension (γ)–log C curves. CMC = critical micelle concentration.

Combining Equations 1.19 and 1.21,

$$-d\gamma = \Gamma^{\sigma}_{2,1} RT d \ln a_2^L \tag{1.22}$$

or

$$\frac{d\gamma}{d \ln a_2^L} = -RT\Gamma^{L}_{2,1} \tag{1.23}$$

In dilute solutions, $f_2 \sim 1$ and,

$$\frac{d\gamma}{d \ln C_2} = -\Gamma_2 RT \tag{1.24}$$

Equations 1.23 and 1.24 are referred to as the Gibbs adsorption equations, and show that Γ_2 can be determined from the experimental results of variation of γ with log C_2, as is illustrated in Figure 1.14 for the A/W and O/W interfaces.

Γ_2 can be calculated from the linear portion of the γ–log C curve just before the critical micelle concentration (CMC):

$$slope = -\frac{d\gamma}{d \log C_2} = -2.303 \Gamma_2 RT \tag{1.25}$$

From Γ_2 the area per molecule of surfactant (or ion) can be calculated,

$$area/molecule = \frac{1}{\Gamma_2 N_{av}} (m^2) = \frac{10^{18}}{\Gamma_2 N_{av}} (nm^2) \tag{1.26}$$

where N_{av} is the Avogadro's constant (equal to 6.023×10^{23}).

The area per surfactant ion or molecule provides information on the orientation of the ion or molecule at the interface. The area depends on whether the molecules lie flat or vertically at the interface, and also on the length of the alkyl chain length (if the molecules lie flat) or the cross-sectional area of the head group (if the molecules lie vertically). For example, for an ionic surfactant such as sodium dodecyl sulfate (SDS), the area per molecule depends on the orientation. If the molecule lies flat, the area is determined by the area occupied by the alky chain and that by the sulfate head group. In this case the area per molecule increases with an increase in the alkyl chain length, and will be in the range 1–2 nm². In contrast, for vertical orientation, the area per molecule is determined by the cross-sectional area of the sulfate group, which is ~0.4 nm², and is virtually independent of the alkyl chain length. The addition of electrolytes screens the charge on the head group, and hence the area per molecule decreases. For nonionic surfactants such as alcohol ethoxylates, the area per molecule for flat orientation is determined by the length of the alkyl chain and the number of ethylene oxide (EO) units. For vertical orientation, the area per molecule is determined by the cross-sectional area of the PEO chain, and this increases with an increase in the number of EO units.

At concentrations just before the break point, the slope of the $\gamma - \log C$ curve is constant,

$$\left(\frac{\partial \gamma}{\partial \log C_2}\right) = \text{constant} \tag{1.27}$$

which indicates that saturation of the interface occurs just below the CMC.

Above the break point (C > CMC), the slope is zero,

$$\left(\frac{\partial \gamma}{\partial \log C_2}\right) = 0 \tag{1.28}$$

or

$$\gamma = \text{constant} \times \log C_2 \tag{1.29}$$

As γ remains constant above the CMC then C_2 or a_2 of the monomer must remain constant.

The addition of surfactant molecules above the CMC must result in an association to form micelles which have low activity, and hence a_2 remains virtually constant.

The hydrophilic head group of the surfactant molecule can also affect its adsorption. These head groups can be unionized (e.g. alcohol or PEO), weakly ionized (e.g. COOH), or strongly ionized (e.g. sulfates $-O-SO_3^-$, sulfonates $-SO_3^-$ or ammonium salts $-N^+(CH_3)_3$). The adsorption of the different surfactants at the A/W and O/W interfaces depends on the nature of the head group. With nonionic surfactants, repulsion between the head groups is smaller than with ionic head groups and adsorption occurs from dilute solutions; the CMC is low, typically 10^{-5} to 10^{-4} mol dm^{-3}. Nonionic surfactants with medium PEO form closely packed layers at C < CMC. Adsorption is slightly affected by the moderate addition of electrolytes or a change in the pH. Nonionic surfactant adsorption is relatively simple and can be described by the Gibbs adsorption equation.

With ionic surfactants, adsorption is more complicated, depending on the repulsion between the head groups and addition of indifferent electrolyte. The Gibbs adsorption equation must be solved to take into account the adsorption of the counterions and any indifferent electrolyte ions.

For a strong surfactant electrolyte such as $R-O-SO_3^- Na^+$ ($R^- Na^+$):

$$\Gamma_2 = -\frac{1}{2RT}\left(\frac{\partial \gamma}{\partial \ln a_\pm}\right) \tag{1.30}$$

The factor 2 in Equation 1.30 arises because both surfactant ion and counterion must be adsorbed to maintain neutrality. $(\partial\gamma/\partial \ln a\pm)$ is twice as large for an unionized surfactant molecule.

For a nonadsorbed electrolyte such as NaCl, any increase in $Na^+ R^-$ concentration produces a negligible increase in Na^+ concentration ($d\mu_{Na^+}$ is negligible; $d\mu_{Cl^-}$ is also negligible).

$$\Gamma_2 = -\frac{1}{RT}\left(\frac{\partial \gamma}{\partial \ln C_{NaR}}\right) \tag{1.31}$$

which is identical to the case of nonionics.

The above analysis shows that many ionic surfactants may behave like nonionics in the presence of a large concentration of an indifferent electrolyte such as NaCl.

1.6.2
Mechanism of Emulsification

As mentioned above, in order to prepare an emulsion, oil, water, a surfactant and energy are required. This can be considered from a consideration of the energy needed to expand the interface, $\Delta A \gamma$ (where ΔA is the increase in interfacial area when the bulk oil with area A_1 produces a large number of droplets with area A_2; $A_2 \gg A_1$, γ is the interfacial tension). Since γ is positive, the energy needed to expand the interface is large and positive; this energy term cannot be compensated by the small entropy of dispersion $T\Delta S$ (which is also positive) and the total free energy of formation of an emulsion, ΔG given by Equation 1.5 is positive. Thus, emulsion formation is nonspontaneous and energy is required to produce the droplets.

The formation of large droplets (a few μm), as is the case for macroemulsions, is fairly easy and hence high-speed stirrers such as the Ultra-Turrax or Silverson Mixer are sufficient to produce the emulsion. In contrast, the formation of small drops (submicron, as is the case with nanoemulsions) is difficult and requires a large amount of surfactant and/or energy. The high energy required for the formation of nanoemulsions can be understood from a consideration of the Laplace pressure Δp (the difference in pressure between inside and outside the droplet), as given by Equations 1.3 and 1.4.

In order for a drop to be broken up into smaller droplets it must be strongly deformed, and this deformation increases Δp. This is illustrated in Figure 1.15, which shows the situation when a spherical drop deforms into a prolate ellipsoid [8].

Near point 1 there is only one radius of curvature R_a, whereas near point 2 there are two radii of curvature $R_{b,1}$ and $R_{b,2}$. Consequently, the stress needed to deform the drop is higher for a smaller drop. Since the stress is generally transmitted by the surrounding liquid via agitation, higher stresses require a more vigorous agitation, and hence more energy is needed to produce smaller drops.

Surfactants play major roles in the formation of emulsions. By lowering the interfacial tension, p is reduced and hence the stress required to break up a drop is reduced. Surfactants also prevent the coalescence of newly formed drops.

1.6 Adsorption of Surfactants at the Liquid/Liquid Interface

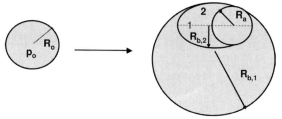

Figure 1.15 Schematic representation of the increase in Laplace pressure when a spherical drop is deformed to a prolate ellipsoid.

Figure 1.16 shows the various processes which occur during emulsification, namely the break up of the droplets, adsorption of the surfactants and droplet collision (which may or may not lead to coalescence) [8].

Each of the represented processes occurs numerous times during emulsification, and the time scale of each process is very short, typically one microsecond. This shows that the emulsification process is a dynamic process and events that occur within a microsecond range may be very important.

In order to describe emulsion formation, two main factors must be considered, namely *hydrodynamics* and *interfacial science*. To assess emulsion formation, the usual approach is to measure the droplet size distribution, using for example laser diffraction techniques. A useful average diameter d is,

$$d_{nm} = \left(\frac{S_m}{S_n}\right)^{1/(n-m)} \tag{1.32}$$

In most cases, d_{32} (the volume/surface average or Sauter mean) is used. The width of the size distribution can be given as the variation coefficient c_m, which is the

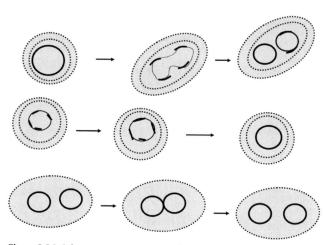

Figure 1.16 Schematic representation of the various processes occurring during emulsion formation. The drops are depicted by thin lines, and the surfactant by heavy lines and dots.

standard deviation of the distribution weighted with d^m divided by the corresponding average d. Generally C_2 will be used which corresponds to d_{32}.

An alternative way to describe the emulsion quality is to use the specific surface area A (the surface area of all emulsion droplets per unit volume of emulsion),

$$A = \pi s^2 = \frac{6\phi}{d_{32}} \tag{1.33}$$

1.6.3
Methods of Emulsification

Several procedures may be applied for emulsion preparation, ranging from simple pipe flow (low agitation energy, L), static mixers and general stirrers (low to medium energy, L–M), high-speed mixers such as the Ultra-Turrax (M), colloid mills and high-pressure homogenizers (high energy, H), and ultrasound generators (M–H).

The method of preparation can be either continuous (C) or batch-wise (B): Pipe flow and static mixers – C; stirrers and Ultra-Turrax – B and C; colloid mill and high-pressure homogenizers – C; Ultrasound – B and C.

In all methods, there is liquid flow and unbounded and strongly confined flow. In unbounded flow, any droplet is surrounded by a large amount of the flowing liquid (the confining walls of the apparatus are far away from most droplets). Thus, the forces can be either frictional (mostly viscous) or inertial. *Viscous forces* cause shear stresses to act on the interface between the droplets and the continuous phase (primarily in the direction of the interface). The *shear stresses* can be generated by either laminar flow (LV) or turbulent flow (TV), and this depends on the Reynolds number, R_e:

$$R_e = \frac{v/\rho}{\eta} \tag{1.34}$$

where v is the linear liquid velocity, ρ is the liquid density and η is its viscosity. The characteristic length l is given by the diameter of flow through a cylindrical tube, and by twice the slit width in a narrow slit.

For laminar flow, $R_e < \sim 1000$, whereas for turbulent flow $R_e > \sim 2000$. Thus, whether the regime is linear or turbulent depends on the scale of the apparatus, the flow rate, and the liquid viscosity [8–11].

If the turbulent eddies are much larger than the droplets they exert shear stresses on the droplets. However, if the turbulent eddies are much smaller than the droplets then the inertial forces will cause disruption (TI).

In bounded flow, other relationships hold. For example, if the smallest dimension of that part of the apparatus where the droplets are disrupted (e.g. a slit) is comparable to the droplet size, then other relationships hold (the flow is always laminar). However, a different regime prevails if the droplets are directly injected through a narrow capillary into the continuous phase (injection regime), and membrane emulsification will occur.

Within each regime, one essential variable is the intensity of the forces acting; thus, the *viscous stress* during laminar flow $\sigma_{viscous}$ is given by:

$$\sigma_{viscous} = \eta G \tag{1.35}$$

where G is the velocity gradient.

Intensity in turbulent flow is expressed by the power density ε (the amount of energy dissipated per unit volume per unit time); thus, for a laminar flow:

$$\varepsilon = \eta G^2 \tag{1.36}$$

The most important regimes are: Laminar/Viscous (LV); Turbulent/Viscous (TV); and Turbulent/Inertial (TI). With water as the continuous phase, the regime is always TI, whereas for a higher viscosity of the continuous phase ($\eta_C = 0.1$ Pa·s), the regime is TV. For still higher viscosity or for a small apparatus (small *l*), the regime is LV, whilst for a very small apparatus (as is the case with most laboratory homogenizers) the regime is nearly always LV.

For the above regimes, a semi-quantitative theory is available that can provide the time scale and magnitude of the local stress σ_{ext}, the droplet diameter *d*, the time scale of droplet deformation τ_{def}, the time scale of surfactant adsorption, τ_{ads}, and the mutual collision of droplets.

One important parameter that describes droplet deformation is the Weber number, W_e (this gives the ratio of the external stress over the Laplace pressure):

$$W_e = \frac{G\eta_C R}{2\gamma} \tag{1.37}$$

The viscosity of the oil plays an important role in the break-up of droplets – that is, the higher the viscosity, the greater the time taken to deform a drop. The deformation time τ_{def} is given by the ratio of oil viscosity to the external stress acting on the drop:

$$\tau_{def} = \frac{\eta_D}{\sigma_{ext}} \tag{1.38}$$

The viscosity of the continuous phase η_C plays an important role in some regimes. For a turbulent inertial regime, η_C has no effect on droplet size, whereas for a turbulent viscous regime a larger η_C leads to smaller droplets. For laminar viscous the effect is even stronger.

1.6.4
Role of Surfactants in Emulsion Formation

Surfactants lower the interfacial tension γ, which in turn causes a reduction in droplet size (the latter decreases with a decrease in γ). For laminar flow the droplet diameter is proportional to γ, whereas for a turbulent inertial regime the droplet diameter is proportional to $\gamma^{3/5}$.

The effect of reducing γ on the droplet size is shown in Figure 1.17, where a droplet surface area A and mean drop size d_{32} are plotted as a function of surfactant concentration m for various systems.

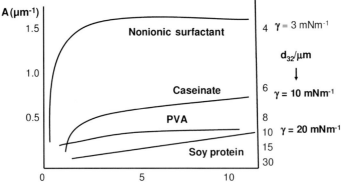

Figure 1.17 Variation of A and d_{32} with m for various surfactant systems.

The amount of surfactant required to produce the smallest drop size will depend on its activity a (concentration) in the bulk, which in turn determines the reduction in γ, as given by the Gibbs adsorption equation:

$$-d\gamma = RT\Gamma d \ln a \qquad (1.39)$$

where R is the gas constant, T is the absolute temperature and Γ is the surface excess (the number of moles adsorbed per unit area of the interface).

Γ increases with an increase in surfactant concentration until it eventually reaches a plateau value (saturation adsorption). This is illustrated in Figure 1.18 for various emulsifiers.

The value of γ obtained depends on the nature of the oil and surfactant used; typically, small molecules such as nonionic surfactants reduce γ to a greater degree than do polymeric surfactants such as polyvinyl alcohol (PVA).

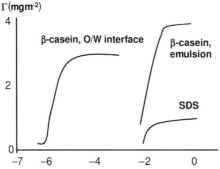

Figure 1.18 Variation of Γ (mg m^{-2}) with log C_{eq} (wt%). The oils are β-casein (O/W interface) toluene, β-casein (emulsions) soybean, and SDS benzene.

1.6 Adsorption of Surfactants at the Liquid/Liquid Interface

Another important role of the surfactant is its effect on the interfacial dilational modulus ε,

$$\varepsilon = \frac{d\gamma}{d \ln A} \tag{1.40}$$

During emulsification, an increase in the interfacial area A occurs which in turn causes a reduction in Γ. The equilibrium is restored by the adsorption of surfactant from the bulk, but this takes time (the time is shorter at a higher surfactant activity). Thus, ε is small at small at small activity and also at large activity. Because of the lack of equilibrium (or the slowness of it being achieved) with polymeric surfactants, ε will not be the same for the expansion and compression of the interface.

In practice, surfactant mixtures are used which have pronounced effects on γ and ε. Some specific surfactant mixtures provide lower γ values than either of the two individual components, and the presence of more than one surfactant molecule at the interface tends to increase ε at high surfactant concentrations. The various components vary in surface activity; for example, those with the lowest γ tend to predominate at the interface, although if they are present at low concentrations it may take a long time before the lowest value is reached. Polymer–surfactant mixtures may in fact demonstrate some synergetic surface activity.

1.6.5
Role of Surfactants in Droplet Deformation

Apart from their effect on reducing γ, surfactants play major roles in the deformation and break up of droplets, and this may be summarized as follows. Surfactants allow the existence of interfacial tension gradients which are crucial for the formation of stable droplets. In the absence of surfactants (clean interface), the interface cannot withstand any tangential stress and the liquid motion will be continuous (Figure 1.19a).

If a liquid flows along the interface with surfactants, the latter will be swept downstream causing an interfacial tension gradient (Figure 1.19b). Hence, a balance of forces will be established:

$$\eta \left[\frac{dV_x}{dy} \right]_{y=0} = -\frac{d\gamma}{dx} \tag{1.41}$$

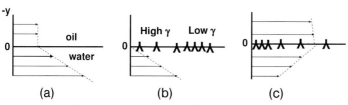

Figure 1.19 Interfacial tension gradients and flow near an oil/water interface. (a) No surfactant; (b) the velocity gradient causes an interfacial tension gradient; (c) the interfacial tension gradient causes flow (Marangoni effect).

If the γ-gradient become sufficiently large it will arrest the interface. However, if the surfactant is applied at one site of the interface, a γ-gradient is formed that will cause the interface to move roughly at a velocity given by

$$v = 1.2[\eta \rho z]^{-1/3} |\Delta \gamma|^{2/3} \tag{1.42}$$

The interface will then drag some of the bordering liquid with it (Figure 1.19c).

Interfacial tension gradients are very important in stabilizing the thin liquid film between the droplets – a step which is very important at the start of the emulsification (films of the continuous phase may be drawn through the disperse phase and the collision is very large). The magnitude of the γ-gradients and of the Marangoni effect depend on the surface dilational modulus ε which, for a plane interface with one surfactant-containing phase, is given by the expression

$$\varepsilon = \frac{-d\gamma/d \ln \Gamma}{(1 + 2\xi + 2\xi^2)^{1/2}} \tag{1.43}$$

$$\xi = \frac{dm_C}{d\Gamma} \left(\frac{D}{2\omega} \right)^{1/2} \tag{1.44}$$

$$\omega = \frac{d \ln A}{dt} \tag{1.45}$$

where D is the diffusion coefficient of the surfactant and ω represents a time scale (the time needed to double the surface area) that is approximately equal to τ_{def}.

During emulsification, ε is dominated by the magnitude of the denominator in Equation 1.43 because ζ remains small. The value of $dm_C/d\Gamma$ tends to go to be very high when Γ reaches its plateau value; ε goes to a maximum when m_C is increased.

For conditions that prevail during emulsification, ε increases with m_C and is given by the relationship

$$\varepsilon = \frac{d\pi}{d \ln \Gamma} \tag{1.46}$$

where π is the surface pressure (π = $\gamma_o - \gamma$). Figure 1.20 shows the variation of π with ln Γ, where ε is given by the slope of the line.

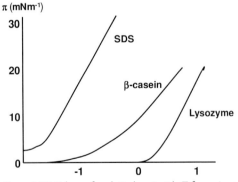

Figure 1.20 Values of π plotted against ln Γ for various emulsifiers.

Sodium dodecyl sulfate (SDS) shows a much higher ε-value when compared with β-casein and lysozyme; this is because the value of Γ is higher for SDS. The two proteins show differences in their ε-values which may be attributed to the conformational change that occur upon adsorption.

The presence of a surfactant means that, during emulsification, the interfacial tension need not be the same everywhere (see Figure 1.19). This has two consequences: (i) the equilibrium shape of the drop is affected; and (ii) any γ-gradient formed will slow down the motion of the liquid inside the drop (this diminishes the amount of energy needed to deform and break up the drop).

Another important role of the emulsifier is to prevent coalescence during emulsification. This is clearly not due to the strong repulsion between the droplets, since the pressure at which the two drops are pressed together is much greater than the repulsive stresses. Hence, the counteracting stress must be due to the formation of γ-gradients. When two drops are pushed together, liquid will flow out from the thin layer between them; such flow will induce a γ-gradient (see Figure 1.19c), producing a counteracting stress given by

$$\tau_{\Delta\gamma} \approx \frac{2|\Delta\gamma|}{(1/2)d} \tag{1.47}$$

Here, the factor of 2 results from there being two interfaces involved. Taking a value of $\Delta\gamma = 10\,\mathrm{mN\,m^{-1}}$, the stress amounts to 40 kPa (which is of the same order of magnitude as the external stress).

Closely related to the above mechanism is the Gibbs–Marangoni effect [13–17], which is represented schematically in Figure 1.21. The depletion of surfactant in the thin film between approaching drops results in a γ-gradient without liquid flow being involved, and in turn an inward flow of liquid that tends to drive the drops apart.

The Gibbs–Marangoni effect also explains the Bancroft rule, which states that the phase in which the surfactant is most soluble forms the continuous phase. If the surfactant is in the droplets, a γ-gradient cannot develop and the drops would be prone to coalescence. Thus, surfactants with a hydrophilic–lipophilic balance (HLB) number > 7 tend to form O/W emulsions, while those with HLB < 7 tend to form W/O emulsions.

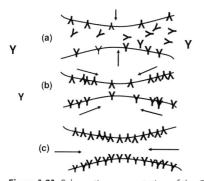

Figure 1.21 Schematic representation of the Gibbs–Marangoni effect for two approaching drops.

The Gibbs–Marangoni effect also explains the difference between surfactants and polymers for emulsification, namely that polymers provide larger drops compared to surfactants. Polymers provide a smaller ε-value at low concentrations when compared to surfactants (see Figure 1.20).

Another factor which should also be considered in relation to emulsification is that of the *disperse phase volume fraction*, ϕ. An increase in ϕ leads to increase in droplet collision and hence coalescence during emulsification. With an increase in ϕ, the viscosity of the emulsion increases and may change the flow from turbulent to laminar (LV regime).

The presence of many particles results in a local increase in velocity gradient, which in turn means that G increases. In turbulent flow, an increase in ϕ will induce turbulence depression and this will result in larger droplets. Turbulence depression by the addition of polymers tends to remove the small eddies, and this results in the formation of larger droplets.

If the mass ratio of surfactant to continuous phase is kept constant, an increase in ϕ results in a decrease in surfactant concentration and hence an increase in γ_{eq}; the result is the formation of larger droplets. However, if the mass ratio of the surfactant to disperse phase is held constant, then the above changes are reversed.

At this point it is impossible to draw any general conclusions as several of the above-mentioned mechanisms may come into play. Experiments using a high-pressure homogenizer at various φ-values at constant initial m_C (with the regime of TI changing to TV at higher φ-values) showed that, with increasing ϕ (>0.1) the resultant droplet diameter increased and the dependence on energy consumption became weaker. A comparison of the average droplet diameter versus power consumption using different emulsifying machines, is shown in Figure 1.22. Here, it can be seen that the smallest droplet diameters were obtained when using the high-pressure homogenizers.

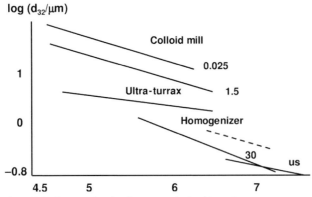

Figure 1.22 Average droplet diameters obtained in various emulsifying machines as a function of energy consumption. Numbers adjacent to the curves denote the viscosity ratio, λ. The solid lines indicate $\phi = 0.04$; the broken line indicates $\phi = 0.3$. us = ultrasonic generator.

1.7
Selection of Emulsifiers

1.7.1
The Hydrophilic-Lipophilic Balance (HLB) Concept

Today, the selection of different surfactants in the preparation of either O/W or W/O emulsions is often made on an empirical basis. One such semi-empirical scale for selecting surfactants is the hydrophilic-lipophilic balance (HLB) number developed by Griffin [18]. This scale is based on the relative percentage of hydrophilic to lipophilic (hydrophobic) groups in the surfactant molecule(s). For an O/W emulsion droplet the hydrophobic chain resides in the oil phase, while the hydrophilic head group resides in the aqueous phase. In contrast, for a W/O emulsion droplet the hydrophilic group(s) reside in the water droplet, whereas the lipophilic groups reside in the hydrocarbon phase.

A guide to the selection of surfactants for particular applications is provided in Table 1.2. Here, the HLB number is seen to depend on the nature of the oil and, as an illustration, the required HLB numbers to emulsify various oils are listed in Table 1.3.

The relative importance of the hydrophilic and lipophilic groups was first recognized when using mixtures of surfactants containing varying proportions of low and high HLB numbers. The efficiency of any combination (as judged by phase separation) was found to pass a maximum when the blend contained a particular proportion of the surfactant with the higher HLB number. This is illustrated in Figure 1.23, which shows the variation of emulsion stability, droplet size and interfacial tension in relation to the percentage of surfactant with a high HLB number.

Table 1.2 A summary of surfactant HLB ranges and their applications.

HLB range	Application
3–6	W/O emulsifier
7–9	Wetting agent
8–18	O/W emulsifier
13–15	Detergent
15–18	Solubilizer

Table 1.3 HLB numbers required for the emulsification of various oils.

Oil	W/O emulsion	O/W emulsion
Paraffin oil	4	10
Beeswax	5	9
Linolin, anhydrous	8	12
Cyclohexane	–	15
Toluene	–	15

The average HLB number may be calculated from additivity:

$$\text{HLB} = x_1 \text{HLB}_1 + x_2 \text{HLB}_2 \tag{1.48}$$

where x_1 and x_2 are the weight fractions of the two surfactants with HLB_1 and HLB_2, respectively.

Griffin developed simple equations for calculating the HLB number of relatively simple nonionic surfactants. For example, in the case of a polyhydroxy fatty acid ester

$$\text{HLB} = 20\left(1 - \frac{S}{A}\right) \tag{1.49}$$

where S is the saponification number of the ester and A is the acid number. For a glyceryl monostearate, $S = 161$ and $A = 198$, and the HLB is 3.8 (suitable for W/O emulsion).

For a simple alcohol ethoxylate, the HLB number can be calculated from the weight percent of ethylene oxide (E) and polyhydric alcohol (P):

$$\text{HLB} = \frac{E + P}{5} \tag{1.50}$$

If the surfactant contains PEO as the only hydrophilic group, then the contribution from one OH group may be neglected:

$$\text{HLB} = \frac{E}{5} \tag{1.51}$$

For a nonionic surfactant $C_{12}H_{25}-O-(CH_2-CH_2-O)_6$, the HLB is 12 (suitable for O/W emulsion).

The above simple equations cannot be used for surfactants containing propylene oxide or butylene oxide; neither can they be applied for ionic surfactants. Davies [19, 20] devised a method for calculating the HLB number for surfactants from their chemical formulae, using empirically determined group numbers. A group number is assigned to various component groups (a summary of the group numbers for some surfactants is shown in Table 1.4).

The HLB is then given by the following empirical equation:

$$\text{HLB} = 7 + \Sigma(\text{hydrophilic group Nos}) - \Sigma(\text{lipophilic group Nos}) \tag{1.52}$$

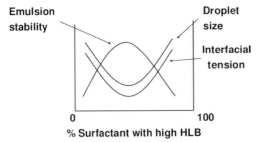

Figure 1.23 Variation of emulsion stability, droplet size and interfacial tension in relation to the percentage of surfactant with a high hydrophilic–lipophilic balance (HLB) number.

Table 1.4 HLB group numbers.

Surfactant type/group	Group number
Hydrophilic	
$-SO_4Na^+$	38.7
$-COO^-$	21.2
$-COONa$	19.1
N (tertiary amine)	9.4
Ester (sorbitan ring)	6.8
$-O-$	1.3
CH–(sorbitan ring)	0.5
Lipophilic	
$(-CH-)$, $(-CH_2-)$, CH_3	0.475
Derived	
$-CH_2-CH_2-O$	0.33
$-CH_2-CH_2-CH_2-O-$	-0.15

Davies has shown that the agreement between HLB numbers calculated from the above-described equation and those determined experimentally is quite satisfactory.

Various other procedures have been developed to obtain a rough estimate of the HLB number. Griffin found a good correlation between the cloud points of 5% solutions of various ethoxylated surfactants and their HLB numbers.

Davies [19, 20] also attempted to relate the HLB values to the selective coalescence rates of emulsions. These correlations were not realized as the emulsion stability and even its type were found to depend to a large extent on the method of dispersing the oil into the water, and *vice versa*. At best, the HLB number can only be used as a guide for selecting the optimum compositions of emulsifying agents.

It is possible to take any pair of emulsifying agents, which fall at opposite ends of the HLB scale, for example Tween 80 (sorbitan mono-oleate with 20 moles EO, HLB = 15) and Span 80 (sorbitan mono-oleate, HLB = 5) and to use them in various proportions to cover a wide range of HLB numbers. The emulsions should be prepared in the same way, with a few percent of the emulsifying blend. The stability of the emulsions is then assessed at each HLB number, either from the rate of coalescence, or qualitatively by measuring the rate of oil separation. In this way it is possible to determine the optimum HLB number for a given oil. Having found the most effective HLB value, various other surfactant pairs are compared at this HLB value, to identify the most effective pair.

1.7.2
The Phase Inversion Temperature (PIT) Concept

Shinoda and coworkers [21, 22] found that many O/W emulsions, when stabilized with nonionic surfactants, undergo a process of inversion at a critical temperature (known as the PIT). The PIT can be determined by following the emulsion conductivity (a small amount of electrolyte is added to increase the sensitivity) as a function of temperature. The conductivity of the O/W emulsion increases with increase of temperature until the PIT is reached, above which there will be a rapid reduction in conductivity (a W/O emulsion is formed). Shinoda and coworkers found that the PIT is influenced by the HLB number of the surfactant. The size of the emulsion droplets was also found to depend on the temperature and HLB number of the emulsifiers, with the droplets being more likely to coalesce when close to the PIT. However, rapid cooling of the emulsion can be used to produce a stable system. Relatively stable O/W emulsions were obtained when the PIT of the system was 20–65 °C higher than the storage temperature. Emulsions prepared at a temperature just below the PIT, followed by rapid cooling, generally have smaller droplet sizes. This is more easily appreciated by considering the change in interfacial tension with temperature, as illustrated in Figure 1.24. The interfacial tension is seen to decrease as the temperature is increased, to reach a minimum close to the PIT, and then to increase again. Thus, droplets prepared close to the PIT are smaller than those prepared at lower temperatures. Such droplets are also relatively unstable towards coalescence near the PIT, although by rapid cooling of the emulsion it is possible to retain the smaller size. This procedure may also be applied in the preparation of mini (nano) emulsions.

The optimum stability of the emulsion was found to be relatively insensitive to changes in the HLB value or the PIT of the emulsifier, although instability was very sensitive to the PIT of the system. It is essential, therefore to measure the PIT of the emulsion as a whole (with all other ingredients).

At a given HLB value, the stability of an emulsion against coalescence increases markedly as the molar mass of both the hydrophilic and lipophilic components increases. The enhanced stability using high-molecular-weight surfactants (polymeric surfactants) may be more easily understood from a consideration of the steric

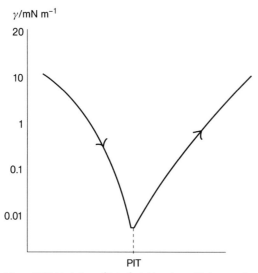

Figure 1.24 Variation of interfacial tension with temperature increase for an O/W emulsion.

repulsion which produces more stable films. Films produced using macromolecular surfactants resist thinning and disruption, thus reducing the possibility of coalescence. The emulsions showed maximum stability when the distribution of the PEO chains was broad; moreover, the cloud point was lower, but the PIT higher, than in the corresponding case for narrow size distributions. Thus, it is clear that the PIT and HLB number are directly related parameters.

The addition of electrolytes reduces the PIT, and hence an emulsifier with a higher PIT value is required when preparing emulsions in the presence of electrolytes. Electrolytes cause dehydration of the PEO chains, and in effect this reduces the cloud point of the nonionic surfactant. It is necessary to compensate for this effect by using a surfactant with a higher HLB. The optimum PIT of the emulsifier is fixed if the storage temperature is fixed.

In view of the above correlation between PIT and HLB, and the possible dependence of the kinetics of droplet coalescence on the HLB number, Sherman and coworkers have suggested the use of PIT measurements as a rapid method for assessing emulsion stability. However, care must be taken when using such methods to assess the long-term stability as the correlations were based on a very limited number of surfactants and oils.

Measurement of the PIT can at best be used as a guide for the preparation of stable emulsions. An assessment of the stability should be made by following the droplet size distribution as a function of time using either a Coulter counter or light-diffraction techniques. An alternative method of assessing stability against coalescence may be to follow the rheology of the emulsion as a function of time and temperature. However, care should be taken when analyzing the rheological data, as coalescence may result in an increase in droplet size which is usually followed by a reduction in the viscosity of the emulsion. This trend is only observed if the coalescence is not accompanied by flocculation of the emulsion droplets (which results in an increase in the viscosity). Ostwald ripening may also complicate the analysis of rheological data.

1.7.3
The Cohesive Energy Ratio (CER) Concept

Beerbower and Hill [23] considered the dispersing tendency on the oil and water interfaces of the surfactant or emulsifier in terms of the ratio of the cohesive energies of the mixtures of oil with the lipophilic portion of the surfactant and the water with the hydrophilic portion. These authors used the Winsor R_o concept, which is the ratio of the intermolecular attraction of oil molecules (O) and the lipophilic portion of surfactant (L), C_{LO}, to that of water (W) and hydrophilic portion (H), C_{HW}

$$R_o = \frac{C_{LO}}{C_{HW}} \quad (1.53)$$

Several interaction parameters may be identified at the oil and water sides of the interface. Typically, at least nine interaction parameters can be identified, as shown schematically in Figure 1.25.

C_{LL}, C_{OO}, C_{LO} (at oil side)
C_{HH}, C_{WW}, C_{HW} (at water side)
C_{LW}, C_{HO}, C_{LH} (at oil interface)

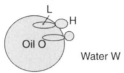

Figure 1.25 The Cohesive Energy Ratio (CER) concept.

In the absence of an emulsifier, there will be only three interaction parameters, namely C_{OO}, C_{WW} and C_{OW}; then, if $C_{OW} \ll C_{WW}$, the emulsion will break.

The above interaction parameters may be related to the Hildebrand solubility parameter [24] δ (at the oil side of the interface) and the Hansen [25] nonpolar, hydrogen bonding and polar contributions to δ at the water side of the interface. The solubility parameter of any component is related to its heat of vaporization ΔH by the expression

$$\delta^2 = \frac{\Delta H - RT}{V_m} \tag{1.54}$$

where V_m is the molar volume.

Hansen considered δ (at the water side of the interface) to consist of three main contributions: a dispersion contribution, δ_d; a polar contribution, δ_p; and a hydrogen bonding contribution, δ_h, each of which have different weighting factors:

$$\delta^2 = \delta_d^2 + 0.25\delta_p^2 + \delta_h^2 \tag{1.55}$$

Beerbower and Hills used the following expression for the HLB number:

$$HLB = 20\left(\frac{M_H}{M_L + M_H}\right) = 20\left(\frac{V_H \rho_H}{V_L \rho_L + V_H \rho_H}\right) \tag{1.56}$$

where M_H and M_L are the molecular weights of the hydrophilic and lipophilic portions of the surfactants, respectively, V_L and V_H are their corresponding molar volumes, respectively, and ρ_H and ρ_L are the densities, respectively.

The cohesive energy ratio was originally defined by Winsor (see Equation 1.53). When $C_{LO} > C_{HW}$, then $R > 1$ and a W/O emulsion forms, but if $C_{LO} < C_{HW}$, then $R < 1$ and an O/W emulsion forms. Yet, if $C_{LO} = C_{HW}$, then $R = 1$ and a planar system results; this denotes the *inversion point*.

R_o can be related to V_L, δ_L and V_H, δ_H by the expression

$$R_o = \frac{V_L \delta_L^2}{V_H \delta_H^2} \tag{1.57}$$

From Equation 1.57,

$$R_o = \frac{V_L(\delta_d^2 + 0.25\delta_p^2 + 0.25\delta_h^2)_L}{V_h(\delta_d^2 + 0.25\delta_p^2 + 0.25\delta_h^2)_H} \tag{1.58}$$

By combining Equations 1.57 and 1.58 one obtains the following general expression for the CER:

$$R_o = \left(\frac{20}{\text{HLB}} - 1\right) \frac{\rho_h(\delta_d^2 + 0.25\delta_p^2 + 0.25\delta_h^2)_L}{\rho_L(\delta_d^2 + 0.25\delta_p^2 + 0.25\delta_p^2)_H} \qquad (1.59)$$

For an O/W system, HLB = 12–15 and R_o = 0.58–0.29 (R_o < 1), while for a W/O system HLB = 5–6 and R_o = 2.3–1.9 (R_o > 1). For a planar system, HLB = 8–10 and R_o = 1.25–0.85 ($R_o \sim 1$).

The R_o equation combines both the HLB and cohesive energy densities, and also provides a more quantitative estimate of emulsifier selection. Moreover, R_o takes into consideration the HLB, molar volume and chemical match, with the success of the approach depending on the availability of data on the solubility parameters of the various surfactant portions. Some values of these parameters are provided elsewhere [26].

1.7.4
The Critical Packing Parameter for Emulsion Selection

The critical packing parameter (CPP) is a geometric expression which relates the hydrocarbon chain volume (v) and length (l), and also the interfacial area occupied by the head group (a) [27]:

$$\text{CPP} = \frac{v}{l_c a_o} \qquad (1.60)$$

where a_o is the optimal surface are per head group and l_c is the critical chain length.

Regardless of the shape of any aggregated structure (spherical or cylindrical micelle, or a bilayer), no point within the structure can be farther from the hydrocarbon–water surface than l_c. The critical chain length, l_c, is roughly equal but less than the fully extended length of the alkyl chain.

The above concept can be applied to predict the shape of an aggregated structure. If we consider a spherical micelle with radius r and aggregation number n, then the volume of the micelle is given by

$$\left(\frac{4}{3}\right)\pi r^3 = nv \qquad (1.61)$$

where v is the volume of a surfactant molecule.

The area of the micelle is then given by

$$4\pi r^2 = n a_o \qquad (1.62)$$

where a_o is the area per surfactant head group. By combining Equations 1.61 and 1.62 we obtain

$$a_o = \frac{3v}{r} \qquad (1.63)$$

where the cross-sectional area of the hydrocarbon chain, a, is given by the ratio of its volume to its extended length l_c:

$$a = \frac{v}{l_c} \quad (1.64)$$

From Equations 1.63 and 1.64,

$$\text{CPP} = \frac{a}{a_o} = \left(\frac{1}{3}\right)\left(\frac{r}{l_c}\right) \quad (1.65)$$

Since $r < l_c$, then CPP $\leq (1/3)$.

For a cylindrical micelle with length d and radius r,

$$\text{Volume of the micelle} = \pi \cdot r^2 \cdot d = nv \quad (1.66)$$

$$\text{Area of the micelle} = 2\pi \cdot r \cdot d = na_o \quad (1.67)$$

By combining Equations 1.66 and 1.67,

$$a_o = \frac{2v}{r} \quad (1.68)$$

$$a = \frac{v}{l_c} \quad (1.69)$$

$$\text{CPP} = \frac{a}{a_o} = \left(\frac{1}{2}\right)\left(\frac{r}{l_c}\right) \quad (1.70)$$

Since $r < l_c$, then $(1/3) < \text{CPP} \leq (1/2)$. For vesicles (liposomes), $1 > \text{CPP} \geq (2/3)$, while for lamellar micelles $P \sim 1$.

Surfactants that make spherical micelles with the above packing constraints [i.e. CPP $\leq (1/3)$] are more suitable for O/W emulsions, whereas surfactants with CPP > 1 (i.e. forming inverted micelles) are suitable for the formation of W/O emulsions.

1.8
Creaming or Sedimentation of Emulsions

This is the result of gravity, when the density of the droplets and the medium are not equal. When the density of the disperse phase is less than that of the medium then creaming will occur; however, if the density of the disperse phase is greater than that of the medium then sedimentation will occur. Figure 1.24 gives A schematic representation of the creaming of emulsions for these three cases is shown in Figure 1.26 [28].

In Figure 1.26, case (a) represents the situation for small droplets (<0.1 μm, i.e. nanoemulsions) whereby the Brownian diffusion kT (where k is the Boltzmann

1.8 Creaming or Sedimentation of Emulsions

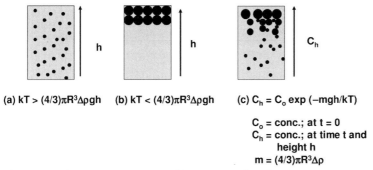

Figure 1.26 Schematic representation of the creaming of emulsions.

constant and T is the absolute temperature) exceeds the force of gravity (mass × acceleration due to gravity, g):

$$kT \gg \frac{4}{3}\pi R^3 \Delta\rho g L \tag{1.71}$$

where R is the droplet radius, $\Delta\rho$ is the density difference between the droplets and the medium, and L is the height of the container.

Case (b) in Figure 1.26 represents emulsions consisting of 'monodisperse' droplets with radius >1 μm. In this case, the emulsion separates into two distinct layers, with the droplets forming a cream or sediment and leaving the clear supernatant liquid. However, this situation is seldom observed in practice.

Case (c) in Figure 1.26 is that for a polydisperse (practical) emulsion, where the droplets will cream or sediment at various rates. In the latter case a concentration gradient build-up will occur, with the larger droplets staying either at the top of the cream layer or at the bottom of the sediment:

$$C(h) = C_o \exp\left(-\frac{mgh}{kT}\right) \tag{1.72}$$

where $C(h)$ is the concentration (or volume fraction ϕ) of the droplets at height h, while C_o is the concentration at zero time, which is the same at all heights.

1.8.1
Creaming or Sedimentation Rates

Here, three situations must be taken into account.

Very Dilute Emulsions ($\phi < 0.01$) In this case the rate could be calculated using Stokes' law, which balances the hydrodynamic force with gravity force

$$\text{Hydrodynamic force} = 6\pi\eta_o R v_o \tag{1.73}$$

$$\text{Gravity force} = \frac{4}{3}\pi R^3 \Delta\rho g \tag{1.74}$$

$$v_o = \frac{2}{9}\frac{\Delta \rho g R^2}{\eta_o} \tag{1.75}$$

where v_o is the Stokes' velocity and η_o is the viscosity of the medium.

For an O/W emulsion with $\Delta \rho = 0.2$ in water ($\eta_o \sim 10^{-3}$ Pa·s), the rate of creaming or sedimentation is $\sim 4.4 \times 10^{-5}$ ms^{-1} for 10 μm droplets, and $\sim 4.4 \times 10^{-7}$ ms^{-1} for 1 μm droplets. This means that, in a 0.1 m container, creaming or sedimentation of the 10 μm droplets is complete in ~ 0.6 h, but for the 1 μm droplets this takes ~ 60 h.

Moderately Concentrated Emulsions ($0.2 < \phi < 0.1$)   In this case one must take into account the hydrodynamic interaction between the droplets, which reduces the Stokes' velocity (v) to a value given by the following expression [29]:

$$v = v_o(1-k\phi) \tag{1.76}$$

where k is a constant that accounts for hydrodynamic interaction, and is of the order of 6.5; this means that the rate of creaming or sedimentation is reduced by about 65%.

Concentrated Emulsions ($\phi > 0.2$)   The rate of creaming or sedimentation becomes a complex function of ϕ, as illustrated in Figure 1.27, which also shows the change of relative viscosity η_r with ϕ. As can also be seen from Figure 1.27, v decreases with increase in ϕ and ultimately approaches zero when ϕ exceeds a critical value, ϕ_p, which is the so-called 'maximum packing fraction'. The value of ϕ_p for monodisperse 'hard-spheres' ranges from 0.64 (for random packing) to 0.74 for hexagonal packing, and exceeds 0.74 for polydisperse systems. Also, for emulsions which are deformable, ϕ_p can be much larger than 0.74.

The data in Figure 1.27 also show that when ϕ approaches ϕ_p, η_r approaches ∞. In practice, most emulsions are prepared at ϕ values well below ϕ_p, usually in the range 0.2–0.5, and under these conditions creaming or sedimentation is the rule rather than the exception. Several procedures may be applied to reduce or eliminate creaming or sedimentation, and these are discussed below.

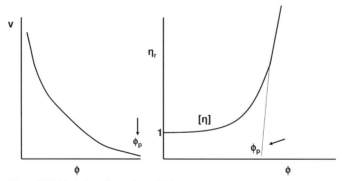

Figure 1.27 Variation of v and η_r with ϕ.

1.8.2
Prevention of Creaming or Sedimentation

Matching Density of Oil and Aqueous Phases Clearly, if $\Delta\rho = 0$, then $v = 0$; however, this method is seldom practical as density matching (if possible) occurs at only one temperature.

Reduction of Droplet Size As the gravity force is proportional to R^3, then if R is reduced by a factor of 10 the gravity force is reduced by 1000. Below a certain droplet size (which also depends on the density difference between oil and water), the Brownian diffusion may exceed gravity, and creaming or sedimentation is prevented. This is the principle of the formulation of nanoemulsions (with size range 20–200 nm) that may show very little or no creaming or sedimentation. The same applies for microemulsions (size range 5–50 nm).

Use of 'Thickeners' These are high-molecular-weight polymers, and may be either natural or synthetic; examples include xanthan gum, hydroxyethyl cellulose, alginates and carragenans. In order to understand the role of these 'thickeners', we should first consider the gravitational stresses exerted during creaming or sedimentation:

$$\text{Stress} = \text{mass of drop} \times \text{acceleration of gravity} = \frac{4}{3}\pi R^3 \Delta\rho g \quad (1.77)$$

In order to overcome such stress, a restoring force is needed:

$$\text{Restoring force} = \text{area of drop} \times \text{stress of drop} = 4\pi R^3 \sigma_p \quad (1.78)$$

Thus, the stress exerted by the droplet σ_p is given by

$$\sigma_p = \frac{\Delta\rho R g}{3} \quad (1.79)$$

Simple calculation shows that σ_p is in the range of 10^{-3} to 10^{-1} Pa, which implies that for the prediction of creaming or sedimentation there is a need to measure the viscosity at such low stresses. This can be achieved by using constant stress or creep measurements.

The above-described 'thickeners' satisfy the criteria for obtaining very high viscosities at low stresses or shear rates; this can be illustrated from plots of shear stress σ and viscosity η versus shear rate $\dot{\gamma}$ (or shear stress), as shown in Figure 1.28. These systems are described as either 'pseudoplastic' or 'shear thinning'. The low-shear (residual or zero shear rate) viscosity $\eta_{(o)}$ can reach several thousand Pas, and such high values prevent either creaming or sedimentation [30, 31].

The behavior described in Figure 1.28 is obtained above a critical polymer concentration (C^*), which can be located from plots of log η versus log C, as shown in Figure 1.29.

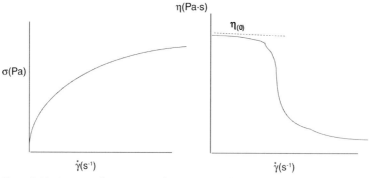

Figure 1.28 Variation of (stress) σ and viscosity η with shear rate $\dot{\gamma}$.

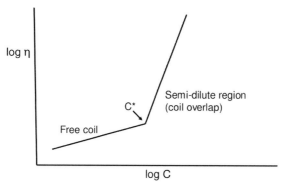

Figure 1.29 Variation of log η with log C for polymer solutions.

Below C^* the log η–log C curve has a slope in the region of 1, whereas above C^* the slope of the line exceeds 3. In most cases a good correlation between the rate of creaming or sedimentation and $\eta_{(o)}$ is obtained.

Controlled Flocculation As discussed earlier, the total energy-distance of the separation curve for electrostatically stabilized shows a shallow minimum (secondary minimum) at relatively long distance of separation between the droplets. By adding small amounts of electrolyte, such a minimum can be made sufficiently deep for weak flocculation to occur. The same applies to sterically stabilized emulsions, which show only one minimum, the depth of which can be controlled by reducing the thickness of the adsorbed layer. This can be achieved by reducing the molecular weight of the stabilizer and/or the addition of a nonsolvent for the chains (e.g. electrolyte).

The above phenomenon of weak flocculation may be applied to reduce creaming or sedimentation, although in practice this is not easy as there is also a need to control the droplet size.

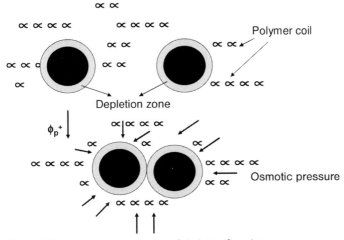

Figure 1.30 Schematic representation of depletion flocculation.

Depletion Flocculation This is achieved by the addition of 'free' (nonadsorbing) polymer in the continuous phase [32]. At a critical concentration, or volume fraction of free polymer, ϕ_p^+, weak flocculation occurs, as the free polymer coils become 'squeezed-out' from between the droplets. This is illustrated in Figure 1.30, which shows the situation when the polymer volume fraction exceeds the critical concentration. Here, the osmotic pressure outside the droplets is higher than in between the droplets, and this results in an attraction the magnitude of which depends on the concentration of the free polymer and its molecular weight, as well as on the droplet size and ϕ. The value of ϕ_p^+ decreases not only with an increase in the molecular weight of the free polymer, but also as the volume fraction of the emulsion increases.

The weak flocculation shown in Figure 1.30 can be applied to reduce creaming or sedimentation, although it suffers from the following drawbacks:

- Temperature dependence: as the temperature increases, the hydrodynamic radius of the free polymer decreases (due to dehydration) and hence more polymer will be required to achieve the same effect at lower temperatures.
- If the free polymer concentration is increased above a certain limit, phase separation may occur and the flocculated emulsion droplets may cream or sediment faster than in the absence of the free polymer.

1.9
Flocculation of Emulsions

Flocculation is the result of van der Waals attraction that is universal for all disperse systems. The van der Waals attraction G_A was described earlier (see Figure 1.5), and is inversely proportional to the droplet–droplet distance of separation h; it also depends on the effective Hamaker constant A of the emulsion system. The primary way of overcoming the van der Waals attraction is by electrostatic stabilization using ionic surfactants; this results in the formation of electrical double layers that introduce a repulsive energy which overcomes the attractive energy. Emulsions stabilized by electrostatic repulsion become flocculated at intermediate electrolyte concentrations (see below). A second – and most effective – method of overcoming flocculation is by 'steric stabilization', using nonionic surfactants or polymers. Here, stability may be maintained in electrolyte solutions (as high as 1 mol dm^{-3}, depending on the nature of the electrolyte) and up to high temperatures (in excess of 50 °C), provided that the stabilizing chains (e.g. PEO) are still under better than θ-conditions ($\chi < 0.5$).

1.9.1
Mechanism of Emulsion Flocculation

This can occur if the energy barrier is small or absent (for electrostatically stabilized emulsions), or when the stabilizing chains reach poor solvency (for sterically stabilized emulsions, $\chi > 0.5$). For convenience, the flocculation of electrostatically and sterically stabilized emulsions will be discussed separately.

1.9.1.1 Flocculation of Electrostatically Stabilized Emulsions
As discussed earlier, the condition for kinetic stability is $G_{max} > 25\,kT$, but when $G_{max} < 5\,kT$, then flocculation will occur. Two types of flocculation kinetics may be distinguished: (i) fast flocculation with no energy barrier; and (ii) slow flocculation, when an energy barrier exists.

Fast Flocculation Kinetics These were treated by Smoluchowski [33], who considered the process to be represented by second-order kinetics and the process to be simply diffusion-controlled. The number of particles n at any time t may be related to the final number (at $t = 0$), n_o, by the following expression:

$$n = \frac{n_o}{1 + k n_o t} \tag{1.80}$$

where k is the rate constant for fast flocculation that is related to the diffusion coefficient of the particles D. That is:

$$k = 8\pi D R \tag{1.81}$$

where D is given by the Stokes–Einstein equation

$$D = \frac{kT}{6\pi \eta R} \tag{1.82}$$

By combining Equations 1.81 and 1.82, we get

$$k = \frac{4\,kT}{3\,\eta} = 5.5 \times 10^{-18}\,\text{m}^3\,\text{s}^{-1} \text{ for water at } 25\,°C \tag{1.83}$$

Table 1.5 Half-life values of emulsion flocculation.

R (μm)	ϕ-Value			
	10^{-5}	10^{-2}	10^{-1}	5×10^{-1}
0.1	765 s	76 ms	7.6 ms	1.5 ms
1.0	21 h	76 s	7.6 s	1.5 s
10.0	4 months	21 h	2 h	25 min

The half life $t_{1/2}$ [$n = (1/2) n_o$] can be calculated at various n_o or volume fractions ϕ, as shown in Table 1.5.

Slow Flocculation Kinetics These were treated by Fuchs [34], who related the rate constant k to the Smoluchowski rate by the stability constant W:

$$W = \frac{k_o}{k} \tag{1.84}$$

W is related to G_{max} by the following expression [35]:

$$W = \frac{1}{2} \exp\left(\frac{G_{max}}{kT}\right) \tag{1.85}$$

As G_{max} is determined by the salt concentration C and valency, it is possible to derive an expression relating W to C and Z:

$$\log W = -2.06 \times 10^9 \left(\frac{R\gamma^2}{Z^2}\right) \log C \tag{1.86}$$

where γ is a function that is determined by the surface potential ψ_o,

$$\gamma = \left[\frac{\exp(Ze\psi_o/kT) - 1}{\exp(ZE\psi_o/kT) + 1}\right] \tag{1.87}$$

Plots of log W versus log C are shown in Figure 1.31. The condition log $W = 0$ ($W = 1$) is the onset of fast flocculation, and the electrolyte concentration at this point

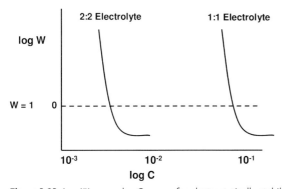

Figure 1.31 Log W versus log C curves for electrostatically stabilized emulsions.

defines the critical flocculation concentration (CFC). Above the CFC, $W<1$ (due to the contribution of van der Waals attraction which accelerates the rate above the Smoluchowski value). Below the CFC, $W>1$ and continues to increase with a decrease of the electrolyte concentration. The data in Figure 1.31 also show that the CFC decreases with increase of valency, in accordance to the Schultze–Hardy rule.

Another mechanism of flocculation is that involving the secondary minimum (G_{min}) which is few kT units. In this case flocculation is weak and reversible, and hence one must consider both the rate of flocculation (forward rate k_f) and deflocculation (backward rate k_b). The rate of decrease of particle number with time is given by the expression

$$-\frac{dn}{dt} = -k_f n^2 + k_b n \qquad (1.88)$$

The backward reaction (break-up of weak flocs) reduces the overall rate of flocculation.

1.9.1.2 Flocculation of Sterically Stabilized Emulsions

This occurs when the solvency of the medium for the chain becomes worse than for a θ-solvent ($\chi > 0.5$). Under these conditions, G_{mix} becomes negative (i.e. attractive) and a deep minimum is produced which results in catastrophic flocculation (referred to as *incipient flocculation*). This situation is shown schematically in Figure 1.32.

With many systems a good correlation between the flocculation point and the θ point is obtained. For example, the emulsion will flocculate at a temperature which is referred to as the *critical flocculation temperature* (CFT) that is equal to the θ-temperature of the stabilizing chain. The emulsion may flocculate at a critical volume fraction of a nonsolvent critical flotation volume (CFV), which is equal to the volume of nonsolvent that brings it to a θ-solvent.

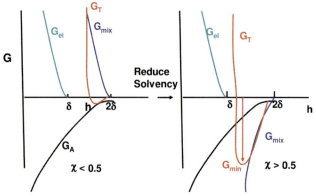

Figure 1.32 Schematic representation of flocculation of sterically stabilized emulsions.

1.9.2
General Rules for Reducing (Eliminating) Flocculation

The rules to be applied for the reduction and/or elimination of flocculation depend on whether charge-stabilized or sterically stabilized emulsions are present.

Charge-Stabilized Emulsions For charge-stabilized emulsions (e.g. using ionic surfactants) the most important criterion is to make G_{max} as high as possible; this is achieved in three main ways: (i) with a high surface or zeta potential; (ii) with a low electrolyte concentration; and (iii) with a low valency of ions.

Sterically Stabilized Emulsions In the case of sterically stabilized emulsions, four main criteria are necessary:

- Complete coverage of the droplets by the stabilizing chains.

- Firm attachment (strong anchoring) of the chains to the droplets. This requires the chains to be insoluble in the medium and soluble in the oil, but it is incompatible with stabilization which requires a chain to be soluble in the medium yet strongly solvated by its molecules. These conflicting requirements are solved by using A–B, A–B–A block or BA_n graft copolymers (where B is the 'anchor' chain and A is the stabilizing chain(s)). Examples of B chains for O/W emulsions are polystyrene, polymethylmethacrylate, polypropylene oxide and alkyl polypropylene oxide. For the A chain(s), PEO or polyvinyl alcohol are good examples. For W/O emulsions, PEO can form the B chain, whereas the A chain(s) may be polyhydroxy stearic acid (PHS), which is strongly solvated by most oils.

- Thick adsorbed layers; the adsorbed layer thickness should be in the region of 5–10 nm. This means that the molecular weight of the stabilizing chains could be in the region of 1000–5000 Da.

- The stabilizing chain should be maintained in good solvent conditions ($\chi < 0.5$), under all conditions of temperature changes on storage.

1.10
Ostwald Ripening

The driving force for Ostwald ripening is the difference in solubility between the small and large droplets (the smaller droplets have higher Laplace pressure and higher solubility than their larger counterparts). This is illustrated in Figure 1.33, where R_1 decreases and R_2 increases as a result of diffusion of molecules from the smaller to the larger droplets.

The difference in chemical potential between different-sized droplets was first proposed by Lord Kelvin [36]:

$$S(r) = S(\infty)\exp\left(\frac{2\gamma V_m}{rRT}\right) \tag{1.89}$$

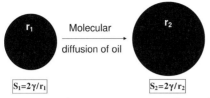

Figure 1.33 Schematic representation of Ostwald ripening.

where $S(r)$ is the solubility surrounding a particle of radius r, $S(\infty)$ is the bulk solubility, V_m is the molar volume of the dispersed phase, R is the gas constant and T is the absolute temperature. The quantity $(2\gamma V_m/RT)$ is termed the *characteristic length*, and has an order of $\sim$1 nm or less, indicating that the difference in solubility of a 1 μm droplet is of the order of 0.1%, or less. In theory, Ostwald ripening should lead to the condensation of all droplets into a single drop, but this does not occur in practice as the rate of growth decreases with an increase of droplet size.

For two droplets with radii r_1 and r_2 ($r_1 < r_2$),

$$\frac{RT}{V_m}\ln\left[\frac{S(r_1)}{S(r_2)}\right] = 2\gamma\left[\frac{1}{r_1} - \frac{1}{r_2}\right] \tag{1.90}$$

From Equation (1.90) it can be shown that the larger the difference between r_1 and r_2, the higher the rate of Ostwald ripening.

Ostwald ripening can be quantitatively assessed from plots of the cube of the radius versus time t [37, 38]:

$$r^3 = \frac{8}{9}\left[\frac{S(\infty)\gamma V_m D}{\rho RT}\right]t \tag{1.91}$$

where D is the diffusion coefficient of the disperse phase in the continuous phase and ρ is the density of the disperse phase.

Several methods may be applied to reduce Ostwald ripening [39–41], including:

- The addition of a second disperse phase component which is insoluble in the continuous medium (e.g. squalane). In this case, partitioning between different droplet sizes occurs, with the component having low solubility expected to be concentrated in the smaller droplets. During Ostwald ripening in a two-component system, equilibrium is established when the difference in chemical potential between different size droplets (which results from curvature effects) is balanced by the difference in chemical potential resulting from partitioning of the two components. This effect reduces further growth of droplets.

- Modification of the interfacial film at the O/W interface. According to Equation 1.91, a reduction in γ results in a reduction of the Ostwald ripening rate. By using surfactants that are strongly adsorbed at the O/W interface (i.e. polymeric surfactants) and which do not desorb during ripening (by choosing a molecule that is insoluble in the continuous phase), the rate could be significantly reduced. An increase in the surface dilational modulus ε ($=d\gamma/d\ln A$) and a decrease in γ would be observed for the shrinking drop, and this tends to reduce further growth.

- A–B–A block copolymers such as PHS-PEO-PHS (which is soluble in the oil droplets but insoluble in water) can be used to achieve the above effect. Similar effects can also be obtained using a graft copolymer of hydrophobically modified inulin, namely INUTEC®SP1 (ORAFTI, Belgium). This polymeric surfactant adsorbs with several alkyl chains (which may dissolve in the oil phase) to leave loops and tails of strongly hydrated inulin (polyfructose) chains. The molecule has limited solubility in water and hence it resides at the O/W interface. These polymeric emulsifiers enhance the Gibbs elasticity, thus significantly reducing the Ostwald ripening rate.

1.11
Emulsion Coalescence

When two emulsion droplets come in close contact in a floc or creamed layer, or during Brownian diffusion, a thinning and disruption of the liquid film may occur that results in eventual rupture. On close approach of the droplets, film thickness fluctuations may occur; alternatively, the liquid surfaces undergo some fluctuations forming surface waves, as illustrated in Figure 1.34.

Figure 1.34 Schematic representation of surface fluctuations.

The surface waves may grow in amplitude and the apices may join as a result of the strong van der Waals attraction (at the apex, the film thickness is the smallest). The same applies if the film thins to a small value (critical thickness for coalescence).

A very useful concept was introduced by Deryaguin and Scherbaker [42], who suggested that a 'disjoining pressure' $\pi(h)$ is produced in the film which balances the excess normal pressure,

$$\pi(h) = P(h) - P_o \tag{1.92}$$

where $P(h)$ is the pressure of a film with thickness h, and P_o is the pressure of a sufficiently thick film such that the net interaction free energy is zero.

$\pi(h)$ may be equated to the net force (or energy) per unit area acting across the film:

$$\pi(h) = -\frac{dG_T}{dh} \tag{1.93}$$

where G_T is the total interaction energy in the film.

$\pi(h)$ is composed of three contributions due to electrostatic repulsion (π_E), steric repulsion (π_s) and van der Waals attraction (π_A):

$$\pi(h) = \pi_E + \pi_s + \pi_A \tag{1.94}$$

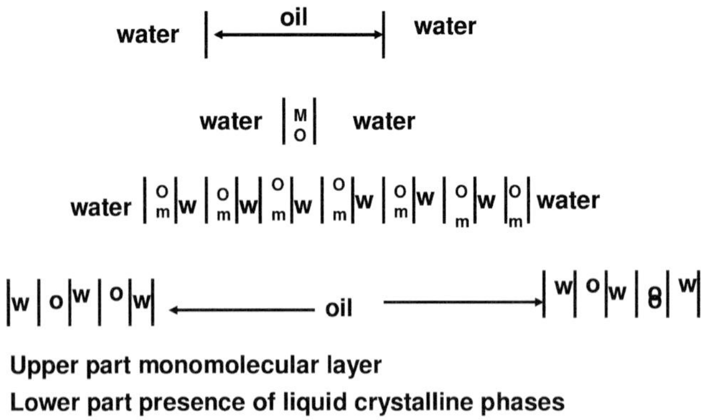

1 Emulsion Science and Technology: A General Introduction

Upper part monomolecular layer
Lower part presence of liquid crystalline phases

Figure 1.35 Schematic representation of the role of liquid crystalline phases.

In order to produce a stable film $\pi_E + \pi_s > \pi_A$, and this is the driving force for the prevention of coalescence that may be achieved by either of two mechanisms (and/or their combination):

- Increased repulsion, both electrostatic and steric.
- Dampening of the fluctuation by enhancing the Gibbs elasticity. In general, smaller droplets are less susceptible to surface fluctuations and hence coalescence is reduced. This explains the high stability of nanoemulsions.

Several methods may be applied to achieve the above effects:

- The use of mixed surfactant films: In many cases the use of mixed surfactants (e.g. anionic and nonionic or long-chain alcohols) can reduce coalescence as a result of various effects such as high Gibbs elasticity, high surface viscosity, and hindered diffusion of the surfactant molecules from the film.
- The formation of lamellar liquid crystalline phases at the O/W interface: This mechanism was suggested by Friberg and coworkers [43], who proposed that the surfactant or mixed surfactant film could produce several bilayers that 'wrapped' the droplets. As a result of these multilayer structures, the potential drop is shifted to longer distances, thus reducing the van der Waals attraction. A schematic representation of the role of liquid crystals is shown in Figure 1.35, which shows the difference between having a monomolecular layer and a multilayer, as is the case with liquid crystals. In order for coalescence to occur, these multilayers must be removed 'two-by-two', and this forms an energy barrier that prevents coalescence.

1.11.1
Rate of Coalescence

Since film drainage and rupture is a kinetic process, coalescence is also a kinetic process. If the number of particles n (flocculated or not) is measured at time t, then

$$n = n_t + n_v m \tag{1.95}$$

where n_t is the number of primary particles remaining, and n is the number of aggregates consisting of m separate particles. When studying emulsion coalescence, it is important to consider the rate constant of flocculation and coalescence; if coalescence is the dominant factor, then the rate K follows a first-order kinetics:

$$n = \frac{n_o}{Kt}[1 + \exp(-Kt)] \tag{1.96}$$

Hence, a plot of log n versus t should give a straight line from which K can be calculated.

1.11.2
Phase Inversion

The phase inversion of emulsions can be one of two types: (i) Transitional inversion, which is induced by changing the facers which affect the HLB of the system (e.g. temperature and/or electrolyte concentration); and (ii) catastrophic inversion, which is induced by increasing the volume fraction of the disperse phase.

Catastrophic inversion is illustrated graphically in Figure 1.36, which shows the variation of viscosity and conductivity with the oil volume fraction ϕ. It can be seen that inversion occurs at a critical ϕ, which may be identified with the maximum packing fraction. At ϕ_{cr}, η suddenly decreases, such that the inverted W/O emulsion has a much lower volume fraction. κ also decreases sharply at the inversion point as the continuous phase is now oil.

Earlier theories of phase inversion were based on packing parameters. When ϕ exceeds the maximum packing ($\sim$0.64 for random packing and $\sim$0.74 for hexagonal packing of monodisperse spheres; for polydisperse systems, the maximum packing exceeds 0.74), inversion occurs. However, these theories are not adequate, as many emulsions invert at φ-values well below the maximum packing as a result of changes in surfactant characteristics with variation of conditions. For example, when using a

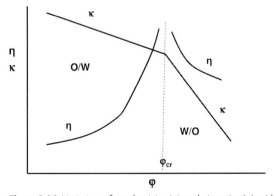

Figure 1.36 Variation of conductivity (κ) and viscosity (η) with the volume fraction of oil, ϕ.

nonionic surfactant based on PEO, the latter chain changes its solvation by an increase of temperature and/or addition of electrolyte. Many emulsions show phase inversion at a critical temperature (the phase inversion temperature) that depends on the HLB number of the surfactant as well as the presence of electrolytes. By increasing the temperature and/or addition of electrolyte, the PEO chains become dehydrated and finally more soluble in the oil phase. Under these conditions, the O/W emulsion will invert to a W/O emulsion. The above dehydration effect amounts to a decrease in the HLB number, and when the latter reaches a value that is more suitable for W/O emulsion, then inversion will occur. At present, there is no quantitative theory that accounts for the phase inversion of emulsions.

1.12
Rheology of Emulsions

Although the rheology of emulsions has many similar features to that of suspensions, there are three main differences in aspect:

- The mobile liquid/liquid interface that contains surfactant or polymer layers introduces a response to deformation; hence, the interfacial rheology must be considered.
- The dispersed phase viscosity relative to that of the medium has an effect on the rheology of the emulsion.
- The deformable nature of the disperse phase droplets, particularly for large droplets, has an effect on the emulsion rheology at high phase volume fraction ϕ.

When the above factors are considered, the bulk rheology of emulsions can be treated in a similar manner as for suspensions, and the same techniques applied.

1.12.1
Interfacial Rheology

A fluid interface in equilibrium exhibits an intrinsic state of tension that is characterized by its interfacial tension γ which is given by the change in free energy with area of the interface, at constant composition n_i and temperature T,

$$\gamma = \left(\frac{\partial G}{\partial A}\right)_{n_i, T} \tag{1.97}$$

The unit for γ is energy per unit area (mJ m^{-1}) or force per unit length (mN m^{-1}), which are dimensionally equivalent.

The adsorption of surfactants or polymers lowers the interfacial tension, and this produces a two-dimensional surface pressure π that is given by

$$\pi = \gamma_o - \gamma \tag{1.98}$$

where γ_o is the interfacial tension of the 'clean' interface (before adsorption) and γ that after adsorption.

The interface is considered to be a macroscopically planar, dynamic fluid interface. Thus, the interface is regarded as a two-dimensional entity independent of the surrounding three-dimensional fluid. The interface is considered to correspond to a highly viscous insoluble monolayer and the interfacial stress σ_s acting within such a monolayer is sufficiently large compared to the bulk-fluid stress acting across the interface. In this way it is possible to define an interfacial shear viscosity η_s,

$$\sigma_s = \eta_s \gamma \tag{1.99}$$

where γ is the shear rate, η_s is given in surface Pa·s (Nm^{-1}s) or surface poise (dyne cm^{-1}s). At this point it should be noted that the surface viscosity of a surfactant-free interface is negligible and can reach high values for adsorbed rigid molecules such as proteins.

1.12.2
Measurement of Interfacial Viscosity

Many surface viscometers utilize torsional stress measurements upon rotating a ring, disk or knife edge (shown schematically in Figure 1.37) within or near to the liquid/liquid interface [44]. This type of viscometer is moderately sensitive. For a disk viscometer the interfacial shear viscosity can be measured in the range $\eta_s \geq 10^{-2}$ surface Pa·s. The disk is rotated within the plane of the interface with angular velocity ω. A torque is then exerted on the disk of radius R by both the surfactant film with surface viscosity η_s and the viscous liquid (with bulk viscosity η) that is given by the expression

$$M = (8/3)R^3\eta\omega + 4\pi R^2\eta_s\omega \tag{1.100}$$

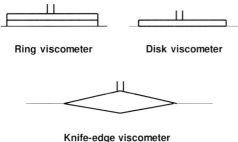

Figure 1.37 Schematic representation of surface viscometers.

1.12.3
Interfacial Dilational Elasticity

The interfacial dilational (Gibbs) elasticity ε, which is an important parameter in determining emulsion stability (reduction of coalescence during formation), is given by the equation:

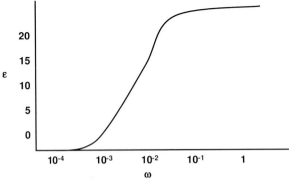

Figure 1.38 Gibbs dilational elasticity versus frequency.

$$\varepsilon = \frac{d\gamma}{d \ln A} \qquad (1.101)$$

where $d\gamma$ is the change in interfacial tension during expansion of the interface by an amount dA (referred to as interfacial tension gradient resulting from nonuniform surfactant adsorption on expansion of the interface).

One of the most convenient methods for measuring ε is to use a Langmuir trough with two moving barriers for expansion and compression of the interface. Another method is to use the oscillating bubble technique, for which instruments are commercially available. A further useful technique for measuring ε is the 'pulsed drop method'. Here, a rapid expansion of a droplet at the end of a capillary from radius r_1 to r_2 is obtained by the application of pressure. The pressure drop within the droplet is measured as a function of time using a sensitive pressure transducer, and from this it is possible to obtain the interfacial tension as a function of time. The Gibbs dilational elasticity is determined from values of the time-dependent interfacial tension. Measurement can be made as a function of frequency, as shown in Figure 1.38 for stearic acid at the decane–water interface at pH 2.5.

1.12.4
Interfacial Dilational Viscosity

Measurement of the dilational viscosity is more difficult than that of the interfacial shear viscosity, due mainly to the coupling between dilational viscous and elastic components. The most convenient method for measuring dilational viscosity is the maximum bubble pressure technique, that can be only applied at the air/water interface [45]. According to this technique, the pressure drop across the bubble surface at the instant when the bubble possesses a hemispherical shape (corresponding to the maximum pressure) is due to a combination of bulk viscous, surface tension and surface dilational viscosity effects; this allows the interfacial dilational viscosity to be obtained.

1.12.5
Non-Newtonian Effects

Most adsorbed surfactant and polymer coils at the O/W interface show non-Newtonian rheological behavior. The surface shear viscosity η_s depends on the applied shear rate, showing shear thinning at high shear rates. Some films also show Bingham plastic behavior with a measurable yield stress. As many adsorbed polymers and proteins demonstrate viscoelastic behavior, it is possible to measure viscous and elastic components using sinusoidally oscillating surface dilation. For example, the complex dilational modulus ε^* which is obtained can be split into 'in-phase' (the elastic component ε') and 'out-of-phase' (the viscous component ε'') components. Creep and stress relaxation methods can be applied to study viscoelasticity.

1.12.6
Correlation of Interfacial Rheology with Emulsion Stability

1.12.6.1 Mixed Surfactant Films

Prince et al. [46] found that emulsions prepared using a mixture of SDS and dodecyl alcohol are more stable than those prepared using SDS alone. This enhanced stability is due to the higher interfacial dilational elasticity ε for the mixture when compared to that of SDS alone. Interfacial dilational viscosity did not play a major role as the emulsions are stable at high temperature, whereby the interfacial viscosity becomes lower.

The above correlation is not general for all surfactant films, since other factors such as thinning of the film between emulsion droplets (which depends on other factors such as repulsive forces) can also play a major role.

1.12.6.2 Protein Films

Biswas and Haydon [47] identified some correlation between the viscoelastic properties of protein (albumin or arabinic acid) films at the O/W interface and the stability of emulsion drops against coalescence. Viscoelastic measurements were carried out using creep and stress relaxation measurements (using a specially designed interfacial rheometer). A constant torque or stress σ (mN m^{-1}) was applied and the deformation γ measured as a function of time for 30 min. After this period the torque was removed and γ (which changes sign) was measured as a function of time to obtain the recovery curve. The results are illustrated graphically in Figure 1.39.

From the creep curves it is possible to obtain the instantaneous modulus G_o ($\sigma/\gamma_{inst.}$) and the surface viscosity η_s from the slope of the straight line (which gives the shear rate) and the applied stress. G_o and η_s are plotted versus pH, as shown in Figure 1.40. Both show an increased creep with increased pH, reaching a maximum at pH ~6 (the isoelectric point of the protein), when the protein molecules show maximum rigidity at the interface.

The stability of the emulsion was assessed by measuring the residence time (t) of several oil droplets at a planar O/W interface containing the adsorbed protein. Figure 1.40 shows the variation of $t_{1/2}$ (the time taken for half the number of oil

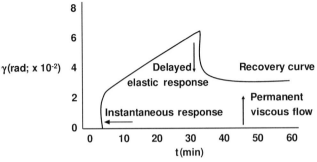

Figure 1.39 Creep curve for a protein film at the oil/water (O/W) interface.

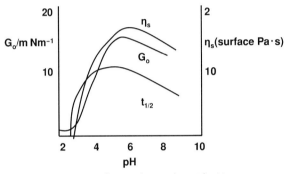

Figure 1.40 Variation of $t_{1/2}$ and G_o and η_s with pH.

droplets to coalesce with the oil at the O/W interface) with pH. A good correlation between $t_{1/2}$ and G_o and η_s was obtained.

Biswas and Haydon subsequently derived a relationship between coalescence time τ and surface viscosity η_s, instantaneous modulus G_o and adsorbed film thickness h:

$$\tau = \eta_s \left[3C' \frac{h^2}{A} - \frac{1}{G_o} - \phi(t) \right] \tag{1.102}$$

where $3C'$ is a critical deformation factor, A is the Hamaker constant and $\phi(t)$ is the elastic deformation per unit stress.

Equation 1.102 shows that τ increases with increase of η_s, but most importantly it is directly proportional to h_2. These results show that viscoelasticity is necessary, but not sufficient, to ensure stability against coalescence. In order to ensure the stability of an emulsion it must be ensured that h is large enough and film drainage is prevented.

1.12.7
Bulk Rheology of Emulsions

For rigid (highly viscous) oil droplets dispersed in a medium of low viscosity such as water, the relative viscosity η_r of a dilute (volume fraction $\phi \leq 0.01$) O/W emulsion of noninteracting droplets behaves as 'hard-spheres' (similar to suspensions).

In the above case, η_r is given by the Einstein equation,

$$\eta_r = 1 + [\eta]\phi \tag{1.103}$$

where $[\eta]$ is the intrinsic viscosity that is equal to 2.5 for hard spheres.

For droplets with low viscosity (comparable to that of the medium), the transmission of tangential stress across the O/W interface, from the continuous phase to the dispersed phase, causes liquid circulation in the droplets. Energy dissipation is less than that for hard spheres, and the relative viscosity is lower than that predicted by the Einstein equation [48–51].

For an emulsion with viscosity η_i for the disperse phase and η_o for the continuous phase

$$[\eta] = 2.5 \left(\frac{\eta_i + 0.4\eta_o}{\eta_i + \eta_o} \right) \tag{1.104}$$

Clearly when $\eta_i \gg \eta_o$, the droplets behave as rigid spheres and $[\eta]$ approaches the Einstein limit of 2.5. In contrast if $\eta_i \ll \eta_o$ (as is the case for foams), $[\eta] = 1$.

In the presence of viscous interfacial layers, Equation 1.104 is modified to take into account the surface shear viscosity η_s and surface dilational viscosity μ_s:

$$[\eta] = 2.5 \left(\frac{\eta_i + 0.4\eta_o + \xi}{\eta_i + \eta_o + \xi} \right) \tag{1.105}$$

$$\xi = \frac{(2\eta_s + 3\mu_s)}{R} \tag{1.106}$$

where R is the droplet radius.

1.12.8
Rheology of Concentrated Emulsions

When the volume fraction of droplets exceed the Einstein limit (i.e. $\phi > 0.01$), it is essential to take into account the effect of Brownian motion and interparticle interactions. The smaller the emulsion droplets, the more important the contribution of Brownian motion and colloidal interactions. Brownian diffusion tends to randomize the position of colloidal particles, leading to the formation of temporary doublets, triplets, and so on. The hydrodynamic interactions are of longer range than the colloidal interactions, and they come into play at relatively low volume fractions ($\phi > 0.01$); this results in an ordering of the particles into layers and tends to destroy the temporary aggregates caused by the Brownian diffusion. This explains the shear thinning behavior of emulsions at high shear rates.

For the volume fraction range $0.01 < \phi < 0.2$, Batchelor derived the following expression for a dispersion of hydrodynamically interacting hard spheres:

$$\eta_r = 1 + 2.5\phi + 6.2\phi^2 + 9\phi^3 \tag{1.107}$$

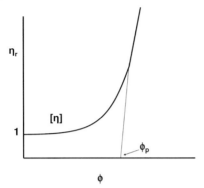

Figure 1.41 The $\eta_r - \phi$ curve.

The second term in Equation 1.107 is the Einstein limit, the third term accounts for the hydrodynamic (two-body) interaction, while the fourth term relates to multibody interaction.

At higher volume fractions ($\phi > 0.2$), η_r is a complex function of ϕ, and the $\eta_r - \phi$ curve is shown in Figure 1.41. This curve is characterized by two asymptotes, namely $[\eta]$ the *intrinsic viscosity* and ϕ_p. A good semi-empirical equation that fits the curve has been provided by Dougherty and Krieger:

$$\eta_r = \left(1 - \frac{\phi}{\phi_p}\right)^{-[\eta]\phi_p} \tag{1.108}$$

Experimental results of $\eta_r - \phi$ curves were obtained for paraffin O/W emulsions stabilized with an A–B–C surfactant consisting of nonyl phenol (B), 13 moles propylene oxide (C) and PEO with 27, 48, 80 and 174 moles EO. As an illustration, Figure 1.42 shows the results for an emulsion stabilized with the surfactant

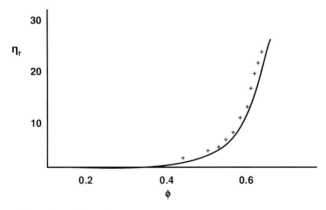

+ Experimental results
Full line calculated curve using Dougherty–Krieger equation

Figure 1.42 Experimental and theoretical $\eta_r - \phi$ curves.

containing 27 moles EO (the volume medium diameter of the droplets is 3.5 (μm). The calculations based on the Dougherty–Krieger equation are also shown in the same figure. In these calculations $[\eta] = 2.5$ and ϕ_p was obtained from a plot of $\eta^{-1/2}$ versus ϕ and extrapolation of the straight line to $\eta^{-1/2} = 0$. The value of ϕ_p was 0.73 (which is higher than the maximum random packing of 0.64, as a result of the polydispersity of the emulsion). The results using the other three surfactants showed the same trend; the experimental $\eta_r - \phi$ curves were close to those calculated using the Dougherty–Krieger equation, indicating that these emulsions were behaving as hard spheres.

1.12.9
Influence of Droplet Deformability on Emulsion Rheology

The influence of droplet deformability on emulsion rheology was investigated by comparing the $\eta_r - \phi$ curves of hard spheres of silica with two polydimethylsiloxane (PDMS) emulsions with low (PDMS 0.3) and high deformability (PDMS 0.45) (by controlling the proportion of crosslinking agent for the droplets; 0.3 low and 0.45 high crosslinking agent). The $\eta_r - \phi$ curves for the three systems are shown in Figure 1.43. The $\eta_r - \phi$ curve for silica can be fitted by the Dougherty–Krieger equation over the whole volume fraction range, indicating typical hard-sphere behavior. The $\eta_r - \phi$ curve for the less deformable PDMS deviates from the hard-sphere curve at $\phi = 0.58$. The $\eta_r - \phi$ curve for the more deformable PDMS deviates from the hard-sphere curve at $\phi = 0.40$, clearly showing the deformation of the 'soft' droplets at a relatively low volume fraction.

1.12.10
Viscoelastic Properties of Concentrated Emulsions

The viscoelastic properties of emulsions can be investigated using dynamic (oscillatory) measurements [49, 50]. A sinusoidal strain with amplitude γ_o is applied to the system at a frequency ω (rad s^{-1}) and the stress σ (with amplitude σ_o) is simultaneously measured. From the time shift Δt between the sine waves of strain and stress

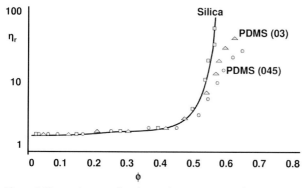

Figure 1.43 $\eta_r - \phi$ curves for silica and two PDMS emulsions.

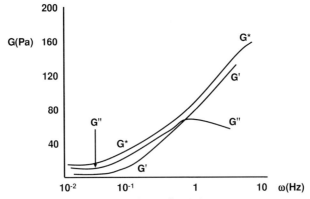

Figure 1.44 Variation of G^*, G' and G'' with frequency ω.

one can measure the phase angle shift δ ($\delta = \Delta t\, \omega$). From σ_o, γ_o and δ it is then possible to obtain the complex modulus G^*, the storage modulus G' (the elastic component) and the loss modulus G'' (the viscous component).

G^*, G' and G'' are measured as a function of strain amplitude to obtain the linear viscoelastic region, and then as a function of frequency (keeping γ_o in the linear region). As an illustration, Figure 1.44 shows the results for an O/W emulsion at $\phi = 0.6$ (the emulsion was prepared using an A–B–A block copolymer of PEO (A) and polypropylene oxide (PPO, B) with an average of 47 PO units and 42 EO units.

The data shown in Figure 1.44 are typical for a viscoelastic liquid. In the low-frequency regime (<1 Hz), $G'' > G'$, but as the frequency ω increases so too does G' increase. At a characteristic frequency ω^* (the crossover point) G' becomes higher than G'', and at high frequency it becomes closer to G^*. G'' then increases with the rise in frequency to reach a maximum at ω^*, after which it decreases with any further increase in frequency.

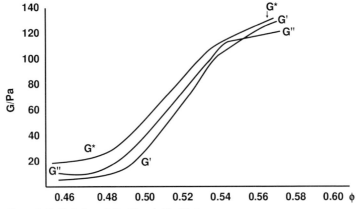

Figure 1.45 Variation of G^*, G' and G'' with ϕ.

From ω^*, the relaxation time t^* can be calculated:

$$t^* = \frac{1}{2\pi\omega^*} \tag{1.109}$$

For the above value of ϕ (=0.6), $t^* = 0.12$ s. In fact, t^* was seen to increase with an increase of ϕ, and this reflected the stronger interaction with an increase of ϕ.

In order to obtain the onset of strong elastic interaction in emulsions, G^*, G' and G'' (obtained in the linear viscoelastic region and high frequency, e.g. 1 Hz) are plotted versus the volume fraction of the emulsion ϕ. Care must be taken to ensure that the droplet size distribution in all emulsions is the same. The most convenient way to do this is to prepare an emulsion at the highest possible ϕ (e.g. 0.6), and then dilute this to obtain a range of ϕ-values. Droplet size analysis should be obtained for each emulsion to ensure that the size distribution is the same. Figure 1.45 shows the plots for G^*, G' and G'' versus ϕ. At $\phi < 0.56$, $G'' > G'$, whereas at $\phi > 0.56$, $G' > G$; $\phi = 0.56$ is the onset of predominantly elastic interaction, and this reflects the small distance of separation between the droplets.

References

1 Tadros, Th.F. and Vincent, B. (1983) in *Encyclopedia of Emulsion Technology* (ed. Becher, P.), Marcel Dekker, New York.

2 Binks, B.P. (ed.) (1998) *Modern Aspects of Emulsion Science*, The Royal Society of Chemistry Publication.

3 Tadros, Th.F. (2005) *Applied Surfactants*, Wiley-VCH, Weinheim.

4 Hamaker, H.C. (1937) *Physica (Utrecht)*, **4**, 1058.

5 Deryaguin, B.V. and Landua, L. (1941) *Acta Physicochem. USSR*, **14**, 633.

6 Verwey, E.J.W. and Overbeek, J.Th.G. (1948) *Theory of Stability of Lyophobic Colloids*, Elsevier, Amsterdam.

7 Napper, D.H. (1983) *Polymeric Stabilisation of Dispersions*, Academic Press, London.

8 Walstra, P. and Smolders, P.E.A. (1998) in *Modern Aspects of Emulsions* (ed. Binks, B.P.), The Royal Society of Chemistry, Cambridge.

9 Stone, H.A. (1994) *Annu. Rev. Fluid Mech.*, **226**, 95.

10 Wierenga, J.A., ven Dieren, F., Janssen, J.J.M. and Agterof, W.G.M. (1996) *Trans. Inst. Chem. Eng.*, **74-A**, 554.

11 Levich, V.G. (1962) *Physicochemical Hydrodynamics*, Prentice-Hall, Englewood Cliffs.

12 Davis, J.T. (1972) *Turbulent Phenomena*, Academic Press, London.

13 Lucasses-Reynders, E.H. (1996) in *Encyclopedia of Emulsion Technology* (ed. Becher, P.), Marcel Dekker, New York.

14 Graham, D.E. and Phillips, M.C. (1979) *J. Colloid Interface Sci.*, **70**, 415.

15 Lucasses-Reynders, E.H. (1994) *Colloids and Surfaces*, **A91**, 79.

16 Lucassen, J. (1981) in *Anionic Surfactants* (ed. E.H. Lucassen-Reynders), Marcel Dekker, New York.

17 van den Tempel, M. (1960) *Proceedings of the International Congress on, Surface Activity*, **2**, 573.

18 (a) Griffin, W.C. (1949) *J. Cosmet. Chemists*, **1**, 311; (b) Griffin, W.C. (1954) *J. Cosmet. Chemists*, **5**, 249.

19 Davies, J.T. (1959) *Proceedings of the International Congress on, Surface Activity*, **1**, 426.

20 Davies, J.T. and Rideal, E.K. (1961) *Interfacial Phenomena*, Academic Press, New York.

21 Shinoda, K. (1967) *J. Colloid Interface Sci.*, **25**, 396.
22 Shinoda, K. and Saito, H. (1969) *J. Colloid Interface Sci.*, **30**, 258.
23 Beerbower, A. and Hill, M.W. (1972) *Amer. Cosmet. Perfum.*, **87**, 85.
24 Hildebrand, J.H. (1936) *Solubility of Non-Electrolytes*, 2nd ed. Reinhold, New York.
25 Hansen, C.M. (1967) *J. Paint Technol.*, **39**, 505.
26 Barton, A.F.M. (1983) *Handbook of Solubility Parameters and Other Cohesive Parameters*, CRC Press, New York.
27 Israelachvili, J.N., Mitchell, J.N. and Ninham, B.W. (1976) *J. Chem. Soc., Faraday Trans. II*, **72**, 1525.
28 Tadros, Th.F. (1967) in *Solid/Liquid Dispersions* (ed. Th.F. Tadros), Academic Press, London.
29 Batchelor, G.K. (1972) *J. Fluid. Mech.*, **52**, 245.
30 Buscall, R., Goodwin, J.W., Ottewill, R.H. and Tadros, Th.F. (1982) *J. Colloid Interface Sci.*, **85**, 78.
31 Krieger, I.M. (1972) *Adv. Colloid Interface Sci.*, **3**, 111.
32 (a) Asakura, S. and Osawa, F. (1954) *J. Phys. Chem.*, **22**, 1255; (b) Asakura, S. and Osawa, F. (1958) *J. Polym. Sci.*, **33**, 183.
33 Smoluchowski, M.V. (1927) *Z. Phys. Chem.* **92**, 129.
34 Fuchs N. (1936) *Z. Physik*, **89** 736.
35 Reerink, H. and Overbeek, J.Th.G. (1954) *Disc. Faraday Soc.*, **18**, 74.
36 Thompson, W. (1871) *(Lord Kelvin), Philosophical Magazine*, **42**, 448.
37 Lifshitz I.M. and Slesov V.V. (1959) *Soviet Physics JETP*, **35**, 331.
38 Wagner C. (1961) *Z. Electrochem.*, **35**, 581.
39 Kabalanov, A.S. and Shchukin, E.D. (1992) *Adv. Colloid Interface Sci.*, **38**, 69.
40 Kabalanov, A.S. (1994) *Langmuir*, **10**, 680.
41 Weers, J.G. (1998) in *Modern Aspects of Emulsion Science* (ed. Binks, B.P.), Royal Society of Chemistry Publication Cambridge.
42 Deryaguin, B.V. and Scherbaker, R.L. (1961) *Kolloidn Zhurnal*, **23**, 33.
43 Friberg, S., Jansson, P.O. and Cederberg, E. (1976) *J. Colloid Interface Sci.*, **55**, 614.
44 Criddle, D.W. (1960) in The viscosity and viscoelasticity of interfaces (ed. Eirich, F.R.), in *Rheology*, Academic Press, New York, Vol. 3, Chapter 11.
45 Edwards, D.A., Brenner, H. and Wasan, D.T. (1991) *Interfacial Transport Processes and Rheology*, Butterworth-Heinemann, Boston, London.
46 Prince, A., Arcuri, C. and van den Tempel, M. (1967) *J. Colloid and Interface Sci.*, **24**, 811.
47 (a) Biswas, B. and Haydon, D.A. (1963) *Proc. Roy. Soc.*, **A271**, 296; (b) Biswas, B. and Haydon, D.A. (1963) *Proc. Roy. Soc.*, **A271**, 317; (c) Biswas, B. and Haydon, D.A. (1962) *Kolloidn Zhurnal*, **185**, 31; (d) Biswas, B. and Haydon, D.A. (1962) *Kolloidn Zhurnal*, **186**, 57.
48 Sherman, P. (1968) Rheology of emulsions (ed. Sherman, P.), in *Emulsion Science*, Academic Press, London, Chapter 4.
49 Tadros, Th.F. (1991) Rheological properties of emulsion systems (ed. Sjoblom, J.), in *Emulsions – A Fundamental and Practical Approach*, NATO ASI Series, Vol. 363, Kluwer Academic Publishers, London.
50 Tadros, Th.F. (1994) *Colloids and Surfaces* **A91**, 215.
51 Dickinson, E. (1998) Rheology of emulsions (ed. Binks, B.P.), in *Modern Aspects of Emulsion Science*, Royal Society of Chemistry Publications, Cambridge.

2
Stabilization of Emulsions, Nanoemulsions and Multiple Emulsions Using Hydrophobically Modified Inulin (Polyfructose)

Tharwat F. Tadros, Elise Vandekerckhove, Martine Lemmens, Bart Levecke, and Karl Booten

2.1
Introduction

The stabilization of emulsions against coalescence requires the presence of an effective energy barrier to prevent thinning and disruption of the liquid film between the droplets. In other words, a high disjoining pressure is required to prevent collapse of the film. The most effective repulsive barrier is produced using polymeric surfactants of the A–B, A–B–A or BA_n graft type. The B ('anchor') chain is chosen to be highly insoluble in the medium and to have a strong affinity to the oil.; this ensures strong adsorption and a lack of displacement of the molecule on close approach. The A ('stabilizing') chain is chosen to be highly soluble in the medium and strongly solvated by its molecules. This provides a strong repulsion between the droplets – a phenomenon referred to as *steric stabilization* [1–4].

Based on the above principle, we have recently designed a graft copolymer based on inulin (linear polyfructose) onto which several alkyl groups have been grafted [5]. This molecule – which is referred to hereafter as 'hydrophobically modified inulin' (HMI; INUTEC®SP1) – can be described as AB_n, where A is the inulin chain and B the alkyl chain. The molecule will adsorb with several alkyl groups at the liquid/liquid interface, leaving loops and tails of polyfructose dangling in solution. This multipoint attachment via several alkyl groups ensures a strong adsorption and a lack of desorption, whereas the strongly hydrated polyfructose loops and tails (up to high temperatures and in the presence of high electrolyte concentrations) ensures effective steric stabilization.

In this chapter we will describe the application of INUTEC® SP1 for the stabilization of emulsions, nanoemulsions, and multiple emulsions.

2.2
Experimental

2.2.1
Materials

In order to prepare the emulsions, two oils were used: Isopar M (Exxon) and silicone oil EU 344, a cyclomethicone (Dow Corning), both of which were used without any further purification. For the nanoemulsions, one-oil and two-oil mixtures were used (using dimethicone):

- Dimethicone DC200/50 cSt (Dow Corning).
- Mixture I (consisting of 10 parts Arlamol HD, isohexadecane, four parts Estol 3606, two parts Sunflower F90 and two parts Avocado oil).
- Mixture II (consisting of one part isopropyl palmitate, one part jojoba oil, one part octyl palmitate and two parts C_{12}–C_{15} alkyl benzoate).

The latter oil mixtures, many of which are used in personal care formulations, were supplied by UNIQEMA (Wilton, UK) and by Gova (Antwerp, Belgium). For the preparation of multiple emulsions, Isopar M was used as the oil.

The HMI (INUTEC®SP1) was synthesized as described previously [5]. The inulin used was INUTEC®N25 with a degree of polymerization (DP) in excess of 23. The polymeric surfactant was purified by solvent precipitation. For nanoemulsions based on the oil mixture, two cosurfactants were used, namely Brij 72 and Brij 721 (both supplied by UNIQEMA).

For the preparation of W/O emulsions (the primary emulsion in W/O/W multiple emulsions), Arlacel P135 (a block copolymer of polyhydroxystearic acid–polyethylene oxide–polyhydroxystearic acid (PHS–PEO–PHS); supplied by UNIQEMA) was used.

2.2.2
Methods

2.2.2.1 Preparation of Emulsions, Nanoemulsions and Multiple Emulsions

- *Emulsions* were prepared by slowly adding the oil to the polymeric surfactant solution while stirring using an Ultra-Turrax homogenizer (Cat. No. X620). Initially, the mixing was carried out at 9500 rpm for 2 min, after which the resultant emulsion was further homogenized in several steps at higher stirrer speeds (1 min at 13 500 rpm; 45 s at 20 500 rpm; and finally 1 min at 24 000 rpm).
- *Nanoemulsions* were created by first preparing an emulsion using the above-described procedure, followed by homogenization at 700 bar for 1 min using a Microfluidizer (Microfluidics, USA).
- *Multiple emulsions* were prepared in a two-step process. For a W/O/W multiple emulsion a primary W/O emulsion was prepared by the addition of 0.1 mol dm^{-3} $MgCl_2$ to an oil solution of Arlacel P135 while stirring using a Ultra-Turrax. This primary emulsion was then added to an aqueous solution of 0.1 mol dm^{-3} $MgCl_2$

while stirring at low speed. The electrolyte is used to balance the osmotic pressure between the internal water droplets and the external continuous phase. For a O/W/O multiple emulsion, an O/W nanoemulsion is used as the primary emulsion, and this is further emulsified in an oil solution of Arlacel P135 at low stirring speed.

2.2.2.2 Investigation of Emulsion Stability

The O/W and multiple emulsions maintained at 50 °C were assessed with optical micrography at various time intervals. The nanoemulsions were investigated by measuring the average droplet radius using dynamic light scattering, namely photon correlation spectroscopy (PCS). For this purpose, a HPPS instrument (supplied by Malvern, UK) was used. The basis of the method is measurement of the intensity fluctuation of scattered (laser) light as the droplets undergo Brownian diffusion. The intensity fluctuation is used to determine the diffusion coefficient D, which in turn allows the z-average radius to be obtained using the Stokes–Einstein equation:

$$D = \frac{kT}{6\Pi\eta R} \tag{2.1}$$

where k is the Boltzmann constant, T is the absolute temperature, η is the viscosity of the medium and R is the droplet radius.

The HPPS instrument provides not only the droplet size distribution and z-average diameter but also the polydispersity index. The latter is a measure of the polydispersity of the nanoemulsion; the higher the number the more polydisperse the nanoemulsion.

2.3
Results and Discussion

2.3.1
Emulsion Stability Using INUTEC®SP1

Stable emulsions may be prepared using 1% INUTEC®SP1 based on the oil phase. As an example, in the case of a 50 : 50 O/W emulsion, 0.5% INUTEC®SP1 was sufficient for the preparation of stable emulsions at 50 °C, both in water and in 2 mol dm^{-3} NaCl. This stability was obtained with several oils, such as paraffinic and silicone oils. The high stability is due to the high affinity of the polymeric emulsifier for the oil droplets (multipoint anchoring) and the strong hydration of the linear polyfructose chain. The situation is illustrated in Figure 2.1, which shows optical micrographs for Isopar O/W emulsions that have been stored at 50 °C for 1.5 and 14 weeks. It is clear from the micrographs that no coalescence has occurred, and that this stability could be maintained for more than a year. The same results were obtained with other oils, and also in the presence of the electrolytes.

As discussed earlier, the polyfructose loops and tails provide a dense, strongly hydrated layer that gives a high free energy of mixing, G_{mix}. Evidence for this has been obtained recently using atomic force microscopy (AFM) measurements [6]. Using a

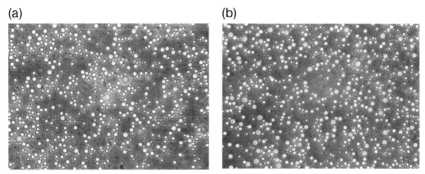

Figure 2.1 Photomicrographs of diluted 50/50 (v/v) Isopar M/water emulsions containing 2% INUTEC® SP1 that were stored at 50 °C for (a) 1.5 weeks and (b) 14 weeks.

hydrophobic glass sphere and a plate onto which INUTEC®SP1 was adsorbed, the force–distance curves showed a steep rise when the distance of separation between the sphere and plate became less than 20 nm. This indicates the presence of a layer thickness of the order of 10 nm, which is obviously greater than the size of the loop; moreover, it also implies a thick hydration layer providing strong repulsion (this is sometimes referred to as 'salvation forces'). This hydration of polyfructose is much stronger than that achieved with polyethylene oxide (PEO), and explains the higher stability obtained with INUTEC®SP1.

Further evidence of film stability and a lack of coalescence was recently obtained by Exerowa *et al.* [7], who measured the disjoining pressure as a function of film thickness both in water and at high NaCl concentration (up to 2 mol dm^{-3}). As the capillary pressure increased, the film thickness was seen to decrease, at which point a sudden jump to a Newton black film occurred. That the film remained stable and did not collapse up to a pressure of 4.5×10^4 Pa provided an explanation for the high stability against coalescence at such high electrolyte concentrations.

2.3.2
Nanoemulsion Stability Using INUTEC®SP1

The small droplets of nanoemulsions (<200 nm diameter) ensures stability against creaming or sedimentation and coalescence [8]. In fact, with such small droplets, the Brownian diffusion is sufficient to overcome the gravity force, thus preventing any separation. In addition, these small droplets are nondeformable, and therefore surface fluctuation is prevented, which in turn prevents coalescence. The thick, strongly hydrated polyfructose loops and tails prevent any thinning or disruption of the liquid film between the droplets, such that coalescence is prevented [8].

The only process of instability seen with nanoemulsions is due to Ostwald ripening. With their high radius of curvature, the small droplets will have a greater solubility than would large drops with a low radius of curvature. This concept was first considered by Lord Kelvin [9], who related the solubility of a drop with radius r, $S(r)$, to that of a drop with infinite radius, $S(\infty)$ by the expression:

$$S(r) = S(\infty)\exp\left(\frac{2\gamma V}{rRT}\right) \qquad (2.2)$$

where γ is the interfacial tension, V is the molar volume of the oil, R is the gas constant and T is the absolute temperature.

For two droplets with radii r_1 and r_2 (where $r_1 < r_2$) $S(r_1) > S(r_2)$ and hence oil molecules will diffuse from the smaller to the larger droplets. Thus, on storage (particularly at high temperature) the droplet size distribution shifts to large sizes and the nanoemulsion becomes increasingly turbid and loses its advantages, such as transparency and the efficient delivery of actives.

Ostwald ripening can be significantly reduced by either of two mechanisms, or by their combination [8]. The addition of a small proportion of a highly insoluble oil (e.g. squalane) will reduce the diffusion of the smaller oil droplets from the small to the large droplets. The most effective method is to use a polymeric surfactant that strongly adsorbs at the interface and enhances the Gibbs elasticity. This effect could be achieved using INUTEC®SP1, as the polymeric surfactant has limited solubility in water and hence adsorbs very strongly at the interface.

The above effect was tested by measuring the rate of Ostwald ripening of nanoemulsions based on silicone oil (Dimethicone DC200/50 cSt), the oil mixture I and oil mixture II. With silicone oil, INUTEC®SP1 was sufficient to produce the nanoemulsion, but with oil mixtures I and II it was necessary to use cosurfactants of Brij 72 and Brij 721, respectively. The rate of Ostwald ripening of the nanoemulsions (which were stored at 50 °C) was investigated by plotting the cube of the radius, R^3, versus time (t) using the Lifshitz–Slesov–Wagner (LSW) theory [10, 11]:

$$R^3 = \frac{8}{9}\left[\frac{S(\infty)\gamma VD}{\rho RT}\right]t \qquad (2.3)$$

Typical plots of R^3 versus t for 20/80 (v/v) silicone O/W nanoemulsions based on 1.6% and 2.4% INUTEC®SP1 are shown in Figure 2.2. The nanoemulsion based on 1.6% INUTEC®SP1 shows a larger Ostwald ripening rate when compared to results obtained with 2.4% INUTEC®SP1. The rate constants obtained from the slope of the lines were 1.1×10^{-29} and 2.4×10^{-30} m^3 s^{-1} for 1.6% and 2.4% INUTEC®SP1, respectively. These rates are very small when compared to nanoemulsions based on conventional surfactants such as alcohol ethoxylates [12], which clearly shows the efficacy of INUTEC®SP1 in reducing Ostwald ripening. The amount of INUTEC®SP1 required to stabilize the nanoemulsion was also relatively smaller, due mainly to the high affinity of the polymeric surfactant to the oil phase, with most molecules being adsorbed but leaving a small amount in bulk solution. With nonionic surfactants of the alcohol ethoxylate type, adsorption is reversible and a significant amount of emulsifier remains in the bulk solution. In addition, such surfactants produce *micelles* with high aggregation numbers, and are able to solubilize oil molecules, thus enhancing their diffusion and obtaining high Ostwald ripening rates. With INUTEC®SP1, aggregation is limited to two or molecules, and hence no oil solubilization occurs; this is another factor responsible for the reduction of Ostwald ripening. The multipoint attachment of the polymeric surfactant

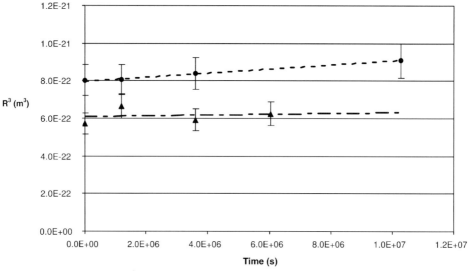

Figure 2.2 Plots of R^3 versus t for nanoemulsions containing 1.6% (upper) and 2.4% (lower) INUTEC®SP1.

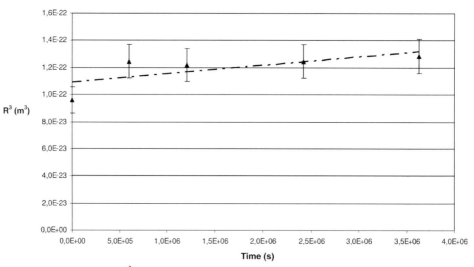

Figure 2.3 Plot of R^3 versus t for nanoemulsion based on oil mixture I.

molecule and the strongly hydrated loops and tails enhances the Gibbs elasticity, and hence the diffusion of oil molecules is very much reduced. Walstra [13] suggested that, by using surfactants which are strongly adsorbed at the O/W interface and which do not desorb during ripening, the rate of Ostwald ripening could be significantly reduced. Indeed, this condition is satisfied using INUTEC®SP1, as observed in the present study.

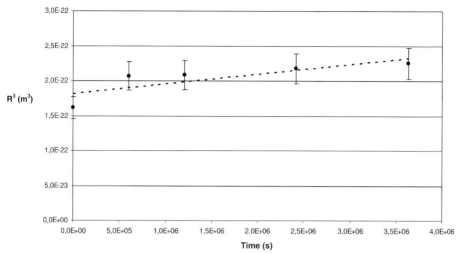

Figure 2.4 Plot of R^3 versus t for nanoemulsion based on oil mixture II.

Figure 2.3 shows the plot of R^3 versus t for the oil mixture I, where Brij 72 and Brij 721 were used as cosurfactants. The 20/80 (v/v) nanoemulsions were prepared using a total surfactant concentration of 6% and with a ratio of Brij 72:Brij 721:INUTEC®SP1 of 1.2:2.8:2. The Ostwald ripening rate constant obtained at 50 °C in this case was $6.0 \times 10^{-30}\,\mathrm{m^3\,s^{-1}}$, which was comparable to that obtained using silicone oil. This low rate, obtained when using more polar oils, confirms the efficacy of INUTEC®SP1 in reducing Ostwald ripening. At this point it should be noted that the concentration of INUTEC®SP1 in the nanoemulsions was only 2%. It is highly likely that the polymeric surfactant displaces the alcohol ethoxylates from the interface due to its much stronger adsorption.

Similar results were obtained using oil mixture II, as illustrated in Figure 2.4, which shows a plot of R^3 versus t for a 20/80 (v/v) nanoemulsion that was stored at 50 °C. In this case the total surfactant concentration was 6%, but the ratio of Brij 72:Brij 721:INUTEC® SP1 was 2:2:2. The rate of Ostwald ripening in this case was $1.0 \times 10^{-29}\,\mathrm{m^3\,s^{-1}}$, which was slightly higher than that of oil mixture I. Although it is likely that oil mixture II is more polar than oil mixture I, the results were still much better than those obtained using alcohol ethoxylates alone, again confirming the efficacy of the polymeric surfactant in reducing Ostwald ripening.

The efficacy of INUTEC® SP1 in reducing Ostwald ripening was best illustrated by comparing the results obtained when using sucrose–cocoate alone (4%) to prepare the nanoemulsion with those using a mixture of sucrose–ester and INUTEC® SP1 (2% each or 3% sucrose–ester and 1% INUTEC® SP1). The results (see Figure 2.5) clearly demonstrated the efficacy of INUTEC®SP1 in reducing the Ostwald ripening rate, with rates of 8×10^{-29}, 1×10^{-29} and $6 \times 10^{-30}\,\mathrm{m^3\,s^{-1}}$ for 4% sucrose–ester, 2% sucrose–ester + 2% INUTEC®SP1, and 3% sucrose–ester + 1% INUTEC®SP1, respectively.

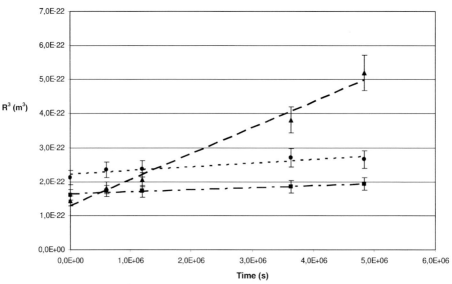

Figure 2.5 Plots of R^3 versus t based on 4% sucrose–ester (▲), 2% sucrose-ester + 2% INUTEC®SP1 (●) and 3% sucrose–ester + 1% INUTEC®SP1 (■).

2.3.3
Multiple Emulsion Stability Using INUTEC® SP1

Optical micrographs of W/O/W and O/W/O multiple emulsions which have been stored at 50 °C for several weeks are shown in Figures 2.6 and 2.7, respectively. The same photographs were obtained directly after preparation of the multiple emulsions, indicating the high stability of the systems based on Arlacel P135 (for W/O) and INUTEC®SP1 (for O/W). Both polymeric surfactants are strongly

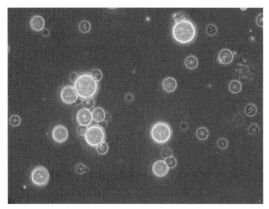

Figure 2.6 Optical micrograph of a W/O/W multiple emulsion.

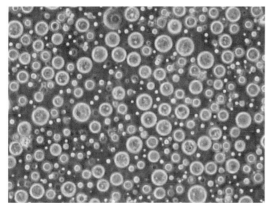
Figure 2.7 Optical micrograph of an O/W/O multiple emulsion.

adsorbed at the O/W interface, and most likely produce a viscoelastic film that prevents any coalescence [14]. For W/O/W multiple emulsions it is necessary to control the osmotic balance between the internal water droplets and the external continuous phase aqueous phase, and this was achieved by using $0.1\,\mathrm{mol\,dm^{-3}}$ $MgCl_2$.

2.4
Conclusions

Hydrophobically modified inulin, INUTEC®SP1, is an effective polymeric surfactant for the stabilization of emulsions, nanoemulsions and multiple emulsions. For O/W emulsions, multipoint attachment of the polymer by several alkyl groups (which may dissolve in the oil phase) and the strongly hydrated polyfructose loops and tails (both at high temperature and in the presence of high electrolyte concentrations) leads to a prevention of coalescence. With nanoemulsions, INUTEC®SP1 is strongly adsorbed at the O/W interface, while the lack of desorption significantly reduces the Ostwald ripening rate as a result of an enhanced Gibbs elasticity. With multiple emulsions, INUTEC®SP1 most likely produces a viscoelastic film that prevents coalescence.

References

1 Tadros, Th.F. and Vincent, B. (1983) in *Encyclopedia of Emulsion Technology* (ed. Becher, P.), Vol. 1, Marcel Dekker, New York.
2 Tadros, Th.F. (1999) Polymeric Surfactants (eds. Goddard, E.D. and Gruber, J.V.), in *Principles of Polymer Science and Technology Cosmetics and Personal Care*, Marcel Dekker, New York.
3 Tadros, Th.F. (2003) in *Novel Surfactants* (ed. Holmberg, K.), Marcel Dekker, New York.

4 Tadros, Th.F., Booten, K. and Levecke, B. (2004) *Cosmetics Toiletries*, **119**, 51.
5 Stevens, C.V., Nerrigi, A., Peristeropoulou, M., Christov, P.P., Booten, K., Levecke, B., Vandamme, A., Pittivils, N. and Tadros, Th.F. (2001) *Biomacromolecules*, **2**, 1256.
6 Nestor, J., Esquena, J., Solans, C., Luckham, P.F., Musoke, M., Levecke, B., Booten, K. and Tadros, Th.F. (2007) *Journal of Colloid and Interface Science*, **34**, 430.
7 Exerowa, D., Gotchev, G., Kolarev, T., Khristove, Khr., Levecke, B. and Tadros, Th.F. (2007) *Langmuir*, **23**, 1711.
8 Tadros, Th.F., Izquierdo, P., Esquena, J. and Solans, C. (2004) *Advances in Colloid Interface Science*, **108–109**, 303.
9 Thompson, W. (Lord Kelvin) (1871) *Phil. Mag.*, **42**, 448.
10 Lifshitz, I.M. and Slesov, V.V. (1959) *Soviet Physics JETP*, **35**, 331.
11 Wagner, C. (1961) *Zeitschrift Electrochemische*, **35**, 581.
12 Izquierdo, P., Esquena, J., Tadros, Th.F., Federen, C., Garcia, M.J., Azemar, N. and Solans, C. (2002) *Langmuir*, **18**, 26.
13 Walstra, P. (1993) *Chemistry and Engineering Science*, **48**, 333.
14 Tadros, Th.F. (1994) *Colloids and Surfaces*, **94**, 39.

3
Interaction Forces in Emulsion Films Stabilized with Hydrophobically Modified Inulin (Polyfructose) and Correlation with Emulsion Stability

Tharwat Tadros, Dotchi Exerowa, Georgi Gotchev, Todor Kolarov, Bart Levecke, and Karl Booten

3.1
Introduction

The stabilization of emulsions against coalescence requires the formation of a stable liquid film between the emulsion droplets. A useful concept to illustrate such stability is the disjoining pressure $\pi(h)$, as introduced by Deryaguin and Scherbaker [1], which balances the excess normal pressure,

$$\pi(h) = P(h) - P_o \tag{3.1}$$

where $P(h)$ is the pressure of a film with thickness h, and P_o is the pressure of a sufficiently thick film such that the interaction free energy is equal to zero.

$\pi(h)$ may be equated to the net force (or energy) per unit area across the film,

$$\pi(h) = -\frac{dG_T}{dh} \tag{3.2}$$

G_T is the total interaction energy in the film.

$\pi(h)$ consists of three contributions, due to electrostatic repulsion (π_E), steric repulsion (π_s) and van der Waals attractions (π_A):

$$\pi(h) = \pi_E + \pi_s + \pi_A \tag{3.3}$$

To produce a stable film, $\pi_E + \pi_s > \pi_A$, and this is the driving force for the prevention of coalescence.

By using the above concepts, Exerowa and Kruglyakov [2] developed a technique for measuring the disjoining pressure of a microscopic horizontal film between two macroscopic emulsion drops. A special measuring cell was used in which the pressure π (that is equal to the disjoining pressure) was measured as a function of film thickness, h.

In this chapter we will describe the results obtained using the above techniques for two isoparaffinic oil (Isopar M) drops that contain adsorbed layers of hydrophobically modified inulin (INUTEC® SP1). This is a graft copolymer consisting of a backbone

Emulsion Science and Technology. Edited by Tharwat F. Tadros
Copyright © 2009 WILEY-VCH Verlag GmbH & Co. KGaA, Weinheim
ISBN: 978-3-527-32525-2

of inulin (linear polyfructose) onto which several alkyl groups have been attached. At an oil-in-water (O/W) interface, the polymeric surfactant adsorbs with multipoint attachment with several alkyl chains, leaving strongly hydrated loops and tails of polyfructose. This polymeric surfactant has proved to be an effective stabilizer against the coalescence of various emulsions, both in water and in high electrolyte concentrations. The π–h isotherms provide a direct measure of the interaction forces between the oil drops, and allow the study of the effect of electrolyte addition on such interactions.

3.2
Materials and Methods

Isopar M (an isoparaffinic oil) was supplied by Exxon (Belgium), and the graft copolymer INUTEC®SP1 by ORAFTI (Belgium); both materials were used as received. The molecule has an average molar mass of 5000 Da, and the inulin backbone a degree of polymerization (DP) greater than 23. The water used was double-distilled and had a conductivity of $10^{-6}\,\Omega^{-1}\,\text{cm}^{-1}$.

The microscopic emulsion films were formed in a specially designed measuring cell, as shown schematically in Figure 3.1. The emulsion films of radius r shown in Figure 3.1a are formed in the middle of a biconcave drop in a glass tube by drawing out the solution from the drop, through the capillary tube.

The emulsion film in the measuring cell (Figure 3.1a) is under a constant capillary pressure $P_c = 2\sigma/R$, where σ is the O/W interfacial tension and r is the radius of curvature of the meniscus. Figure 3.1b represents the effect of increasing the capillary pressure to a maximum value of about 10^5 Pa that is determined by the minimum radius of the porous plate.

The measuring cell is immersed in a thermostat bath that contains a microscope and a special optical and electronic system for observation of the film and determining its thickness. The film thickness is determined using a microinterferometric technique as described previously [2].

The above method allows the pressure (that is equal to the disjoining pressure π) to be obtained as a function of film thickness, h. This method is referred to as the 'film

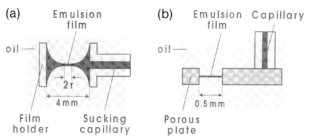

Figure 3.1 Schematic representation of the measuring cell for the formation of a horizontal emulsion film.

pressure balance technique'; isotherms of π–h provide a direct measure of the interaction forces in the liquid film between emulsion drops.

3.3
Results and Discussion

Measurement of the interfacial tension σ as a function of INUTEC® SP1 concentration in the presence of 2×10^{-4} and $0.5\,\mathrm{mol\,dm^{-3}}$ NaCl showed a gradual reduction in σ with increasing INUTEC® SP1 concentration; in time, σ reached a value of $\sim 18\,\mathrm{mN\,m^{-1}}$ at $\sim 5 \times 10^{-5}\,\mathrm{mol\,dm^{-3}}$ INUTEC® SP1, above which it remained constant. This concentration of INUTEC® SP1 at the 'break point' may be identified as the 'critical association concentration' (or 'critical micelle concentration'; CMC). The value of σ at the plateau showed no significant change with increasing NaCl concentration (up to $1\,\mathrm{mol\,dm^{-3}}$), indicating that the adsorption of INUTEC® SP1 is not affected by the addition of electrolyte.

Figure 3.2 shows, graphically, the variation of equivalent film thickness in relation to NaCl concentration at a constant INUTEC® SP1 concentration of $2 \times 10^{-5}\,\mathrm{mol\,dm^{-3}}$, at a constant capillary pressure P_c of 18 Pa, a constant film radius of 100 μm and a constant temperature of 22 °C. The error bar in the film thickness is larger at low electrolyte concentrations. However, the results of Figure 3.2 show a gradual decrease in film thickness up to a NaCl concentration of $5 \times 10^{-2}\,\mathrm{mol\,dm^{-3}}$, after which it remained constant (11 nm) above this concentration.

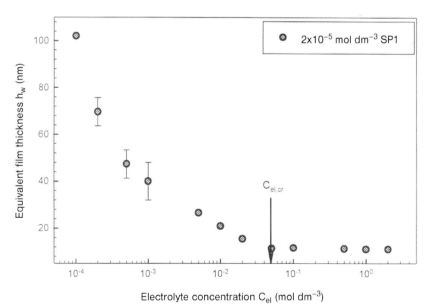

Figure 3.2 Variation of equivalent film thickness with NaCl concentration.

The large thicknesses obtained at low NaCl concentration are due to the presence of residual double-layer repulsion and the presence of extended double layers (which have a thickness of ~33 nm in 10^{-4} mol dm^{-3} NaCl). Thus, a thickness of ~100 nm in 10^{-4} mol dm^{-3} NaCl is a reasonable value (note that the film thickness is approximately twice the double layer extension). With an increase in electrolyte concentration, compression of the double layers occurs and this results in a decrease in the film thickness. For example, at 10^{-2} mol dm^{-3} NaCl, the film thickness is ~20 nm; this is a reasonable value considering the contribution from the adsorbed polymer layer. However, above 5×10^{-2} mol dm^{-3} NaCl the film thickness remains constant, giving a value of 11 nm which is the contribution from the adsorbed polymer layers (that give a thickness of ~5.5 nm).

The above change in film thickness at and above 5×10^{-2} mol dm^{-3} NaCl represents the transition from electrostatic to steric repulsion. This transition can be best investigated by considering the π–h isotherms at constant electrolyte concentration, as shown in Figure 3.3 at 2×10^{-4} mol dm^{-3} NaCl.

In Figure 3.3, the details of three independent experiments are shown, using different symbols. The trend in the π–h isotherms is the same. Initially, the film thickness shows a gradual decrease with increase in capillary pressure; however, when the thickness reaches ~30 nm a jump occurs to ~7.2 nm, after which it remains constant with further increase in capillary pressure up to 4.5×10^4 Pa. The (jump) transition in film thickness takes place over a wide pressure range, namely

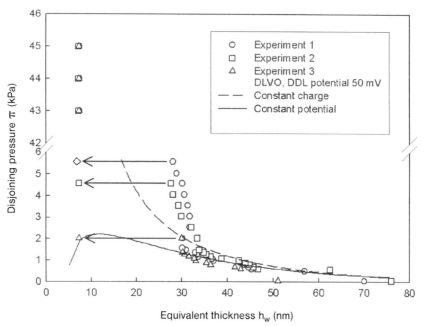

Figure 3.3 π–h isotherms at C_{NaCl} of 2×10^{-4} mol dm^{-3}.

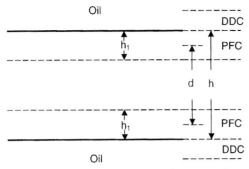

Figure 3.4 Schematic representation of the model for the DLVO calculation.

2×10^3 to 5.5×10^3 Pa. The jump can be accounted for by the formation of a Newton Black film (NBF).

The above trends can be quantitatively accounted for by consideration of the double-layer repulsion before the jump (i.e. in the thick film region) and the steric repulsion that accounts for the NBF. The results before the jump can be analyzed using the Deyaguin–Landua–Verwey–Overbook (DLVO) theory [3, 4]. The adsorption layer thickness h_1 is considered to be equal to half of the NBF thickness. A model for the DLVO calculation is shown in Figure 3.4.

In the model shown in Figure 3.4, d is the distance between the planes of origin of the electrostatic disjoining pressure π_{el} – that is, those of the electric diffuse double layers in the film. As the thickness of the NBF is 7.2 nm, then $h_1 = 3.6$ nm and $d = h - 3.6$ nm. π_{el} is computed using a numerical solution of the complete Poisson–Boltzmann equation [5]. The van der Waals contribution for the disjoining pressure π_A is calculated using a value for the Hamaker constant of 5.26×10^{-21} J.

The results for the DLVO calculations for the limiting cases of constant potential and constant charge are shown in Figure 3.3 using a diffuse double-layer potential of 50 mV. As can be seen, the experimental results up to a thickness of 30 nm are between the theoretical curves for constant potential and constant charge.

Similar results are obtained at a NaCl concentration of 10^{-3} mol dm^{-3} as shown in Figure 3.5 where in this case four independent experiments have been carried out. In the DLVO calculations a diffuse double layer potential of 20 mV was used. In this case also, the transition area is shifted to smaller thickness and smaller pressure.

π–h isotherms were obtained at higher electrolyte concentrations corresponding to the plateau value of the thickness shown in Figure 3.2 – that is, in the range 5×10^{-2} to 2 mol dm^{-3} NaCl. The results are shown in Figure 3.6, where the initial thicknesses are within the range 9–11 nm, after which the transition zone starts, corresponding to a capillary pressure of 0.6–1×10^3 Pa. In this zone, all films transform to a NBF with a jump, after which the pressure increases but the film thickness remains constant. This indicates that the barrier in the π–h isotherm is relatively low, and this allows for the jump to occur. Such behavior is consistent with the absence of electrostatic repulsion at these high electrolyte concentrations.

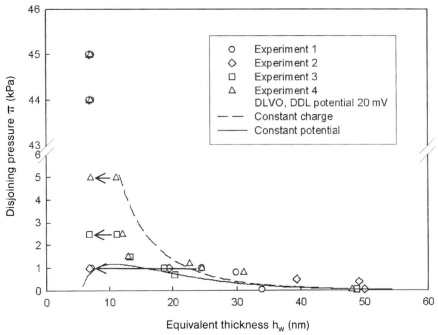

Figure 3.5 π–h isotherms at C_{NaCl} of 10^{-3} mol dm^{-3}.

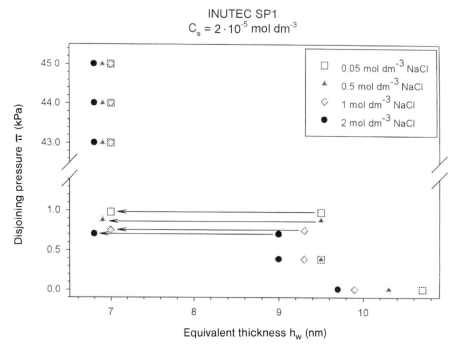

Figure 3.6 π–h isotherms at various NaCl concentrations.

The lack of rupture of the NBF up to the highest pressure applied (namely 4.5×10^4 Pa) clearly indicates the high stability of the liquid film in the presence of high NaCl concentrations (up to 2 mol dm^{-3}). This result is consistent with the high emulsion stability obtained at high electrolyte concentrations and high temperature [6, 7]. Emulsions of Isopar M-in-water are very stable under such conditions, and this may be accounted for by the high stability of the NBF. The droplet size of 50:50 O/W emulsions prepared using 2% INUTEC®SP1 is in the region of 1 to 10 μm. This corresponds to a capillary pressure of $\sim 3 \times 10^4$ Pa for the 1 μm drops and $\sim 3 \times 10^3$ Pa for the 10 μm drops. These capillary pressures are lower than those to which the NBF have been subjected, and this clearly indicates the high stability obtained against coalescence in these emulsions.

3.4
Conclusions

By measuring the disjoining pressure, π, as a function of film thickness, h, for emulsion films stabilized using hydrophobically modified inulin, INUTEC®SP1, it is possible to obtain information on the interaction forces operating in the film. At low NaCl concentrations (<5×10^{-2} mol dm^{-3}) a residual electrostatic repulsion is observed which may be analyzed using the DLVO theory. The polymer-stabilized films showed transition from electrostatic to steric repulsion when the capillary pressure was increased beyond a certain range, and this caused a 'jump' to a NBF which remained stable up to high capillary pressure (up to 4.5×10^4 Pa). These stable NBFs persisted in high NaCl concentrations (up to 2 mol dm^{-3}), and this accounted for the high stability of emulsions prepared at high electrolyte concentrations and high temperatures.

References

1 Deryaguin, B.V. and Scherbaker, R.L. (1961) *Kolloid Zh.*, **23**, 33.
2 Exerowa, D. and Kruglyakov, P.M. (1998) *Foam and Foam Films*, Elsevier, Amsterdam.
3 Deryaguin, B.V. and Landau, L. (1941) *Acta Physicochem. USSR*, **14**, 633.
4 Verwey, E.J.W. and Overbeek, J.Th.G. (1948) *Theory of Stability of Lyophobic Colloids*, Amsterdam.
5 Lyklema, J. (2005) *Fundamentals of Interface and Colloid Science*, Elsevier, Amsterdam.
6 Tadros, Th.F., Vandamme, A., Booten, K., Levecke, B. and Stevens, C. (2004) *Advances Colloid and Interface Sci.*, **108/109**, 207.
7 Tadros, Th.F., Vandamme, A., Booten, K., Levecke, B. and Stevens, C. (2004) *Colloids and Surfaces*, **250**, 133.

4
Enhancement of Stabilization and Performance of Personal Care Formulations Using Polymeric Surfactants

Tharwat F. Tadros, Martine Lemmens, Bart Levecke, and Karl Booten

4.1
Introduction

Most personal-care formulations are complex multiphase systems that include solid/liquid dispersions (suspensions), liquid/liquid dispersions (emulsions), mixtures of suspensions and emulsions (suspoemulsions), multiple emulsions (W/O/W or O/W/O), nanoemulsions (covering the size range 50–200 nm) and microemulsions (covering the size range 5–50 nm) [1]. All of these systems are formulated using complex mixtures of surfactants and, apart from the microemulsions, they are thermodynamically unstable. This situation can be better understood by considering the free energy of formation ΔG of these disperse systems, which consists of an energy term $\Delta A \gamma$ (where ΔA is the increase in surface area accompanied by the formation of particles or droplets with a high surface area from the bulk phase with much lower surface area, and γ is the interfacial tension) and an entropy term $-T\Delta S$ (on the formation of a large number of particles or droplets from the bulk phase there is an increase in entropy). In all disperse systems $\Delta A\gamma \gg -T\Delta S$ and hence ΔG is positive. Thus, the formation of disperse systems is nonspontaneous and energy is required to form the particles or droplets. The added surfactants reduce γ and produce an energy barrier to provide the system with kinetic stability.

When storing these formulations certain breakdown processes may take place, such as *flocculation, Ostwald ripening, coalescence* and *phase inversion*, but the correct choice of surfactant system can lead to these breakdown processes being reduced and the shelf life of the product extended. Unfortunately, most surfactant systems used in personal-care formulations are reversibly adsorbed at the interface, and consequently a high concentration is required to maintain the long-term physical stability. However, such high surfactant concentrations may be undesirable in cosmetics and personal-care formulations as they may cause skin irritation.

The above problems can be solved to a large extent by the use of polymeric surfactants of the A–B, A–B–A block and BA_n graft types [2, 3]. The B chain (sometimes referred to as the 'anchor' chain) is chosen to have a high affinity to the surface of particles or droplets, and this ensures a lack of desorption. The A chain (the 'stabilizing' chain) is chosen to be soluble in the medium and strongly solvated by its molecules. This provides an effective stabilization of the dispersion against flocculation, Ostwald ripening, coalescence and phase inversion. Due to the high affinity of the polymeric surfactant to the surface, the concentration required to prepare the dispersion is much lower than that used with conventional surfactants. This means that the surfactant concentration left in bulk solution is small, which in turn should limit the possibility of skin irritation. The latter effect is also minimized as the molecule has a high molecular weight, such that its penetration through the skin is largely prevented.

By applying the above principles, we have recently developed a graft copolymer based on inulin (linear polyfructose) onto which several alkyl groups have been grafted. This polymeric surfactant – which hereafter is referred to as INUTEC®SP1 – has been applied for the stabilization of emulsions [4, 5], nanoemulsions [6, 7] and multiple emulsions [8]. In this chapter we will describe the applications of INUTEC®SP1 in the enhancement of performance and stability of some personal-care formulations.

4.2
Experimental

Four different personal-care formulations have been prepared into which INUTEC®SP1 has been incorporated to enhance their stability and performance. These include a massage lotion formulation, a hydrating shower cream, a soft conditioner and a sun spray (SPF19). The composition of each formulation is provided in Table 4.1.

The above formulations were prepared using standard procedures. The oil phase was added to the aqueous phase while stirring at 10 000 rpm using a high-speed stirrer (e.g. Ultra-Turrax). Occasionally, the thickener was included in the aqueous phase and with a cross-polymer of polyacrylate, and the pH adjusted with NaOH to produce the microgel. With solid materials heating was necessary to melt the solid before emulsification. All formulations were maintained at room temperature and their stability was assessed by visual inspection. In addition, the performance of the formulation on application, such as its 'stickiness' and 'skin feel', was also assessed.

4.3
Results and Discussion

4.3.1
Massage Lotion

This formulation contains 52.5% oil-phase composed of five different materials that vary in their polarity. The addition of 1% INUTEC®SP1 was sufficient to stabilize this formulation against any strong flocculation, coalescence and phase-inversion. When

Table 4.1 Compositions of the massage lotion, hydrating shower cream, soft conditioner and sun spray (SPF 19) formulations.

Phase	Ingredient	Concentration (%, w/w)
Massage lotion		
A	Parafinium liquidum	42.00
	Helianthus annuus	5.00
	Cetearyl ethylhexanoate	2.50
	Cetearyl isononoate	2.50
	Perfume	0.30
B	Water	36.68
	INUTEC®SP1	1.00
	Vitis vinifira (grape) skin extract	0.02
	Preservatives	
	Lactic acid to pH 4–5	
C	C_{10-30} alkyl acrylate crosspolymer	3.00
	Xanthan gum	3.00
D	10% NaOH to pH 5–6	3.00
Hydrating shower cream		
A	Helianthus annuus (vegetable oil)	9.00
	Questamix (blend skin lipids)	0.10
	Pentaerythrityl (substantia oil) Tetracaprylate/caprate	1.00
	Ammonium lauryl sulfate (primary surfactant)	9.00
	Perfume	
B	Water	
	INUTEC®SP1 (emulsion stabilizer)	0.20
	Xanthan gum (viscosity modifier)	0.50
	C_{10-30} alkyl acrylate crosspolymer (viscosity modifier)	0.70
	Preservatives	
C	Cocoamidopropyl betaine (secondary emulsifier)	3.50
D	10% NaOH to pH 4.7–5.2	
Soft conditioner		
A	Cetearyl alcohol (conditioning)	4.00
	Cyclopentamethicone (for shine)	0.60
	Bishydroxyethyl biscetyl (color)	0.05
	Malonamide (maintenance, repairing, strength)	

Table 4.1 (Continued)

Phase	Ingredient	Concentration (%, w/w)
B	Water	
	Hydroxyethylcellulose (2%) viscosity modifier	25.00
	Polyquaternium-10	0.10
	INUTEC®SP1 (emulsion stabilizer)	0.05
	Preservatives	
C	Cetrimonium chloride	1.50
D	Perfume	0.20
Sunspray SPF19		
A	C_{10-30} alkyl benzoate (dry oil)	6.00
	Jojoba oil pressed (vegetable oil)	2.00
	Isoamyl *p*-methoxycinnamate (UVB-filter)	10.00
	Ethylhexyl dimethyl PABA (UVB-filter)	7.00
	Cyclopentasiloxane, C_{30-45} alkyl cetearyl dimethicone crosspolymer (structuring agent)	1.00
	Ethylhexylpalmitate (spreading oil)	1.00
	Sorbitan isostearate (coemulsifier)	0.50
B	Water	
	Xanthan gum	0.10
	INUTEC®SP1 (emulsion stabilizer)	0.75
	Glycerin	3.00
	Preservatives	
C	Aqua, Galactobrabinan (stabilizing)	5.00
D	10% NaOH to pH 5–6	

stored at ambient temperature the formulation showed no separation for more than one year, this high stability against coalescence being the result of the adsorption and conformation of the polymer at the oil/water interface [4]. The polymer molecule adsorbs with multipoint attachment with several alkyl chains (that may be soluble in the oil-phase), leaving loops and tails of the linear polyfructose chain dangling in solution. A schematic representation of the adsorption conformation of the polymer chain at the oil/water interface is shown in Figure 4.1.

The multipoint attachments of the chains at the oil/water interface prevents any desorption on approach of the oil droplets [4, 5]. The strongly hydrated polyfructose loops and tails (which have a thickness in the region of 10 nm) [6] provide strong steric repulsion as a result of the unfavorable mixing of the polyfructose chain and loss in

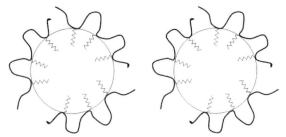

Figure 4.1 Schematic presentation of the adsorption and conformation of INUTEC®SP1 at the oil/water interface.

configurational entropy of the chains on considerable overlap [7]. Thus, thinning and disruption of the aqueous film between the oil droplets is prevented, which eliminates coalescence [8]. The polymer-surfactant molecule at the oil/water interface enhances the Gibbs dilational elasticity, and this also prevents any coalescence.

To prevent creaming of the emulsion, two rheology modifiers – namely crosslinked polyacrylate and xanthan gum – were added to the aqueous phase. The crosslinked acrylate is neutralized with NaOH, and the resulting electrolyte in the system does not affect the stability of the formulation. As discussed previously [4, 5], INUTEC®SP1 can be used to stabilize emulsions in high electrolyte concentrations.

An additional advantage of using INUTEC®SP1 is the excellent performance on application. The formulation proved to be nonsticky, light and nongreasy, and also showed an excellent skin-feel. However, this was not surprising as the polymeric surfactant helps to lubricate the skin surface.

4.3.2
Hydrating Shower Gel

This formulation contains 11% oil-phase and 9% ammonium lauryl sulfate, which is necessary to produce a stable foam on application. Unfortunately, the presence of oil droplets in a foam is known to cause its destabilization, as a result of the following mechanism [9].

Undissolved oil droplets form in the surface of the film, and this can lead to film rupture. One widely accepted mechanism for the destabilization considers two steps: the oil droplets enter the air/water interface, and then spread over the film, causing its rupture. Now, the destabilizing action can be rationalized [10] in terms of the balance between the entry coefficient E and the Harkens [11] spreading coefficient S:

$$E = \gamma_{W/A} + \gamma_{W/O} - \gamma_{O/A} \tag{4.1}$$

$$S = \gamma_{W/A} - \gamma_{W/O} - \gamma_{O/A} \tag{4.2}$$

where $\gamma_{W/A}$, $\gamma_{O/A}$ and $\gamma_{W/O}$ are the macroscopic interfacial tensions of the aqueous phase, oil phase and interfacial tension of the oil/water interface, respectively.

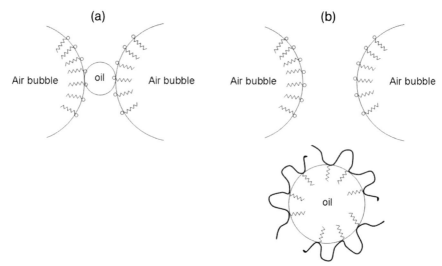

Figure 4.2 (a) Destabilization of the foam film by entering oil droplets; (b) the oil droplets which are stabilized by INUTEC®SP1 do not enter the foam film, which remains stable.

Ross and McBain [12] suggested that destabilization occurs when E and S are both greater than zero for entering and spreading. This leads to a displacement of the original film, leaving an unstable oil film that can easily be broken. For the above hydrating shower gel, without any added INUTEC®SP1 the oil droplets can enter the liquid film between the air bubbles, adsorbing some of the ammonium lauryl sulfate surfactant and causing destabilization of the foam film. In the presence of INUTEC®SP1, which is strongly and preferentially adsorbed at the oil/water interface, the oil droplets cannot enter the aqueous film between the air bubbles and hence the film remains stable. A schematic representation of the destabilization and stabilization mechanism of the foam is shown in Figure 4.2.

At this point it should be mentioned that the concentration of INUTEC®SP1 required to stabilize the system is 0.2% (2% based on the oil phase), and comparable to that used with the massage lotion. The formulation also has good skin-feel on application.

4.3.3
Soft Conditioner

This formulation contains 4.6% (w/w) hydrophobic components (cetearyl alcohol and cyclopentamethicone), and in this case 0.05% INIUTEC®SP1 was sufficient for its stabilization. This shows that the polymer surfactant is also effective in the stabilization of relatively more polar oils. The formulation also has a smooth skin-feel on application.

4.3.4
Sun Spray SPF19

One of the most useful applications of INUTEC®SP1 is with sprayable formulations, as the polymeric surfactant does not cause any increase in the viscosity of the system on application. The total oil content of this formulation was about 30% (w/w), and 0.75% INUTEC®SP1 was sufficient for its stabilization. This sprayable formulation is nonsticky and gives a pleasant skin-feel.

4.4
Conclusions

Hydrophobically modified inulin (INUTEC®SP1) can be applied in a wide variety of personal-care and cosmetic formulations. Its role is to enhance the stabilization of the system against strong flocculation, Ostwald ripening, coalescence and phase-inversion. The effect is due to the strong adsorption of the polymer surfactant at the oil/water interface (with multipoint attachment with several alkyl groups) and the strong hydration of the linear polyfructose loops and tails (enhanced steric stabilization). With shower gels, which contain oils, the addition of INUTEC®SP1 prevents foam destabilization by blocking entry of the oil droplets into the aqueous foam film between the air bubbles. In addition, INUTEC®SP1 enhances the performance of formulations on application, such as a lack of stickiness and greasiness, thus producing a good skin-feel.

References

1 Tadros, Th. F. (2005) *Applied Surfactants, Principles and Application*, Wiley-VCH, Germany.
2 Tadros, Th. F. (1999) Polymeric surfactants (eds. Goddard, E.D. and Gruber, J.V.), in *Principles of Polymer Science and Technology in Cosmetics and Personal Care*, Marcel Dekker, New York.
3 Tadros, Th. F. (2003) Polymeric surfactants (ed. Holmberg, K.), in *Novel Surfactants*, Marcel Dekker, New York.
4 Tadros, Th. F., Vandamme, A. Booten, K., Levecke, B. and Stevens, C.V. (2004) *Advances in Colloid and Interface Science*, **108–109**, 207.
5 Tadros, Th. F., Vandamme, A., Booten, K., Levecke, B. and Stevens, C.V. (2004) *Colloids and Surfaces*, **250**, 133.
6 Nestor, J., Esquena, J., Solans, C., Luckham, P.F., Musoke, M., Levecke, B., Booten, K. and Tadros, Th.F. (2007) *Journal of Colloid Interface Science*, **311**, 430.
7 Napper, D.H. (1983) *Polymeric stabilisation of colloidal dispersions*, Academic Press, London.
8 Binks, B.P. (ed.) (1998) *Modern aspects of emulsions*, The Royal Society of Chemistry, Cambridge.
9 Garett, P.R. (ed.) (1993) *Defoaming*, Marcel Dekker, New York.
10 Robinson, J.V. and Woods, W.W. (1948) *J. Soc. Chem. Ind.*, **67**, 361.
11 Harkins, W.D.(1941) *J. Phys. Chem.*, **9**, 552.
12 Ross, S. and McBain, J.W. (1944) *Ind. Chem. Eng.*, **36**, 560.

5
Effect of an External Force Field on Self-Ordering of Three-Phase Cellular Fluids in Two Dimensions

Waldemar Nowicki and Grażyna Nowicka

5.1
Introduction

Ordered cellular materials characterized by a periodic unit cell or by a repeating structure topology [1] are subjects of growing scientific interest because of their potential prognostic significance (see, for example, [2–18]). A variety of templating methods that use self-organization have been developed to produce two-dimensional (2-D) and three-dimensional (3-D) ordered structures [4, 19, 20]. Some of the methods utilize cellular fluids, represented by drained foams, high-internal-phase emulsions or foamed concentrated emulsions, as precursors of highly advanced materials, such as photonic crystals [20], well-ordered porous materials [18, 21–23], particles of special shape and properties [24, 25], composite polymers [26–33] as well as membranes for separation processes [29].

The cellular fluids have volume fractions of the dispersed phase, φ, higher than 0.74 [34, 35], which is the maximum volume fraction of close-packed monodisperse hard spheres; with increasing φ the bubbles/droplets deform to a polyhedral shape which enables their denser packing. Normally, dry foams or high-internal-phase emulsions have φ higher than 0.8 [36], and on occasion it may be as high as 0.99 [35, 37]. The topology of these structures is governed by the interfacial energy minimization. Two-dimensional monodisperse systems form highly ordered structures, the ideal case being the hexagonal regular honeycomb [38]. Recently, it has been shown that bidisperse 2-D systems may also organize into ordered cellular structures [39–43].

The ability of mono- and bidisperse cellular fluids to form highly organized structures has been studied mainly with two-phase systems. Recently we have shown that, under certain conditions, the mixing of two different cellular fluids, confined between parallel plates, may also lead to the cell arrangement into ordered patterns [43]. These 2-D patterns demonstrate a very interesting pseudo-phase behavior and anisotropic properties. However, the pattern formation requires some energy to be supplied to the system in order to overcome the energy barriers between different

Emulsion Science and Technology. Edited by Tharwat F. Tadros
Copyright © 2009 WILEY-VCH Verlag GmbH & Co. KGaA, Weinheim
ISBN: 978-3-527-32525-2

cell arrangements. The mixing of different cellular fluids is one of the possible pathways of composite materials formation [26–31, 33].

The problem addressed in this chapter is to examine possibilities of formation of ordered 2-D structures due to the gravity-driven mixing of cellular fluids. Here, we have used a modification of an earlier model [43] which consists of accounting for different fluid densities and the formation of cell clusters.

5.2
The Model

The modeling starts with the insertion of cells A into a network consisting of equal-volume fluid cells B dispersed in a liquid C, the content of which in the system is so low that cells B form space-filling polyhedra. The inserted A cells are assumed to consist of a fluid which is immiscible both with B and C phases, to be of the same volume and small as compared with the size of B cells. In order to force the 2-D structure the system is confined between two parallel plates separated by a distance d which is smaller than the criterion of Raleigh–Plateau instability [44] and allows the assumption that both A and B cells are in the shape of prisms, the bases of which are at surfaces of the plates. Thus, in the initial B/C dispersion, liquid C films in equilibrium make a hexagonal, regular honeycomb. In the mixed network, because of different interfacial tensions at the A/C and B/C interfaces, the Plateau angles can deviate from 120° [45, 46]. Attention is focused on the time scales that would allow neglect of the film breakage, and also on network sizes that would allow neglect of the influence of boundary conditions on the network topology.

Initially, the inserted cells make a cluster situated at a node of the network formed by B–C–B films or a wall of cell B. The cluster is surrounded by asymmetrical A–C–B films, whereas its inside-cells are separated by symmetrical A–C–A films. The behavior of the system after A cell insertion is governed by the tendency to reduce both the interfacial energy and the potential energy due to gravity. It is assumed that the gravitational force does not affect the cell shapes; such an assumption allows estimation of the two effects on the cell arrangement separately.

The energy of cluster insertion is defined as a difference between the interfacial energies of systems with and without the cluster (denoted as E_{ABC} and E_{BC}, respectively) in the absence of the gravitational field. The value of E_{ABC} was calculated by minimizing the interfacial energies of all films; film tensions were assumed as equal to the sum of interfacial tensions of film-constituting interfaces. In the energy minimization procedure, the cluster's size and position were assumed to be unchangeable; the optimization concerned only the lengths, curvatures and positions of the films. The angles between films meeting at the nodes were determined assuming the mechanical equilibrium and fitted in the course of energy minimization. As the cluster insertion involves a simultaneous addition to and elimination of network films, the expression for the energy of insertion of a cluster composed of N cells, $E(N)$, takes the form:

$$E(N) = E_{ABC} - E_{BC} = d[L_{AB}(\gamma_{AC} + \gamma_{BC}) + 2L_A\gamma_{AC} - 2L_B\gamma_{BC}] \quad (5.1)$$

where L_{AB}, L_A and L_B denote the lengths of A–C–B, A–C–A and B–C–B films, whereas γ_{AC} and γ_{BC} describe the interfacial tensions at the A/C and B/C interfaces, respectively. Further on in the studies (for the sake of convenience) we use the values of $E(N)$ calculated for a single A cell denoted as $\varepsilon(N_n)$ or $\varepsilon(N_w)$, where the subscripts n and w correspond to the node- and wall-locations of clusters, respectively. The Surface Evolver [47, 48] was used to compute all $E(N)$ values.

In order to study the evolution of the cluster inserted, we introduced a concept of the energy of cluster transformation which corresponds to a change in the system's energy due to the cluster dislocation and/or fragmentation. Because of topological limitations only the following cluster transformations are considered:

$$N_n \rightarrow N_w \quad (5.2)$$

$$N_w \rightarrow M_w + (N-M)_w \quad (5.3)$$

$$N_n \rightarrow M_n + (N-M)_w \quad (5.4)$$

where N indicates the initial size of the cluster, M is the size of the cluster residue, whereas (M–N) corresponds to the size of the dislocated cluster. The first process represents the cluster dislocation from the node to the wall, the second the cluster fragmentation on the wall, while the third process describes the fragmentation of node-located clusters accompanied by the dislocation of fragmentation product to the wall location. The transformation energies of these processes, denoted as E_d, E_f and E_{fd} respectively, were calculated from the following equations:

$$E_d = N\varepsilon(N_w) - N\varepsilon(N_n) \quad (5.5)$$

$$E_f = M\varepsilon(M_w) + (N-M)\varepsilon[(N-M)_w] - N\varepsilon(N_w) \quad (5.6)$$

$$E_{fd} = M\varepsilon(M_n) + (N-M)\varepsilon[(N-M)_w] - N\varepsilon(N_n) \quad (5.7)$$

The effect of gravity on cluster evolution was studied under assumptions that the gravitational force is parallel to plates confining the system, and that phases A and B differ in their densities.

5.3
Results and Discussion

5.3.1
Energies of Cluster Insertion and Transformation

In the first stage of the study we collected values of $\varepsilon(N_n)$ and $\varepsilon(N_w)$ calculated for different cluster sizes (up to $N=6$) and also for different values of the relative interfacial tension, μ, defined as:

$$\mu = \frac{\gamma_{AC}}{\gamma_{BC}} \tag{5.8}$$

Some exemplary results visualizing the effect of μ on the insertion energy are represented in Figure 5.1; the proportionality between these two quantities is observed for both node- and wall-cluster locations. The influence of cluster size on the values of $\varepsilon(N_n)$ and $\varepsilon(N_w)$ at different μ values is illustrated in Figure 5.2; for a better visibility of the effect the relative increments in the insertion energy of N-size cluster and that of a single A cell are presented. As a result, for the node-located clusters the lowest energy species is the singlet (1_n), except for extremely small μ values. The increase in N involves non-monotonic increase in $\varepsilon(N_n)$. It is noteworthy that the minima observed occur at N values being multiples of three; the clusters comprising such numbers of cells fit better the nodes having a three-fold symmetry axis and, hence, can have lower energy. On the other hand, for the wall-located clusters $\varepsilon(N_w)$ decreases monotonically with N in the whole range of the examined μ values; the decreasing trend of the dependencies points to the singlets as being the most labile species.

The effect of cluster location on the insertion energy is apparent in Figure 5.3. The left-hand part of the figure shows the dependencies of $\varepsilon(N_n)$ and $\varepsilon(N_w)$ on N, determined for $\mu = 1$. As can be seen, with increasing N the curves tend to a common asymptote, the position of which coincides with the energy of insertion of an infinite cluster ($\varepsilon(\infty)$). This observation allows it to be concluded that the energies of insertion of clusters comprising larger cell numbers than those studied here (i.e. $N > 6$) are in between those obtained for small clusters and $\varepsilon(\infty)$. The results presented in the right-hand part of Figure 5.3 allow an extension of the conclusion

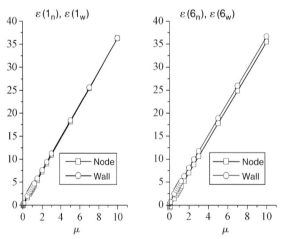

Figure 5.1 The influence of the relative interfacial tension, μ, on the insertion energies of both node- and wall-located clusters, $\varepsilon(N_n)$ and $\varepsilon(N_w)$ respectively, of two different sizes: $N = 1$ (left) and $N = 6$ (right).

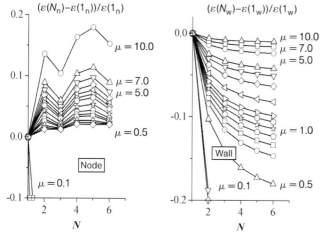

Figure 5.2 The influence of cluster size on the relative increments in insertion energies in the case of node-location (left) and wall-location (right) at different μ values (as indicated in the figures). The corresponding relative increments in insertion energies are defined as: $[\varepsilon(N_n) - \varepsilon(1_n)]/\varepsilon(1_n)$ and $[\varepsilon(N_w) - \varepsilon(1_w)]/\varepsilon(1_w)$.

over other μ values studied because, as shown in Figure 5.4, ordinate values of the asymptotes increase proportionally to μ, as should be expected for incompressible cellular fluids. A slight discrepancy from linearity, seen at small μ values, can be attributed to the symmetry compatibility between the nodes and the clusters where N is a multiple of 3.

The values of $\varepsilon(N_n)$ and $\varepsilon(N_w)$ collected for different N and μ were next used to calculate the transformation energies of various processes. Some exemplary results,

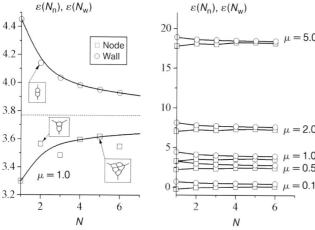

Figure 5.3 Plots of $\varepsilon(N_n)$ and $\varepsilon(N_w)$ versus N for $\mu = 1$ (left) and different values of μ (right).

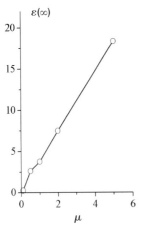

Figure 5.4 The dependence of $\varepsilon(\infty)$ on μ.

presented as the ratios of the transformation energy of a given process to that of the process: $2_n \rightarrow 1_n + 1_w$, are listed in Tables 5.1 to 5.3. These tables also provide the transformation energies corresponding to the detachment of small A clusters from the infinite cluster, followed by their dislocation to the wall. These energies, denoted as E_{dd}, were calculated assuming that the detachment of a small cluster does not influence the insertion energy of the infinite one. This process is represented schematically as:

$$(A/C) \rightarrow (A/C) + M_w \tag{5.9}$$

Table 5.1 Transformation energies E_d, E_{fd}, E_f and E_{dd} calculated for $\mu = 0.10$ ($E_0 = E_{fd}(2_n \rightarrow 1_n + 1_w)$).

Process	E_d/E_0	Process	E_{fd}/E_0	Process	E_f/E_0	Process	E_{dd}/E_0
$6_n \rightarrow 6_w$	3.955	$6_n \rightarrow 5_n + 1_w$	1.002	$6_w \rightarrow 5_w + 1_w$	0.562	$(A/C) \rightarrow (A/C) + 1_w$	0.799
$5_n \rightarrow 5_w$	3.514	$6_n \rightarrow 4_n + 2_w$	1.457	$6_w \rightarrow 4_w + 2_w$	0.669	$(A/C) \rightarrow (A/C) + 2_w$	1.143
$4_n \rightarrow 4_w$	3.166	$6_n \rightarrow 3_n + 3_w$	1.716	$6_w \rightarrow 3_w + 3_w$	0.652	$(A/C) \rightarrow (A/C) + 3_w$	1.371
$3_n \rightarrow 3_w$	2.890	$6_n \rightarrow 2_n + 4_w$	2.307	$5_w \rightarrow 4_w + 1_w$	0.560	$(A/C) \rightarrow (A/C) + 4_w$	1.615
$2_n \rightarrow 2_w$	2.317	$6_n \rightarrow 1_n + 5_w$	2.746	$5_w \rightarrow 3_w + 2_w$	0.661	$(A/C) \rightarrow (A/C) + 5_w$	1.853
$1_n \rightarrow 1_w$	1.770	$5_n \rightarrow 4_n + 1_w$	0.909	$4_w \rightarrow 3_w + 1_w$	0.555	$(A/C) \rightarrow (A/C) + 6_w$	2.090
		$5_n \rightarrow 3_n + 2_w$	1.285	$4_w \rightarrow 2_w + 2_w$	0.672		
		$5_n \rightarrow 2_n + 3_w$	1.859	$3_w \rightarrow 2_w + 1_w$	0.572		
		$5_n \rightarrow 1_n + 4_w$	2.304	$2_w \rightarrow 1_w + 1_w$	0.454		
		$4_n \rightarrow 3_n + 1_w$	0.830				
		$4_n \rightarrow 2_n + 2_w$	1.522				
		$4_n \rightarrow 1_n + 3_w$	1.950				
		$3_n \rightarrow 2_n + 1_w$	1.145				
		$3_n \rightarrow 1_n + 2_w$	1.691				
		$2_n \rightarrow 1_n + 1_w$	1.000				

Table 5.2 Transformation energies E_d, E_{fd}, E_f and E_{dd} calculated for $\mu = 0.50$ ($E_0 = E_{fd}(2_n \rightarrow 1_n + 1_w)$).

Process	E_d/E_0	Process	E_{fd}/E_0	Process	E_f/E_0	Process	E_{dd}/E_0
$6_n \rightarrow 6_w$	3.445	$6_n \rightarrow 5_n + 1_w$	1.709	$6_w \rightarrow 5_w + 1_w$	0.877	$(A/C) \rightarrow (A/C) + 1_w$	0.954
$5_n \rightarrow 5_w$	2.613	$6_n \rightarrow 4_n + 2_w$	1.953	$6_w \rightarrow 4_w + 2_w$	0.921	$(A/C) \rightarrow (A/C) + 2_w$	1.073
$4_n \rightarrow 4_w$	2.413	$6_n \rightarrow 3_n + 3_w$	1.892	$6_w \rightarrow 3_w + 3_w$	0.925	$(A/C) \rightarrow (A/C) + 3_w$	1.153
$3_n \rightarrow 3_w$	2.478	$6_n \rightarrow 2_n + 4_w$	2.603	$5_w \rightarrow 4_w + 1_w$	0.878	$(A/C) \rightarrow (A/C) + 4_w$	1.230
$2_n \rightarrow 2_w$	1.763	$6_n \rightarrow 1_n + 5_w$	2.724	$5_w \rightarrow 3_w + 2_w$	0.922	$(A/C) \rightarrow (A/C) + 5_w$	1.305
$1_n \rightarrow 1_w$	1.597	$5_n \rightarrow 4_n + 1_w$	1.078	$4_w \rightarrow 3_w + 1_w$	0.878	$(A/C) \rightarrow (A/C) + 6_w$	1.382
		$5_n \rightarrow 3_n + 2_w$	1.057	$4_w \rightarrow 2_w + 2_w$	0.917		
		$5_n \rightarrow 2_n + 3_w$	1.772	$3_w \rightarrow 2_w + 1_w$	0.874		
		$5_n \rightarrow 1_n + 4_w$	1.894	$2_w \rightarrow 1_w + 1_w$	0.834		
		$4_n \rightarrow 3_n + 1_w$	0.813				
		$4_n \rightarrow 2_n + 2_w$	1.567				
		$4_n \rightarrow 1_n + 3_w$	1.694				
		$3_n \rightarrow 2_n + 1_w$	1.588				
		$3_n \rightarrow 1_n + 2_w$	1.754				
		$2_n \rightarrow 1_n + 1_w$	1.000				

All processes quoted in Tables 5.1 to 5.3 are accompanied by positive values of transformation energy, which means that they are not spontaneous, although the opposite processes may proceed spontaneously. An analysis of the results presented in Tables 5.1 to 5.3 indicates that it is energetically favorable for the system to contain large, node-located clusters. Thus, when considering a system consisting of a B/C

Table 5.3 Transformation energies E_d, E_{fd}, E_f and E_{dd} calculated for $\mu = 1.00$ ($E_0 = E_{fd}(2_n \rightarrow 1_n + 1_w)$).

Process	E_d/E_0	Process	E_{fd}/E_0	Process	E_f/E_0	Process	E_{dd}/E_0
$6_n \rightarrow 6_w$	3.463	$6_n \rightarrow 5_n + 1_w$	1.868	$6_w \rightarrow 5_w + 1_w$	0.917	$(A/C) \rightarrow (A/C) + 1_w$	0.984
$5_n \rightarrow 5_w$	2.513	$6_n \rightarrow 4_n + 2_w$	2.078	$6_w \rightarrow 4_w + 2_w$	0.953	$(A/C) \rightarrow (A/C) + 2_w$	1.026
$4_n \rightarrow 4_w$	2.338	$6_n \rightarrow 3_n + 3_w$	1.925	$6_w \rightarrow 3_w + 3_w$	0.956	$(A/C) \rightarrow (A/C) + 3_w$	1.154
$3_n \rightarrow 3_w$	2.494	$6_n \rightarrow 2_n + 4_w$	2.679	$5_w \rightarrow 4_w + 1_w$	0.918	$(A/C) \rightarrow (A/C) + 4_w$	1.220
$2_n \rightarrow 2_w$	1.737	$6_n \rightarrow 1_n + 5_w$	2.761	$5_w \rightarrow 3_w + 2_w$	0.954	$(A/C) \rightarrow (A/C) + 5_w$	1.286
$1_n \rightarrow 1_w$	1.620	$5_n \rightarrow 4_n + 1_w$	1.093	$4_w \rightarrow 3_w + 1_w$	0.919	$(A/C) \rightarrow (A/C) + 6_w$	1.353
		$5_n \rightarrow 3_n + 2_w$	0.973	$4_w \rightarrow 2_w + 2_w$	0.952		
		$5_n \rightarrow 2_n + 3_w$	1.731	$3_w \rightarrow 2_w + 1_w$	0.916		
		$5_n \rightarrow 1_n + 4_w$	1.811	$2_w \rightarrow 1_w + 1_w$	0.883		
		$4_n \rightarrow 3_n + 1_w$	0.763				
		$4_n \rightarrow 2_n + 2_w$	1.554				
		$4_n \rightarrow 1_n + 3_w$	1.638				
		$3_n \rightarrow 2_n + 1_w$	1.673				
		$3_n \rightarrow 1_n + 2_w$	1.790				
		$2_n \rightarrow 1_n + 1_w$	1.000				

network in which a number of clusters of A cells is randomly distributed, the only processes that can be expected to occur are shifts of the wall-located clusters towards the nearest nodes, possibly accompanied by cluster aggregations. In order to distribute uniformly cells A between the nodes of the B/C network, some extra energy is required for the fragmentation of clusters into smaller species, and their migration. The necessary energy can be provided by numerous means, including gravity-driven mixing. The effect of gravity on the system evolution is discussed in the next section.

5.3.2
Evolution of the System in a Gravitational Field

Here, it is necessary to consider A/C and B/C systems in direct contact with each other, situated in the gravitational field and oriented in a way that allows the percolation of single A cells or their clusters through the B/C network. The component of the gravitational force acting on a single A cell along the wall, F_g, can be expressed as:

$$F_g = Vg|\rho_A - \rho_B|\cos\alpha \tag{5.10}$$

where V denotes the volume of A cell, g is the acceleration due to gravity, ρ_A and ρ_B are the densities of A and B phases, respectively, and α represents the angle between a wall of cell B and the direction of the gravitational field. Above a certain value of F_g the percolation of A cells through the B/C system will be complete, resulting in the reseparation of A and B cells. Also, there is a low limit of F_g beyond which no transformation in the system will occur. The question to be answered here is whether there is a force which can initiate a set of processes, the only end products of which are the node-located A clusters of the same size.

In the simplest case – that is, decoration of the nodes with A singlets – the applied force should be strong enough to cause fragmentation of cluster of any size but, at the same time, not be strong enough to move the singlet from the node. The force does not need to induce every possible fragmentation process – it must be sufficient only for the detachment of the single A cell from clusters of any size. This means that, for the decoration of nodes in the B/C network with A singlets, the following relationship should be fulfilled:

$$E_d(1_n \rightarrow 1_w) > E_{fd}[N_n \rightarrow (N-1)_n + 1_w] \quad \text{for} \quad N > 1 \tag{5.11}$$

This method of production of ordered mixed systems is further referred to as 'strategy I'. An analysis of the results presented in Tables 5.1 to 5.3 indicates that the criterion of Equation 5.11 is satisfied only at the lowest μ value – that is, when $\mu = 0.10$. Under the action of F_g that fulfils the requirements of strategy I, the only clusters infiltrating into the B/C system are those of $N \leq 4$, as their transformation energies $[E_{dd}((A/C) \rightarrow (A/C) + N_w)]$ are below the upper limit set by $E_d(1_n \rightarrow 1_w)$ value (see Table 5.1).

At higher μ the E_{fd} values for N being multiples of 3 exceed $E_d(1_n \rightarrow 1_w)$ and the larger the cluster size the higher the E_{fd}. This means that, during the course of the percolation process, in addition to A singles, also triplets, sextuplets and so on will be trapped in the nodes of B/C network. The results presented in Tables 5.1 to 5.3 allow a supposition that there is a certain limited maximum size of clusters, $N^{\lim}$, above which the transformation energy decreases with increasing N, as the energy of detachment of a single cell from clusters consisting of three or six cells is greater than that of the singlet detachment from the infinite cluster:

$$E_{fd}(3_n \rightarrow 2_n + 1_w) > E_{dd}[(A/C) \rightarrow (A/C) + 1_w] \tag{5.12}$$

and

$$E_{fd}(6_n \rightarrow 5_n + 1_w) > E_{dd}[(A/C) \rightarrow (A/C) + 1_w] \tag{5.13}$$

Thus, the application of a force which is sufficient for the fragmentation of any cluster except this with $N = N^{\lim}$ may be expected to result in the formation of a well-organized A/B/C system. This pathway of production of ordered mixed systems is further referred to as 'strategy II'.

A closer analysis of the results gathered in Tables 5.1 to 5.3 reveals yet another possible mode of decoration of the nodes in the B/C network with A singlets, referred to as 'strategy III'. This possible scenario assumes the singlets to be the only species percolating through the B/C system. To this end, the force F_g should be adjusted to a value which permits the detachment of singlets from the A/C cluster but is insufficient for the detachment of doublets or larger clusters. This means, that the following inequality should be fulfilled:

$$E_{dd}[(A/C) \rightarrow (A/C) + 2_w] > E_{dd}[(A/C) \rightarrow (A/C) + 1_w] \tag{5.14}$$

As results from the data given in Tables 5.1 to 5.3 show, the force F_g fulfilling the above requirement is also large enough for the fragmentation of doublets that can be formed in the nodes during the percolation, but insufficient for the decomposition of larger, node-located clusters. Thus, this pathway of production of ordered mixed cellular fluids might be effective provided that the process of doublet fragmentation would be fast enough to prevent the formation of irremovable triplets.

It follows from the above discussion that the forces applied to produce well-organized mixed cellular fluids should be carefully adjusted. In general, these forces should be within a certain interval, the boundaries of which are determined by the energies of two processes: on the one side (the lower boundary) it is the process of the largest energy from among all processes needed to ensure the system ordering (E_{min}), whereas on the other side (the upper boundary) – the process of the lowest energy (E_{max}) from among these that can contribute to the disorder.

When strategy I is applied the force used should be from the interval (F_{min}, F_{max}). Assuming the proportionality between the transformation energies and the forces needed for the corresponding processes to occur, it is possible to write:

$$\frac{E_{max}}{E_{min}} = \frac{F_{max}}{F_{min}} = Z_n \tag{5.15}$$

where $E_{max} = E_d(1_n \rightarrow 1_w)$ and $E_{min} = E_{fd}(N_n \rightarrow (N-1)_n + 1_w)$. The ratio Z_n for the other strategies can be obtained from Equation 5.15, when adequate values of E_{max} and E_{min} are substituted.

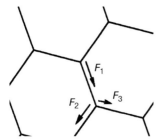

Figure 5.5 Schematic representation of F_g components acting along walls in the B/C system.

In the B/C system three different wall orientations in relation to the direction of gravitational field are possible and hence, as results from Equation 5.10, three different components of F_g. The percolation of A singlets through the system will occur only due to the action of the two highest components, F_1 and F_2, as shown schematically in Figure 5.5; usually, the two components are not the same ($F_1 > F_2$ in the situation depicted). In order to obtain a mixed system with an ordered arrangement of cells, the ratio of these two forces, $Z_g = F_1/F_2$, cannot exceed Z_n ($Z_g \leq Z_n$). Because of the network geometry, Z_g depends on α, and for α from 0 to $\pi/6$ it fulfils the following relationship:

$$Z_g = \max\left(\frac{\cos(\alpha)}{\sin(\alpha + \frac{\pi}{6})}, \frac{\sin(\alpha + \frac{\pi}{6})}{\cos(\alpha)}\right) \tag{5.16}$$

Because of the network symmetry, Z_g is a periodic function of α with a period $\pi/6$ (Figure 5.6), which also gives the Z_n values corresponding to different strategies and different values of μ. As seen, strategy I is effective only when α takes values close

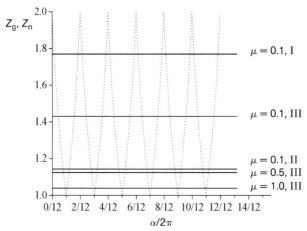

Figure 5.6 The dependence of Z_g on α (dotted line). The values of Z_n calculated for different μ and various strategies (I–III) are also indicated (solid lines).

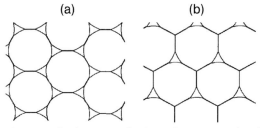

Figure 5.7 The decoration of nodes in the B/C network with A singlets when $\alpha = \pi/12$ (a) and $\alpha = 0$ (b) ($\rho_A > \rho_B$, $\mu = 0.10$).

to $\pi/12$ (where $Z_g < Z_n$). When the smaller force component, F_2, is close to F_{min} the gravity-caused percolation results in the decoration of all nodes with A singlets (Figure 5.7a). On the other hand, when $F_g \geq F_{min}$, and if the B/C walls are oriented parallel to the gravitational force, some singlets can escape from the nodes, since $Z_g = 2 > Z_n$. In consequence, only every second node can be decorated, as illustrated in Figure 5.7b.

Similar considerations can be made about the allowed values of F_g that can be applied for strategies II and III. As the respective interval boundaries for F_g change, Z_n also changes, as illustrated in Figure 5.6. In strategy II, the estimation of Z_n is possible only at $\mu = 0.1$, since at higher μ values we do not know the energy of singlet detachment from the cluster composed of N^{lim} cells. On the other hand, in strategy III it is interesting to note, that the width of the range of α values for which the requirement that $Z_g < Z_n$ is satisfied, increases with decreasing μ.

5.4 Conclusions

The results obtained in these studies indicate that the production of mixed 2-D cellular fluids having an ordered arrangement of cells requires very specific conditions which are difficult to realize. The applied intensity of gravitational field should fulfill certain criteria which depend on both the properties of the fluids (i.e. their densities) and the strategy used. The gravity force can be replaced by a centrifugal force produced by rotation; however, as the shape of the centrifugal force field is radially oriented and divergent, the ordered cell arrangement can be obtained only in a limited region of the system.

The above-presented approach to the behavior of multiphase cellular fluids in a gravitational field is very simplified. In particular, it neglects the energetic effects connected with the deformation of cells and the kinetics of processes involved in the formation of mixed systems. It cannot be excluded that these factors may significantly influence the behavior of the system under consideration. Nevertheless, the results obtained point to the possibility of exploiting the spontaneous mixing of cellular fluids in an external force field for the production of complex materials exhibiting both structural and compositional order.

References

1 Williams, C.B., Mistree, F. and Rosen, D.W., Proceedings of the DETC'05: ASME 2005 Design Engineering Technical Conference and Computer and Information in Engineering Conference, paper no DFMLC- 84832.
2 Priest, C., Herminghaus, S. and Seemann, R. (2006) *Appl. Phys. Lett.*, **88**, 024106.
3 Zsoldos, I., Réti, T. and Szasz, A. (2004) *Comput. Mater. Sci.*, **29**, 119–130.
4 Mezzenga, R., Ruokolainen, J., Fredrickson, G.H., Kramer, E.J., Moses, D., Heeger, A.J. and Ikkala, O. (2003) *Science*, **299**, 1827–1874.
5 Wang, A.-J. and McDowell, D.L. (2005) *Int. J. Plasticity*, **21**, 285–328.
6 Dempsey, B.M., Eisele, S. and McDowell, D.L. (2005) *Int. J. Heat Mass Transf.*, **48**, 527–535.
7 Miyoshi, H., Matsuo, H., Oku, Y., Tanaka, H., Yamada, K., Mikami, N., Takada, S., Hata, N. and Kikkawa, T. (2004) *Jap. J. Appl. Phys. Part 1*, **43**, 498–503.
8 Wang, B. and Cheng, G.D. (2005) *Struct. Multidiscip. Optim.*, **30**, 447–458.
9 Sharafat, S., Ghoniem, N., Williams. B. and Babcock, J. (2005) *Fusion Sci. Technol.*, **47**, 886–890.
10 Gibson, L.J. (2005) *J. Biomech.*, **38**, 377–399.
11 Seepersad, C.C., Dempsey, B.M., Allen, J.K., Mistree, F. and McDowell, D.L. (2004) *American Institute of Aeronautics and Astronautics Journal*, **42**, 1025–1033.
12 Wu, J., Abu-Omar, M.M. and Tolbert, S.H. (2001) *Nano Lett.*, **1**, 27–31.
13 Widawski, G., Rawiso, M. and François, B. (1994) *Nature*, **369**, 387–389.
14 Antonietti, M. and Ozin, G.A. (2004) *Chem. Eur. J.*, **10**, 28–41.
15 Yabu, H., Takebayashi, M., Tanaka, M. and Shimomura, M. (2005) *Langmuir*, **21**, 3235–3237.
16 Yu, C., Zhai, J., Gao, X., Wan, M., Jiang, L., Li, T. and Li, Z. (2004) *J. Phys. Chem. B*, **108**, 4586–4589.
17 Fazekas, A., Dendievel, R., Salvo, L. and Brecht, Y. (2002) *Int. J. Mech. Sci.*, **44**, 2047–2066.
18 Imhof, A. and Pine, D.J. (1997) *Nature*, **389**, 948–951.
19 Srinvasarao, M., Collings, D., Philips, A. and Patel, S. (2001) *Science*, **292**, 79–83 and references therein.
20 Li, J., Peng, J., Huang, W., Wu, Y., Fu, J., Cong, Y., Xue, L. and Han, Y. (2005) *Langmuir*, **21**, 2017–2021 and references therein.
21 Yi, G.R. and Yang, S.M. (1999) *Chem. Mater.*, **11**, 2322–2325.
22 Cai, M.M., Shen, J.H. and Ji, Z. (2003) *Rare Metal Mater. Engin.*, **32**, 752–754.
23 Carnachan, R.J., Bokhari, M., Przyborski, S.A. and Cameron, N.R. (2006) *Soft Matter*, **2**, 608–616.
24 Xiang, A., Du, Z., Zeng, Q., Zhang, C. and Li, H. (2005) *Polym. Int.*, **54**, 1366–1370.
25 Yun, Y., Li, H.Q. and Ruckenstein, E. (2001) *J. Colloid Interf. Sci.*, **238**, 414–419.
26 Ruckenstein, E. and Li, H. (1995) *Polym. Bull.*, **35**, 517–524.
27 Ruckenstein, E. and Li, H. (1996) *Polymer*, **37**, 3373–3378.
28 Ruckenstein, E. and Li, H. (1997) *Polym. Composites*, **18**, 320–331.
29 Ruckenstein, E. (1997) *Adv. Polym. Sci.*, **127**, 1–58.
30 Li, H., Huang, H. and Ruckenstein, E. (1999) *J. Polym. Sci., Part A. Polym. Chem.*, **37**, 4233–4240.
31 Li, H., Huang, H. and Ruckenstein, E. (2001) *J. Polym. Sci., Part A. Polym. Chem.*, **39**, 757–764.
32 Mezzenga, R., Ruokolainen, J., Fredrickson, G.H. and Kramer, E.J. (2003) *Macromolecules*, **36**, 4466–4471.
33 Du, Z., Zhang, C. and Li, H. (2005) *Polym. Int.*, **54**, 1377–1383.
34 Solans, C., Esquena, J. and Azemar, N. (2003) *Curr. Opin. Colloid Interf. Sci.*, **8**, 156–163.

35 Cameron, N.R. (2005) *Polymer*, **46**, 1439–1449.
36 Bergeron, V. and Walstra, P., Foams (2005) in *Fundamentals of Interface and Colloid Science, Volume V: Soft Matter* (ed. J. Lyklema), Elsevier Academic Press, Amsterdam, p. 74.
37 Richez, A. Deleuze, H., Veddrenne, P. and Collier, R. (2005) *J. Appl. Polym. Sci.*, **96**, 2053–2063.
38 Hales, T.C. (2001) *Discrete Comput. Geom.*, **25**, 122.
39 Fortes, M.A. and Teixeira, P.I.C. (2001) *The European Physical Journal E – Soft Matter*, **6**, 131–137.
40 Teixeira, P.I.C., Graner, F. and Fortes, M.A. (2002) *The European Physical Journal E – Soft Matter*, **9**, 447–452.
41 Li, J.R., Cheng, H.F., Yu, J.L. and Han, F.S. (2003) *Mater. Sci. Eng.*, **A362**, 240–248.
42 Seo, M., Nie, Z., Xu, S., Lewis, P. and Kumacheva, E. (2005) *Langmuir*, **11**, 4773–4775.
43 Nowicki, W. and Nowicka, G. (2004) *The European Physical Journal E – Soft Matter*, **13**, 409–415.
44 Isenberg, C., (1992) *The Science of Soap Films and Soap Bubbles*, New York, Dover.
45 Han, G.B., Dussaud, A., Prunet-Foch, B., Neimark, A.V. and Vignes-Adler, M. (2000) *J. Non-Equilib. Thermodyn.*, **25**, 325–335.
46 Neimark, V. and Vignes-Adler, M. (1995) *Phys. Rev. E*, **51**, 788–791.
47 Brakke, K. (1992) *Exp. Math.*, **1**, 141–165.
48 http://www.susqu.edu/facstaff/b/brakke/evolver/evolver.html. Accessed Dec. 11 (2006).

6
The Physical Chemistry and Sensory Properties of Cosmetic Emulsions: Application to Face Make-Up Foundations

Frédéric Auguste and Florence Levy

6.1
Introduction

The purpose of any cosmetic (e.g. skin care, make-up or hairstyling), regardless of the surface to which it is applied – whether skin, hair, eyelashes or lips – is to improve our perception of that surface and, in some circumstances, also to protect it. When the product is intended to improve the perception of the surface, this may be achieved through touch (by modifying the texture or roughness) or through sight (by smoothing out skin contours, by contributing substances with specific optical properties). Moreover, the effect must be as long-lasting as possible.

In this respect, cosmetic emulsions differ minimally from those found in paints or bitumens, except that they must take into account the living tissue of the application surface. In addition to the classic physico-chemical problems of colloids associated with the study of emulsions (preparation, stability, conveyance of active ingredients, etc.), any cosmetic emulsion when applied to a living surface must also feel comfortable and maintain the desired properties for as long as possible. This must be achieved despite assaults emanating not only from the surface itself, such as movement, perspiration, sebum, saliva and tears, but also from the outside, such as mechanical aggression and contact with liquids (e.g. water, coffee, salad dressing).

In a facial make-up such as an emulsified fluid foundation, a major problem is to identify the correct compromise between comfort and the cosmetic result. Traditionally, we refer to the application comfort parameter of 'play-time'; this is defined as the length of time that the user can apply the product to the face before it becomes uncomfortable or is transformed drastically into an undesirable effect (such as 'balling'). In the past, the play-time has always been an empirical, subjective factor associated with comfort, and until recently had never been studied with the aim of achieving a better definition through improved understanding. All that was known was that the so-called 'long-lasting' foundations generally had a short play-time and low comfort level, whilst 'comfort' foundations had a long play-time but came off too easily. Between these two types there were the 'medium' foundations, with middle-level play-time, durability and comfort level – in other words of good, average quality. Not only did the notion of play-time elude the understanding of the physical chemist but the notions of durability and comfort were every bit as ill-defined.

It was this ignorance of the connection between cosmetic properties and formula composition that motivated the present studies. Here, for the first time, we have concentrated – through the physico-chemical characterization of various commercial formulas – on correlating the composition of the formulas with sensory properties such as play-time and durability, rather than on the formulation and stability of the emulsions.

6.2
Materials and Methods

6.2.1
Selection of the Foundations to be Studied

Most liquid foundations are inverse emulsions containing colored or uncolored solid particles (pigments, fillers), an aqueous phase (water, glycols such as propylene glycol or butylene glycol) dispersed in an oily phase (hydrocarbonated oils or silicones, either volatile or not), one or several (surfactants), possibly 1–5% of polymers, and additives (preservatives, fragrances). The magnitude of the average diameter of the emulsion droplets is 1 to 5 μm.

The present study was carried out on commercial inverse emulsion foundations selected partly according to composition (formulas with compositions as different as possible in terms of volume fractions of dispersed solids, volatile and nonvolatile liquids, and polymers) and partly on the basis of their sensory properties [1]. The decision was taken to use the volume fractions of different families of constituents as the criteria upon which to compare the formulas. This was because, on the one hand, the large number of components (up to 30) in the formulas makes any comparison extremely difficult and, on the other hand, because these parameters are of prime importance in understanding how the properties of the products evolve upon application and while drying.

6.2.2
Characterization Methods

When applied to a surface an emulsion undergoes a shear, which is characterized by rheology, and also shows a drift in composition resulting from evaporation, which is characterized by weighing and quantitative analysis of the constituents after application on a model surface.

Drying of the Foundations Samples (50 g) of each foundation were subjected to slow stirring for 24 h, under permanent ventilation and controlled temperature. Preliminary microscopic and granulometric observations established that this shearing brought about no change in the size of the droplets of nonvolatile liquids. Samples were taken at regular intervals, and their composition was determined using various techniques. Water was titrated using the Karl Fischer method [2], concentrations of glycol (propylene glycol, butylene glycol, pentylene glycol, glycerol) and ethanol were monitored by gas chromatography (GC), and the total concentration of nonvolatile components by gravimetry.

Flow Rheology The samples taken at different drying times were characterized in terms of their flow rheology by using a Haake Rheostress RS600 rheometer. The information acquired was the value of the viscosity at $1000\,s^{-1}$, assimilated to the shear gradient while a layer of foundation 10 µm thick was being applied to the skin at a rate of $1\,cm\,s^{-1}$.

Determination of Drying Rate During Application A specific device to deposit the foundation onto a model skin surface was engineered to reproduce coating conditions similar to those of real skin. The initial quantity deposited was $1\,mg\,cm^{-2}$, and the mass of foundation before and after application was monitored using a Mettler Toledo PR503 precision balance. A study carried out simultaneously on the coalescence of water droplets on the skin during shearing showed that, for the formulas studied, no such coalescence occurred ('breakdown' in cosmetic parlance).

Observation of the Deposits Replicas of trapped foundation applications to the inner wrist were cast using a low-surface-tension crosslinkable resin and observed in reflection under a Leica DMLB optical microscope. The images were then processed using a software program (Multifocus, Microvision Instruments, France) to correct the blurred effects caused by the contours.

6.3
Experimental Results and Discussion

6.3.1
Drying of the Foundation Bulk and Drift in Composition During Drying

The drift in composition undergone by foundation D over time during drying under slow blade stirring is shown in Figure 6.1. The raw results (mass loss over time) are shown in Figure 6.1a, while processed results (evolution of the volume fraction of dispersed phase during drying) are shown in Figure 6.1b.

These results, which can be extended to the other foundations under study, concur with previously published data on the drying of emulsions [3–5], and show simultaneous evaporation of the dispersed and continuous phases (Figure 6.1a). Complementary tests have demonstrated that the evaporation of the dispersed

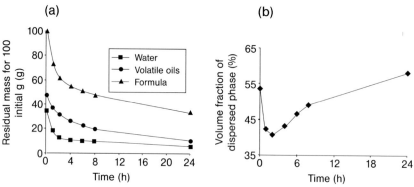

Figure 6.1 Drying of foundation D.

aqueous phase also depends on the nature of the continuous phase, in line with hypotheses proposed elsewhere on the possible diffusion of the dispersed phase through the continuous phase [3].

This simultaneous evaporation of the aqueous and volatile oily phases is faster for the aqueous phase in the present case, and this appears as a two-stage evolution of the volume fraction of the dispersed phase (Figure 6.1b). Initially, there is a decrease in the dispersed phase – that is, there is a faster evaporation of the aqueous dispersed phase than of the oily continuous phase – and this is followed by concentration of the dispersed phase (pigments, fillers, droplets of residual aqueous phase).

6.3.2
Evolution of Viscosity During Drying

Figure 6.2 illustrates the evolution of viscosity at $1000\,s^{-1}$ as a function of drying time for several foundations. A variety of behaviors was observed: for certain foundations the viscosity increased very strongly as drying began (A), whereas for others the increase was much slower (C, D, E, F), while in some cases there is no increase at all (B).

In an attempt to understand these different behaviors, three parameters that can affect viscosity during drying have been considered:

- Evolution of the viscosity of the continuous phase (evaporation of certain solvents, polymer concentration).
- Evolution of the volume fraction of the dispersed phase (according to the evaporation rate of the continuous and dispersed phases, cf. Figure 6.1b).
- Evolution of the aggregation of the dispersed particles [6].

The evolution of viscosity of the oily phase during drying, linked to evaporation of the oils with a lower molar mass, does not suffice to explain the measured evolutions, as in general the viscosities of the oils used range from 5 to $50\,mPa\cdot s$.

Likewise, evolution of the viscosity linked to polymer concentration cannot explain all of these results, since certain foundations become more viscous as the polymers

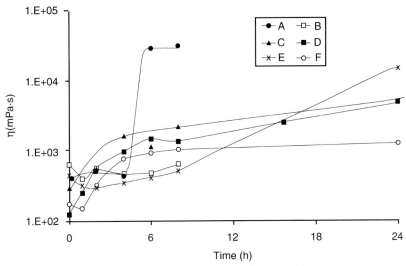

Figure 6.2 Evolution of the viscosity of foundations as a function of drying time.

are concentrated (A and C), while others show an increase in viscosity even though they contain no polymers (B and D), and the viscosity of foundation F remains unchanged despite a notable rise in polymer concentration.

As for the change in viscosity as the volume fraction of the dispersed phase evolves (cf. the curves in Figure 6.3 obtained from those in Figures 6.1b and 6.2), all foundations yielded a viscosity plateau or quasi-plateau at the start of drying

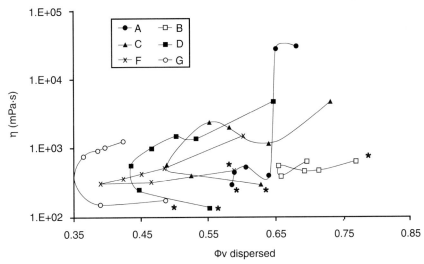

Figure 6.3 Evolution of the viscosity of foundations as a function of the volume fraction of the dispersed phase. The asterisks correspond to the start of drying.

(the points marked with an asterisk) that end only when evaporation of the internal aqueous phase ends. The evaporation of the continuous phase entails an increase in the concentration of the dispersed phase. In this way, evolution of the viscosity can be linked with evolution of the volume fraction of the dispersed phase.

This argument is, however, also insufficient to explain all of the present observations. If evolution of the viscosity were driven only by the volume fraction of the dispersed phase, not only would the viscosity begin by dropping as the dispersed phase decreases but a single viscosity should also characterize a given volume fraction of the dispersed phase for any given foundation, which is not the case.

Particle aggregation in the dispersed phase, taken together (according to the foundation) with a simultaneous decrease of the dispersed phase (lower viscosity) and with a more viscous continuous phase (polymer concentration or nonvolatile oils viscosity), would then be the important parameter in explaining the observations.

Inversely, if there were no decrease in the volume fraction of the dispersed phase, we would see an initial rise in viscosity rather than a plateau.

Thus, it is concluded that whilst the volume fraction of the dispersed phase is important in explaining the plateaus observed, particle aggregation during drying and the increase in polymer concentration in some foundations are also important parameters driving the change in rheology as the formulas dry.

A final significant factor is the presence of small spherical particles that act as lubricants, even when there is practically no liquid left in the formulas (B and F) and even when polymer is present, as in foundation B.

6.3.3
Play-Time and Disposition of Foundation on the Skin

The loss of mass measured during a 10 s application indicates evaporation rates which range from 27% to 41% of the foundations' initial mass, as follows:

	Foundation				
	A	B	D	E	F
Mass loss (%)	30.3	34.7	27.5	23.6	40.6

From these results it is possible to evaluate the drift in composition that takes place during the 10 s period of application. Assuming that this type of shearing has a negligible effect on the formulas' rheological properties, the results given in Figures 6.1, 6.2 and 6.3 can be combined to obtain information on how viscosity evolves during application. By way of example, the evaporated mass during evaporation is plotted in Figure 6.4 on the viscosity/evaporated mass curve for foundation D.

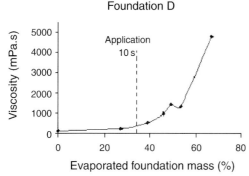

Figure 6.4 Plot of the composition drift at the end of application on the viscosity/composition drift curve.

Although it rests on numerous hypotheses and approximations that remain to be verified, this method can be applied to all of the foundations examined in these studies and those to be developed later to provide qualitative information on how play-time correlates with the sensory evaluations. (Typically, a small increase in viscosity during application = a long play-time and good application comfort; a large increase in viscosity during application = a short play-time and application discomfort.) The evolution of viscosity during drying, driven by the drift in composition which also occurs, is thus the key factor in explaining and eventually modifying the play-time.

A more thorough study of the impact of composition drift on the properties of foundations has shown that it affects not only the play-time but also the way the foundation is dispersed on the skin. The play-time and composition drift determine how the foundation is placed (Figure 6.5):

- When the viscosity is strongly altered during application (A), the foundation 'jams' – that is, it cannot be spread into the hollows of the skin contours and ultimately is mainly applied over the plateaus of the contours. Too short a play-time therefore equals an uneven distribution of the foundation on the skin. The advantage of this type of formula is the durability provided by the polymer, which sticks the particles to the skin.

- When the viscosity of the foundation changes little during application (B, F), the way in which it is finally dispersed depends on how the hollows of the skin contours are filled with the remaining product. If there is only a small volume left, then only the hollows of the skin contours will be filled (B); however, when the leftover volume of make-up surpasses the volume of the skin contour hollows, then the plateaus will also be covered (F).

This type of observation of how foundations are dispersed on the skin also provides information on the different factors which drive durability, such as protection of the foundation in the hollows of the contours (Figure 6.5, foundation B) or how

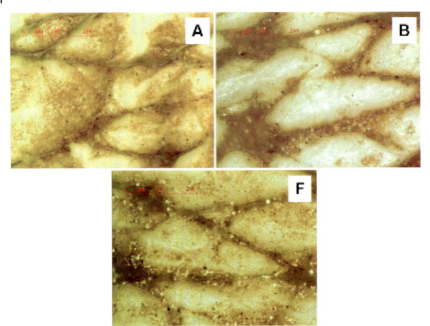

Figure 6.5 Observations of how the foundation is deposited on the skin.

the polymers act to jam and thus protect the particles deposited on the plateaus of the contours (foundation A).

6.4
Conclusions

From the partial understanding acquired to date we are able to link subjective values such as play-time both to measurable values such as the durability of the product on the skin, the viscosity or the drift in composition, and to composition parameters (a choice of solvents of known volatility, polymers, volume fraction of solids, etc.). This in turn makes it possible to vary the composition parameters in many ways to produce foundations that are not only comfortable (i.e. a good play-time) but are also durable and show application comfort, by fine-tuning the evaporation rates of the dispersed and continuous phases of the emulsions, the compatibility between these two phases, and the fillers.

References

1 Barthelemy, J. (1998) Evaluation d'une grandeur sensorielle complexe: description quantifiée. In: Depledt,E. and Société Scientifique d'hygiène Alimentaire (eds.), *Evaluation sensorielle, manuel méthodologique*, 2nd edition, Lavoisier, Technique & Documentation, Paris, pp. 149–162.

2 Charlot, G. (1966), *Les méthodes de la chimie analytique*, 5th edition, Masson & Cie, Paris, p. 834.

3 Aranberri, I., Beverley, K.J., Binks, B.P., Clint, J.H. and Fletcher, P.D.I. (2002) *Langmuir*, **19**, 3471–3475.

4 Friberg, S.E. and Langlois, R.C. (1992) *J. Dispersion Science and Technology*, **13**(2) 223–243.

5 Langlois, R.C. and Friberg, S.E. (1993) *J. Soc. Cosmet. Chem*, **44**, 23–34.

6 Leal-Calderon, F., Mondain-Monval, O., Pays, K., Royer, N. and Bibette, J. (1997) *Langmuir*, **13**(26), 7008–7011.

7
Nanoparticle Preparation by Miniemulsion Polymerization
Man Wu, Elise Rotureau, Emmanuelle Marie, Edith Dellacherie, and Alain Durand

7.1
Introduction

Today, submicronic colloidal systems are to be found over a wide range of applications, including domains such as medicine, cosmetics and food. The properties of these colloid systems are largely controlled by their surface characteristics and are of fundamental importance for their potential applications. The preparation of polymeric nanoparticles with well-defined surface properties can be carried out by several processes, many of which involve an initial oil-in-water (O/W) emulsion in which the stabilizer is a polymeric surfactant. Miniemulsion polymerization represents a convenient method to control the surface properties of colloidal systems. Aqueous miniemulsions are obtained via a high energy input to a mixture of monomer, water, stabilizer and a highly water-insoluble compound (added in the monomer), the so-called 'hydrophobe'. The hydrophobe slows down the spontaneous mass exchange between oil droplets (known as 'Ostwald ripening'), while the stabilizer prevents droplet coalescence [1, 2]. The polymerization of these droplets leads to particles which ideally keep their size (with only a slight reduction in size because of density variation) [3]. A wide range of ionic [4, 5] and nonionic [6, 7] surfactants have been used, leading to lattices with different surface charges and colloidal stability. To the best of our knowledge, although relatively few previous publications have dealt with the use of macromolecular surfactants in miniemulsion polymerization, various studies involving both random [8–12] and block [13–17] copolymers have been conducted. Nevertheless, the use of polymeric surfactants prepared from polysaccharides is a relatively recent innovation in miniemulsion polymerization [9, 18, 19], although several investigations have been conducted on the subject of emulsion polymerization [18, 20].

In these studies, biocompatible polymeric surfactants based on dextran have been used as stabilizers for O/W emulsions. These polysaccharidic surfactants have been prepared by reacting dextran with aliphatic or aromatic epoxides, as described elsewhere [21]. Anionic groups were also attached to the polysaccharide chains by reaction with propanesultone. The stability of the prepared emulsions was analyzed and the main aging processes were identified. Polymeric nanoparticles were prepared by miniemulsion polymerization using two different monomers, namely *styrene* and *butylcyanoacrylate*. Control of the surface properties of the resultant nanoparticles was verified, and the colloidal stability of the lattices examined as a function of ionic strength and temperature.

7.2 Experimental

7.2.1 Materials

The native dextran was obtained from Pharmacia (Uppsala, Sweden). All other materials were from Aldrich (St. Quentin Fallavier, France) and used as received. MilliQ water was used for all of the experiments.

7.2.2 Emulsion Preparation

Oil-in-water emulsions were prepared by sonication (pulsed mode) using a Vibracell model 600 W (Sonics & Materials Inc., Danbury, CT, USA). Dodecane was used without further purification. The total volume of oil and aqueous phase was kept equal at 10 ml for all of the emulsions prepared. The polymer was previously dissolved in the aqueous phase over a 20 h period.

7.2.3 Polymerization

The amphiphilic derivatives of dextran were dissolved in MilliQ water. The organic phase is composed of styrene, hexadecane (5 vol% related to styrene) and azoisobutyronitrile (AIBN) (1.5 wt% related to styrene). The volume fraction of the organic phase was varied from 5 to 20 vol%. After stirring for 1 h, emulsification was achieved via sonication (pulsed mode, 10 W, 3 min; Vibracell model 600W). In order to avoid polymerization due to heating the mixture was ice-cooled during sonication. The polymerization was subsequently performed at 75 °C for 24 h.

7.2.4 Size Measurement of the Emulsion Droplets

Droplet sizes were measured by dynamic light scattering at low concentration using a HPPS-ET® (Malvern). Although this apparatus is able to measure relatively concentrated samples, the emulsions were diluted. As dilution with oil-saturated water or with the polymer solution yielded the same results, pure water was subsequently used for diluting the direct emulsions.

7.2.5
Particle Characterization

Nanosphere size distribution was studied by photon correlation spectroscopy experiments using a HPPS-ET®.

The colloidal stability of latex suspensions in the presence of added electrolyte or at increasing temperature was assessed with turbidimetry. Typically, 20 µl of each dispersion was added to 3 ml of NaCl solutions (ranging from 10^{-4} to 4 mol l^{-1}). The samples were allowed to stand for 40 min, after which their absorbance was measured over the range 450–700 nm, at 50 nm intervals. The slope n of the straight line log (optical density) versus log (wavelength) was taken as an indication of particle size. The presence of flocculation was evidenced by a sharp decrease in n-value [22].

7.3
Results and Discussion

7.3.1
Synthesis of Hydrophobically Modified Dextrans

Polysaccharide surfactants have been synthesized from dextran, a neutral bacterial polysaccharide composed of glucose units with α 1,6 linkages. A commercial dextran sample T40® (average M_n = 26 000 g mol^{-1}; average M_w = 40 000 g mol^{-1}, as determined by size-exclusion chromatography) was used as the precursor.

The hydrocarbon moieties were introduced within the polysaccharide chains by reacting dextran with aliphatic or aromatic epoxides. This synthesis relies on the ability of hydroxyl groups of the polysaccharide to react with the epoxide. A first condition is to carry out the reaction in an alkaline medium, so as to obtain a basic catalysis. This has been used for hydrophobic modification of other polysaccharides such as cellulose or inulin [23]. A second requirement for the reaction is the availability of a solvent which is common to the polysaccharide and to the epoxide.

With an aromatic epoxide such as phenylglycidylether, the solubility in water even low is enough to carry out the modification in an aqueous medium. Phenylglycidylether is dispersed under vigorous stirring in an aqueous alkaline solution of dextran. The epoxide is then progressively dissolved in the aqueous phase and reacts with hydroxyl groups of dextran. In that process, a significant part of the epoxide molecules is in fact consumed by reaction with hydroxide ions, which leads to a yield of epoxide grafting close to 12% [18].

With aliphatic epoxides (epoxyoctane will be used here), the previous procedure leads to a limited dextran modification, partly because of the very low solubility of these epoxides in water. Two possibilities have been examined – either to increase the epoxide transfer into the aqueous phase by adding a cationic surfactant, or to replace water by a solvent which is common to both dextran and aliphatic epoxides, namely dimethylsulfoxide (DMSO) [24, 25]. In the work reported here the latter procedure was used.

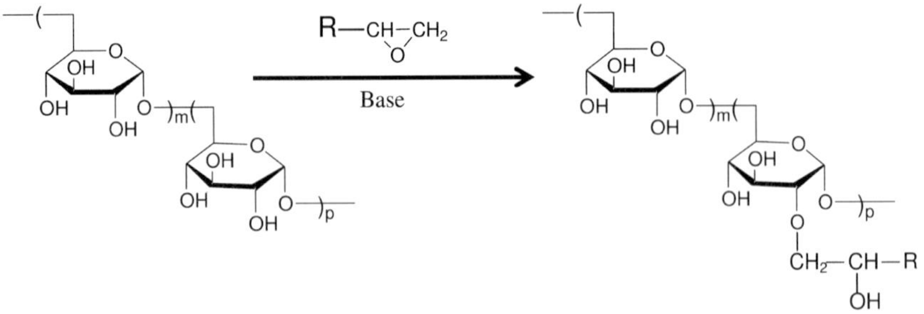

Scheme 7.1 Chemical modification of dextran by epoxides. In an aqueous medium the base is sodium hydroxide; in dimethylsulfoxide, tetrabutylammonium hydroxide is used. R may be either an aromatic group or an aliphatic chain. Although the hydrocarbon groups are represented after reaction with one particular hydroxyl group, it is clear that several positions could be etherified in one glucose unit.

For the two modification procedures, the number of hydrocarbon groups attached onto glucose units was varied by changing either the amount of added epoxide in the feed or the duration of the reaction.

The percentage of glucose units modified with a hydrocarbon group is called the 'degree of hydrophobic modification': $\tau = 100 \times p/(m + p)$ (see Scheme 7.1). The polymers created are named $DexP_\tau$ and $DexC6_\tau$ according to the nature of the hydrocarbon groups attached: phenoxy (P) or n-C_6H_{13}–(C6).

Limited amounts of anionic groups (sodium propylsulfonate) have also been introduced within the polysaccharide chains. This was achieved by reacting dextran with various amounts of 1,3-propanesultone, a reactant which has also been used for other polysaccharides [26, 27]. The reaction conditions were identical to those depicted above for the reaction with aliphatic epoxides.

Further details concerning the chemical modification of dextran can be found elsewhere [18, 28]. Following such modification procedure, amphiphilic dextrans were obtained in which the hydrocarbon groups were linked to the backbone by ether functions. These groups exhibited a better chemical stability against hydrolysis than, for example, ester functions. This may be an important property for some miniemulsion polymerizations, as detailed below (see Section 7.3.4.2).

The amount of anionic groups grafted was expressed by the degree of sulfopropyl-substitution, σ (%), defined by: $\sigma = 100 \times \frac{y}{x+y+z}$ (see Scheme 7.2). The polymers were named $DexP_\tau S_\sigma$ or $DexC6_\tau S_\sigma$, according to the nature of the hydrocarbon moieties. The degrees of hydrophobic and anionic modification were calculated from integrations in ^{1}H NMR spectra carried out in deuterated DMSO. Essentially, three signals were used in the spectra: (1) the peak corresponding to the anomeric proton of glucose units; (2) the peak of the aliphatic chains of epoxides; and (3) the peak of the methylene groups of propanesultone.

The solution and interfacial properties have been reported elsewhere and correlated to the chemical characteristics of the polymers [29]. These data have proved to be fundamental for a convenient design of miniemulsion process, and

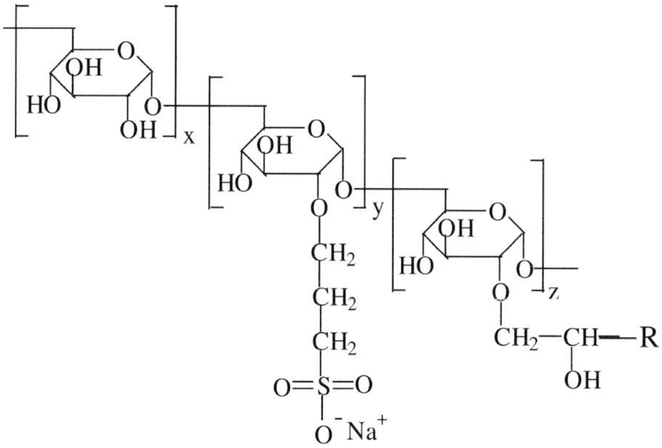

Scheme 7.2 Repeat units of DexP$_\tau$S$_\sigma$ and DexC6$_\tau$S$_\sigma$. The sub-stituent group R holds for —CH$_2$—O—C$_6$H$_6$ and —(CH$_2$)$_5$—CH$_3$, respectively. As for the position of the substituents within one glucose unit, see the footnote of Scheme 7.1.

hence will not be described in detail at this point. Only the emulsifying properties will be discussed in the following sections.

7.3.2
Preparation of O/W Miniemulsions

Here, the emulsifying properties of amphiphilic dextrans will be examined through the preparation of O/W miniemulsions, in which the oil is an aliphatic hydrocarbon compound. Two types of parameter will be distinguished, namely those related to the process conditions and those linked with the chemical structure of the polymers.

7.3.2.1 Control of Initial Droplet Size by Process Variables

The energy required for emulsion preparation is provided by ultrasound, using a sonication process. An initial verification was conducted to determine that, under the process conditions being used, the dextran macromolecules would not be degraded; the results obtained were consistent with data published elsewhere [30]. The influence of process variables on initial droplet size has been examined in detail, with three process parameters being considered: (1) power input; (2) the duration of the sonication stage; and (3) the mean of application of the ultrasound (the ultrasound stages were alternated with rest stages of different duration). Within the ranges examined, the relevant parameter was shown to be the duration of sonication.

For given process conditions, the initial droplet size appeared to depend on the weight ratio of the surfactant polymer-to-oil, denoted α (Figure 7.1). Below a value of approximately 0.05, the initial droplet size was seen to vary significantly with α. Within that range of α, the droplet size was limited by the amount of polymer available to cover the oil droplets with a minimum surface coverage that was high enough to prevent droplet coalescence. For α values above 0.05, the initial droplet size

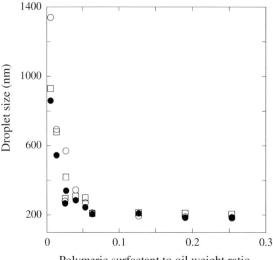

Figure 7.1 Initial droplet diameter as a function of polymer-to-oil weight ratio. The oil is dodecane. ● DexC6$_{12}$; ○ DexC6$_{38}$; □ DexC10$_{11}$.

did not vary significantly, and for this two explanations were proposed. First; the droplet size may have been limited by the emulsification process, independently of the polymeric surfactant itself. Second, the rate of diffusion of the polymer to the newly created interfaces may not have been fast enough to stabilize smaller droplets; consequently, small droplets were able to grow (by coalescence or Ostwald ripening) up to approximately 150 nm before the polymer covered the interface.

7.3.2.2 Influence of Polymer Structure on Initial Droplet Size

With native dextran, the initial droplet size remained close to 800 nm, irrespective of the concentration. For neutral dextran derivatives, the nature of the hydrocarbon groups attached along the polysaccharide backbone did not have any significant effect on the initial droplet size (see Figure 7.1). Similarly, the number of attached hydrocarbon groups was not a relevant parameter for the initial droplet size, at least within the range of degrees of substitution considered in these studies. Nevertheless, the presence of anionic groups along the polysaccharide backbone had a significant effect on the initial droplet size for α values below 0.05. Over that range, the electrostatic repulsions contributed to the prevention of coalescence phenomena and allowed smaller droplet sizes to be obtained. For higher α values, the droplet size was the same for neutral and anionic dextran derivatives (Figure 7.2). The ionic strength of the aqueous phase had a strong influence on the initial droplet size, especially for low values of α. When anionic groups were present within the polysaccharide backbone, the effect of ionic strength was significantly reduced as compared to neutral dextran derivatives.

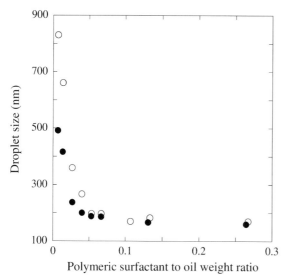

Figure 7.2 Droplet size of dodecane-in-water emulsions as a function of polymeric surfactant to oil weight ratio for DexP$_{23}$ (○) and DexP$_{23}$S$_7$ (●).

Following the emulsion preparation, the amount of polymer remaining in the aqueous phase was determined using the anthrone sugar titration method. The amount of adsorbed polymer per surface units is known to vary with the equilibrium bulk concentration, and follows a curve which is similar to an adsorption isotherm (Figure 7.3). The maximum surface coverage, Γ_{max}, as determined by application of the Scatchard equation, was consistent with the experimental values (Table 7.1). The

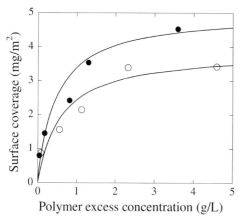

Figure 7.3 Surface coverage as a function of polymer excess concentration in the aqueous phase. Dodecane-in-water emulsions, 5% oil volume fraction. ● DexC6$_{12}$; ○ DexC6$_{38}$. The lines are curve fitted according to the Langmuir equation.

Table 7.1 Maximum surface coverage of dodecane droplets immediately after emulsification. Different polymeric surfactants were used.

Polymeric surfactant	Continuous phase	ϕ^a (%)	Γ_{max} (mg m^{-2})
DexC6$_{12}$	Water	5.0	4.4
DexC6$_{21}$	Water	5.0	4.8
DexP$_{23}$	Water	9.1	4.1
DexC6$_{38}$	Water	5.0	3.2
DexC10$_{11}$	Water	5.0	3.9
DexC6$_{12}$S$_{50}$	Water	5.0	$\approx$0.8
	0.1 M NaCl	5.0	4.0

^{a}Oil volume fraction in the emulsion.

found values were close to those reported for the adsorption of amphiphilic dextrans onto polymeric nanoparticles [31]. For neutral dextran derivatives, the adsorption curves were similar, irrespective of the hydrophobic substitution or the nature of the hydrocarbon group. Dextran derivatives prepared from dextran precursors with different molar masses led to identical results.

In contrast, amphiphilic polymers carrying anionic groups within the polysaccharide chains exhibit a much lower surface coverage. If the ionic strength of the aqueous phase is increased up to 0.1 M, the found values become similar to those of neutral dextran derivatives. Sodium propylsulfonate groups strongly increase the hydrophilicity of polymeric surfactants and limit their adsorption at a hydrophobic surface. However, this effect is cancelled out by an increase in the ionic strength of the aqueous phase.

Polymer surface coverage is never lower than 1 mg m^{-2} when neutral amphiphilic dextrans are used. This value seems to be a minimum surface coverage required for preventing droplet coalescence. With anionic dextran derivatives, the contribution of electrostatic repulsions to the prevention of coalescence allows obtaining stable droplets with a lower surface coverage (0.5 mg m^{-2}).

7.3.3
Stability of Miniemulsions within Polymerization Duration

7.3.3.1 Mechanism and Kinetics of Miniemulsion Polymerization

As the mechanism of miniemulsion polymerization has been detailed in many reviews, only a brief outline will be provided at this point. The description will be restricted to the case of direct miniemulsions – that is, where the monomer droplets are dispersed into a continuous aqueous phase.

In such a polymerization process, the water-insoluble monomer is dispersed into a continuous aqueous phase in the form of submicronic droplets (typical sizes are <500 nm). The amount of surfactant is adjusted so that no micelles remain in the continuous phase. The initiator used for miniemulsion polymerization can be solubilized either in the aqueous phase or in the monomer droplets. If the initiator is in the continuous aqueous phase, radicals are first produced in water but the presence of submicronic oil droplets together with an absence of micelles in the aqueous phase, leads to the complete capture of radicals by droplets. In contrast, if the initiator is solubilized in the monomer droplets, then miniemulsion polymerization is directly initiated in the droplets. Consequently, in any case, polymerization occurs within each individual droplet and leads to polymeric nanoparticles with well-controlled characteristics (size and surface coverage) corresponding to those of the initial miniemulsion, provided that its stability is convenient.

In the case of dextran derivatives, the use of persulfate salts is not convenient because of side reactions with polysaccharides such as cellulose or dextran. Consequently, only AIBN will be used in such situations (i.e. an initiator which is completely solubilized in the monomer droplets).

Although miniemulsion polymerization has been well-studied with molecular surfactants, the use of polymeric surfactants has been much less documented, especially in the case of randomly modified amphiphilic polymers such as polysaccharide derivatives.

The design of convenient conditions for the control of particle properties requires knowledge of the rate of emulsion aging, as well as that of polymerization reaction.

During recent years many studies have been conducted regarding the kinetics of monomer consumption in miniemulsion polymerization. Apart from the initiator concentration, solubilization (either in the continuous aqueous phase or in the monomer droplets) and polymerization temperature, the influence of surfactant concentration has been demonstrated, whichever the localization of the initiator. At this point, data available for the free-radical miniemulsion polymerization of styrene initiated by AIBN will be used as an illustrative example (Table 7.2). When the weight ratio of polymeric surfactant to styrene is increased, the rate of monomer

Table 7.2 Kinetic data for miniemulsion of styrene initiated by AIBN. All miniemulsion polymerizations involved polymeric surfactants. The time required for half-conversion of styrene ($t_{1/2}$) and the maximum monomer conversion obtained are given as a function of reaction conditions.

$[AIBN]_0^a$ (mol l^{-1})	$T(°C)$	α^b	$t_{1/2}$ (h)	Maximum conversion reached^c	Reference
0.0379	70	0.01	4.0	0.71 (8 h)	
0.0379	70	0.02	2.8	0.85 (8 h)	[16]
0.0379	70	0.04	2.1	0.98 (8 h)	
0.0529	65	0.02	4.8	0.62 (6.7 h)	
0.0529	65	0.05	2.9	0.84 (6.7 h)	[11]
0.0529	65	0.08	1.4	0.91 (6.7 h)	
0.0785	75	0.07	0.8	0.88 (5.8 h)	[19]

^{a} Initial AIBN concentration relative to styrene volume.
^{b} Weight ratio of polymeric surfactant to styrene (see Section 7.3.2.1).
^{c} The corresponding time is given in parentheses.

consumption is also increased. Furthermore, the maximum degree of monomer conversion reached is also higher in the presence of larger amounts of the polymeric surfactant. Finally, for common polymerization conditions (temperature and initiator concentration), the time required for the half-consumption of styrene ($t_{1/2}$) is of the order of 1 to 5 h [11, 16, 19].

7.3.3.2 Mechanism and Rate of Emulsion Aging

Theoretical Considerations In the case of submicronic emulsions stabilized with polymeric surfactants, the main aging process is generally that of Ostwald ripening. Creaming also occurs with miniemulsions, but on timescales (days) that are much longer than those corresponding to miniemulsion reactions (hours) and so will not be considered here.

The driving force of Ostwald ripening is the difference of chemical potential of the oil contained in droplets with different sizes. There is a spontaneous diffusion of the oil from small droplets to larger droplets, and this leads to a displacement of droplet size distribution toward higher diameters. Provided that the oil volume fraction is not too high, Ostwald ripening is limited by the diffusion of the oil in the bulk phase. Consequently, the main characteristics of emulsion aging by Ostwald ripening are: (i) the linear variation of the cube of the average droplet radius with time; (ii) the overall displacement of the droplet size distribution; and (iii) the direct dependence of the aging rate on the oil solubility. A theoretical equation describing the droplet radius variation by Ostwald ripening has been derived by Lifshitz and Slyozov [32] and independently by Wagner [33]:

$$\bar{R}^3(t) = \bar{R}^3(0) + \omega t \qquad (7.1)$$

where $\bar{R}$ is the average radius of the droplets (m), t is the time (s) evolved since emulsion preparation, and ω (m^3 s^{-1}) is called the 'rate' of Ostwald ripening. This latter parameter is expressed as a function of the physico-chemical properties of oil and interface:

$$\omega = k(\phi) \frac{8\gamma_i D V_m^2 C_\infty}{9RT} \qquad (7.2)$$

where γ_i is the interfacial tension (N m^{-1}), D is the diffusion coefficient of the oil in water (m^2 s^{-1}), V_m is the molar volume of the oil (m^3 mol^{-1}), C_∞ is the solubility of the oil in pure water (mol m^{-3}), R is the gas constant (8.314 J mol^{-1} K), T is the absolute temperature (K) and $k(\phi)$ is a correcting factor which takes into account the influence of oil volume fraction. The correcting factor $k(\phi)$ was added in more recent approaches as the initial treatment corresponded to the limit of zero volume fraction [34].

Results Obtained with Amphiphilic Dextrans With dextran derivatives, it has been reported that the normalized droplet size distribution is invariant with time [35]. Moreover, the cube of the average droplet radius increases linearly with time for all of

Table 7.3 Ostwald ripening rates of dodecane-in-water emulsions stabilized by neutral and anionic amphiphilic dextrans.

Polymer	C (g l^{-1})	ϕ (%)	γ_i (mN m^{-1})^a	$\omega_{exp}^{\ c}$ (m^3 s^{-1})	ω_{calc} (m^3 s^{-1})	$\omega_{exp}/\omega_{calc}$
T40©	5	5.0	52.8^b	8.0×10^{-26}	19.0×10^{-27}	4.2
DexC6$_{12}$	5	5.0	14.0	2.2×10^{-26}	5.7×10^{-27}	3.9
DexC6$_{21}$	5	5.0	8.0	1.1×10^{-26}	3.1×10^{-27}	3.6
DexC6$_{38}$	5	5.0	3.0	0.6×10^{-26}	1.1×10^{-27}	5.5
DexC10$_{11}$	5	5.0	15.5	1.5×10^{-26}	6.2×10^{-27}	2.4
DexP$_7$	10	9.1	20.0	3.1×10^{-26}	8.8×10^{-27}	3.5
DexP$_{23}$	10	9.1	6.9	1.5×10^{-26}	3.0×10^{-27}	5.0
DexP$_{23}$S$_7$	20	9.1	9.6	3.6×10^{-26}	4.2×10^{-27}	8.5
DexC6$_{12}$S$_{50}$	10	5.0	25.0	5.8×10^{-26}	1.0×10^{-26}	5.8

^{a} Value measured at a polymer concentration of 1 g l^{-1} (for details, see Section 7.3.3.2).
^{b} Tabulated value of dodecane/water interfacial tension from Ref. [65].
^{c} The error was estimated to be ±0. × 10^{-26} m^3 s^{-1} from the linearized plots.

the polymers [21]. The calculated aging rates are consistent with the experimental rates, even if they are systematically lower, and this is also observed with molecular surfactants (Table 7.3). The aging rate is controlled by the chemical structure of the amphiphilic dextran derivative, and increasing the number of attached hydrocarbon groups leads to a lower aging rate. This effect is consistent with the prediction of Equation 7.2, as the interfacial tension is lower with more-substituted dextran derivatives. Although a linear dependence of aging rate on interfacial tension has been identified (Figure 7.4), the experimental slope is much higher than the theoretical prediction (as in Equation 7.2). For these curves, the interfacial tension has been measured independently with polymer aqueous solutions containing similar polymer concentrations.

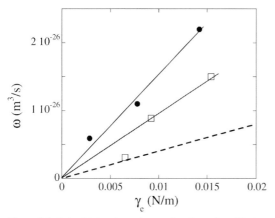

Figure 7.4 Ostwald ripening rate as a function of equilibrium interfacial tension for dodecane-in-water emulsions. ● DexC6$_\tau$; □ DexC10$_\tau$. The dashed line is the prediction of Equation 7.2.

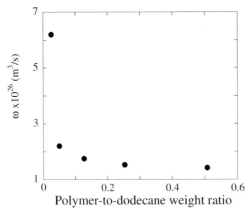

Figure 7.5 Ostwald ripening rate of emulsions as a function of polymeric surfactant: dodecane weight ratio. ● DexC6$_{12}$.

For a given degree of substitution, the aging rate is lower when more-hydrophobic hydrocarbon groups are used, even if the interfacial tension is approximately the same. Such an effect cannot be explained by Equation 7.2, and may result from interfacial dynamics. With C$_{10}$ groups, the rearrangement of polymeric chains at the interface during emulsion aging may be slower than with C$_6$ groups, and this may slow down the aging process. Recently, theoretical calculations have suggested that interfacial elasticity can change the rate of Ostwald ripening [37–39]. Alternatively, it has been suggested that, with sufficiently long aliphatic chains, the hydrocarbon groups partly dissolve into the oil and act as additives, which in turn lowers the rate of Ostwald ripening. Recent results with miniemulsion polymerization stabilized by polymeric surfactants are consistent with that assumption [40].

When anionic dextran derivatives are used, greater aging rates are observed as compared to neutral analogues, even with similar interfacial tensions (see Table 7.3). This result can be related to the previous findings concerning the influence of the nature of hydrocarbon groups.

The concentration of amphiphilic dextran has a strong effect on aging rate below a certain value of the polymer-to-oil weight ratio (Figure 7.5). This can be explained by a default of polymer in the aqueous phase leading to an increase of interfacial tension, and thus of the aging rate of the emulsion.

The effect of oil volume fraction on emulsion aging rate is much more limited than the predictions of several theoretical treatments. A similar discrepancy has been reported in the case of emulsions stabilized with molecular surfactants such as sodium dodecyl sulfate [41].

7.3.3.3 Variation of the Rate of Emulsion Aging with Polymerization Conditions

Taking once again the example of miniemulsion polymerization of styrene initiated by AIBN (see above), the time for half-consumption of the monomer is generally not more than 5 h. It is expected that the initial emulsion will undergo no significant droplet size variation during that time (see Section 7.3.3.1). By using Equation 7.1, it can be easily established that the time required for an increase in the average droplet radius from $\bar{R}(0)$ to $1.5\bar{R}(0)$ is $t_{3/2} = 2.375\frac{\bar{R}^3(0)}{\omega}$ This time should be much greater than 5 h. Consequently, for an initial styrene emulsion such that $\bar{R}(0) = 100$ nm we should have $\omega < 10^{-25}$ m^3 s^{-1} to fulfill the preceding conditions. Clearly, this calculation is not expected to provide more than an order of magnitude. In the following section we will detail how ω varies with the composition of droplets, and how it could be adjusted in a convenient range of values.

Effect of Monomer Droplets Composition Current monomers such as styrene or methyl methacrylate have a much higher solubility in water than in hydrocarbon oils (Table 7.4) [42, 43]. Consequently, miniemulsions of pure monomers in water exhibit aging rates which are much too high as compared to the polymerization rate, even in the presence of efficient surfactants. The final particles would be much bigger than the initial monomer droplets and, on occasion, the initial miniemulsion would not even be stable enough to start the polymerization reaction.

A common method of adjusting the emulsion aging rate is to add in the monomer a small amount of a compound with ultra-low water solubility. This additive is sometimes called the 'costabilizer', even if its main effect is not related to the interfacial tension but rather with the chemical potential of the monomer. Indeed, while the monomer diffuses out of small droplets into bigger ones, the concentration of the highly water-insoluble oil gives rise to an increase in the chemical potential of the monomer in the droplet and thus limits its diffusion [41].

Frequently, the added compound is a highly hydrophobic, nonreactive molecule; hexadecane is a very common example, although the use of other organic compounds such as triglycerides or dyes has been proposed [49, 50].

Table 7.4 Some physical properties of various compounds involved in miniemulsion polymerization.

Component	Molecule	Water solubility[a] (mol m^{-3})	Ref.	Diffusion coefficient[b] (m^2 s^{-1})	Molar volume (m^3 mol^{-1})
Monomer	Styrene	1.7	[66]	8.4×10^{-10}	1.15×10^{-4}
	Methyl methacrylate	150	[66]	9.4×10^{-10}	1.06×10^{-4}
	Lauryl methacrylate	10^{-4}	[66]	4.8×10^{-10}	2.93×10^{-4}
Initiator	AIBN	2.4	[46]	6.5×10^{-10}	–
	Lauroyl peroxide	8×10^{-3}	[46]	3.8×10^{-10}	–
Chain-transfer agent	1-dodecanethiol	3×10^{-2}	[46]	5.5×10^{-10}	–
Costabilizer	Hexadecane	9.3×10^{-8}	[67]	4.9×10^{-10}	2.92×10^{-4}
	Dodecane	2.4×10^{-5}	[67]	5.2×10^{-10}	2.27×10^{-4}
	Decane	3.6×10^{-4}	[67]	5.8×10^{-10}	1.94×10^{-4}
	Hexadecanol	33×10^{-6}	[68]	4.8×10^{-10}	2.96×10^{-4}

[a] All values were given at 25 °C except for hexadecane (20 °C) and hexadecanol (0 °C).
[b] Calculated values using the correlation of Wilke and Chang [69] and taking the temperature equal to 25 °C.

The quantitative treatment of this effect has been proposed by Kabal'nov et al. [51], who derived an equation for the aging rate of binary oil mixtures as a function of the composition:

$$\omega_{mixture} = \left[\frac{\varphi_1}{\omega_1} + \frac{\varphi_2}{\omega_2}\right]^{-1} \quad (7.3)$$

By using Equation 7.3 we can compare the aging rate of the mixture to that of the pure oil 1 by the retardation factor $F_{2,1}$, defined by:

$$F_{2,1} = \omega_1/\omega_{mixture} = 1 + \left[\frac{\omega_1}{\omega_2} - 1\right]\varphi_2 \quad (7.4)$$

In Equation 7.4, the ratio ω_1/ω_2 depends essentially on the respective water solubilities of oils 1 and 2. The greater the difference, the greater is the efficiency of compound 2 in lowering the emulsion aging of oil 1.

Those molecules taking part in the miniemulsion polymerization can also change the rate of Ostwald ripening and even act as costabilizers. The main advantage of such reactive compounds is that their incorporation into macromolecular chain prevents their diffusion out of the latex particles after polymerization. Although, for some applications this may be a major advantage, this incorporation could nevertheless induce changes in the properties of the macromolecules.

For example, the initiator is commonly solubilized in the monomer. Thus, if a highly hydrophobic initiator is used (e.g. lauroyl peroxide) it can contribute to a lowering of the aging rate in a similar way as a hydrophobic additive [45, 52]. In contrast, an initiator such as AIBN exhibits a much too-high solubility in water (see Table 7.4). The same conclusion holds true for benzoylperoxide [45].

Small amounts of a very hydrophobic monomer (e.g. alkyl methacrylates with long alkyl chains) are sometimes added in the droplets (see Table 7.4) [53, 54]. Depending on the reactivity ratios, this monomer may be incorporated into the macromolecules, and its influence on the final properties may be minimal as long as it remains present in small amounts. Apart from monomers, highly hydrophobic chain-transfer agents such as dodecanethiol have also been used as costabilizers in miniemulsions (see Table 7.4) [55]. Once again, these compounds modify the structure of the synthesized macromolecules by reducing their length through transfer reactions.

In addition, in the case of controlled free-radical polymerization, control agents are present in monomer droplets (their nature depends on the type of controlled polymerization). Recently, Qi and Schork proposed an estimation of the effect of a third component (added to the mixture of monomer and hydrophobe) on the aging rate of miniemulsions [56]. These authors expressed the ratio of the rates of miniemulsion aging without and with the third component, and obtained equations similar to Equation 7.4. Although their treatment was focused on the example of reversible addition-fragmentation chain transfer (RAFT) agents, it can apply to any type of additive.

Calculated curves from the results reported by Chern and colleagues [50, 57] concerning the effect of hydrophobic monomers on the rate of Ostwald ripening are consistent with Equation 7.4, and show a linear variation of the retardation factor $F_{2,1}$

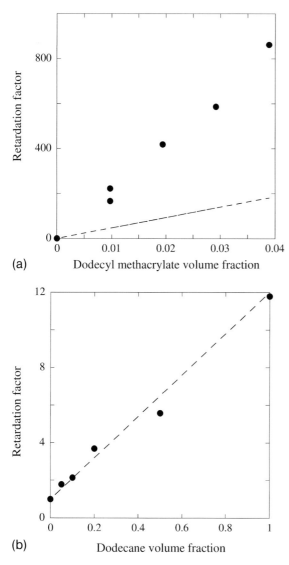

Figure 7.6 (a) Retardation factor calculated for styrene emulsions as a function of the volume fraction of dodecyl methacrylate in styrene. Calculated curve using data from Ref. [52]. The dotted line is the calculated variation. (b) Retardation factor for decane emulsions as a function of the volume fraction of dodecane in decane. The dotted line is the calculated variation.

with the volume fraction of the hydrophobic additive (Figure 7.6a). Although the experimental slope is much higher than the calculated version, the approximated relationship for $\omega_{mixture}$ (see Equation 7.3) is only valid when the molar volumes are almost equal.

It should be noted at this point that the existence of different migrations of polymerization reactants between droplets is not only important for miniemulsion stability. In fact, the variation of droplet composition according to size might induce different reaction conditions and hence a less efficient control of the polymerization reaction [56].

Rather than small molecules, polymers are sometimes added in the monomer droplets so as to slow down Ostwald ripening. These macromolecules may be of the same chemical nature as the polymer to be synthesized (e.g. poly(styrene) dissolved in styrene droplets) or of a different nature (e.g. macromolecules containing very hydrophobic units dissolved in a more polar monomer). In that case, the macromolecules would be expected to have the effect of a highly water-insoluble additive with safety acceptability.

Alternatively, some groups have used polymeric surfactants as the sole stabilizer against Ostwald ripening. In some instances, the polymer is initially dissolved into the monomer phase and reorganizes at the interface during miniemulsion preparation [12]. However, relatively high polymer concentrations are required in this situation. More recently, cationic amphiphilic comb-like copolymers were reported as being capable of stabilizing miniemulsions of styrene, without the use of any other hydrophobic additive [39]. It should be noted that, in both examples, the polymeric surfactants carried highly hydrophobic groups (octadecyl) as sticking units. Their reported efficiency against Ostwald ripening can be compared to the results of the present studies, which showed that increasing the hydrophobicity of the hydrocarbon units leads to a lowering of the rate of Ostwald ripening independently of the interfacial tension itself. It is suggested that, in addition to their effect on the interfacial tension, octadecyl chains partly dissolve into the oil droplets and act as hydrophobic additives.

Effect of Temperature The effect of temperature on miniemulsion aging rate is the last to be considered, as many reactions are carried out at temperatures above 70 °C.

As the continuous phase contains a limited amount of nonadsorbed polymer, the viscosity of this aqueous phase would be expected to vary with temperature. Nevertheless, as the polymer concentration in the aqueous phase is expected to be limited, it is reasonable to assume that the viscosity of this continuous phase will vary with temperature in a similar way as pure water. So, the main effects to be observed derive from the variation of emulsion aging rate and the modification of colloidal interactions between oil droplets.

In a first approach, the effect of temperature on emulsion aging rate can be restricted to the diffusion coefficient and the solubility of oil in water. The variation of interfacial tension with temperature can be neglected [58]. The variation of the rate of Ostwald ripening rate with temperature can be expressed in the following form:

$$\omega = \left(\frac{a}{T}\right)\exp\left(-\frac{E_{app}}{RT}\right) \quad (7.5)$$

where a (m$^3 \cdot$K s^{-1}) is a pre-exponential factor, E_{app} (J mol^{-1}) is the apparent activation energy of the molecular diffusion process, and R (J mol^{-1} K) is the gas

Table 7.5 Apparent activation energy for oil-in-water emulsions of various alkanes stabilized by DexP$_{23}$ (Equation 7.4). (Data from [19])

Oil	$E_{app}^{experimental}$ (kJ mol^{-1})
Decane	47.5
Dodecane	57.1
Hexadecane	72.5

constant. The pre-exponential term inversely proportional to T was introduced because ω is by itself inversely proportional to T (cf. Equation 7.2).

Results with Amphiphilic Dextrans For mixtures of decane and dodecane emulsified in water in the presence of amphiphilic dextran DexP$_{23}$, the variation of aging rate with the proportion of dodecane in the oil is very close to that predicted by the equation of Kabal'nov (Figure 7.6b).

The effect of temperature should also be considered, since many radical polymerization reactions are carried out at temperatures around 70 °C. The apparent activation energy E_{app} of Equation 7.4 was determined for emulsions of various oils stabilized with DexP$_{23}$ (Table 7.5) [19].

7.3.4
Preparation of Defined Nanoparticles with Various Monomers

7.3.4.1 Poly(styrene) Nanoparticles Covered by Dextran

At this point, previous considerations about miniemulsions will be applied to the preparation of styrene miniemulsions stabilized by amphiphilic dextrans. These miniemulsions will be used to obtain dextran-covered poly(styrene) nanoparticles by free-radical miniemulsion polymerization.

When the Ostwald ripening rate of styrene-in-water emulsions is estimated on the basis of the results of Chern and Chen [57], the value of ω obtained is equal to 3.3×10^{-23} m^3 s^{-1}, and is too high (see Section 7.3.3.3). The addition of 5% of hexadecane in the monomer reduces ω by a factor 10 000 (ω = 2.4×10^{-27} m^3 s^{-1})! An increase from 250 nm for the initial droplet size to 350 nm for the final particle size would last for about 16 days, while polymerization would be completed within a few hours (Figure 7.7). Experimentally, no variation in the droplet size of the styrene-in-water emulsions was observed within that time, and therefore Ostwald ripening cannot be the destabilization process of these miniemulsions.

Over the range of the examined degrees of modification, drastic differences were observed during the polymerization step for the different dextran derivatives. The best results were obtained for those derivatives with degree of modification higher than 15% and for a styrene volume fraction ϕ of 10%. In these cases, when varying the polymeric surfactant: oil ratio between 0.015 and 0.15, the particle size was close to the initial droplet size and the coagulate amount was below 5 wt% (Figure 7.8). Styrene particles covered by a hydrophilic shell composed of amphiphilic polysaccharide

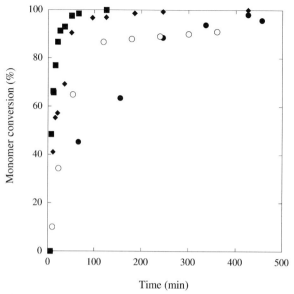

Figure 7.7 Conversion as a function of time for miniemulsion polymerization stabilized by amphiphilic derivatives of dextran. ○, Radical polymerization of styrene. Anionic polymerization of butyl cyanoacrylate at pH 1 (●), pH 1.5 (◆) and pH 2 (■).

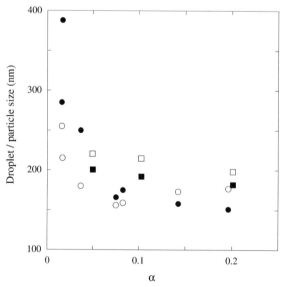

Figure 7.8 Size of the initial emulsion droplets and of the final latex particles as a function of [dextran]:[styrene] mass ratio. The miniemulsions were stabilized by DexP$_{18}$ (○ emulsion; ● latex) and DexP$_{23}$S$_{25}$ (□ emulsion; ■ latex).

could be synthesized using this method, although a marked increase in particle size was observed for low weight ratio of polymeric surfactant to oil.

The particle size and amount of coagulum was also strongly increased by decreasing the modification degree. In the case of DexP$_5$, no latex was obtained, but this could be explained by a limited protection against coalescence.

Finally, charged derivatives of amphiphilic dextran were tested (Figure 7.8). For α-values above 0.05, the particle size plateaus at about the same value as for uncharged derivatives. In contrast, latex prepared with an α-value <0.05 had a lower average particle diameter when the density of charges along the polysaccharidic backbone was increased. The presence of ionizable groups on the polysaccharidic backbone favors the formation of smaller particle sizes by decreasing the initial droplet size and maintaining the stability of the miniemulsion during the polymerization step.

The amount of adsorbed polymer was deduced by titration of the polymer remaining in solution after particle removal, following the anthrone method [59]. When plotting the particle surface coverage (Γ, mg m^{-2}) as a function of the concentration of nonadsorbed polymer remaining in the aqueous phase (the polymer 'in excess'), a curve was obtained which had the shape of an adsorption isotherm (Figure 7.9). Furthermore, the curve could be linearized using the Scatchard method [60], and this provided maximum surface coverages ranging from 3 to 5 mg m^{-2} when increasing the modification degree. In contrast, the amount of adsorbed polymer clearly decreased with the amount of ionizable groups along the polysaccharidic backbone. These results were in total accordance with the

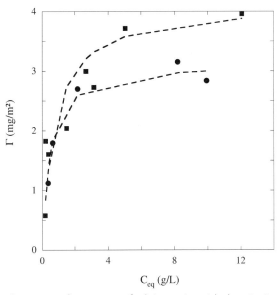

Figure 7.9 Surface coverage of poly(styrene) particles by ● DexP$_{13}$ and ■ DexP$_{28}$ for particles obtained by miniemulsion polymerization. The lines are curve fittings made by using the Scatchard equation.

experimental data obtained in the case of emulsions [61], and proved that no desorption of the dextran derivatives occurred during the polymerization step. Taken together, these results show clearly that the surface coverage of the nanoparticles is controlled by the emulsification step of the process.

For all of the latexes synthesized in these studies, the electrokinetic layer thickness was found to be close to 6 nm. Variations due to polysaccharide surfactant structure or concentration were within experimental errors.

In the case of the adsorption of amphiphilic derivatives of dextran onto preformed nanoparticles, a decrease in the electrokinetic layer thickness was observed when the degree of phenoxy substitution was increased [62]. Indeed, the increase of anchoring groups on the polysaccharidic chain reduced the length of loops and tails on the surface.

However, when amphiphilic derivatives of dextran were used during the miniemulsion polymerization, the OH-groups of the macromolecular polysaccharidic chain were accessible during the polymerization step for the transfer reaction, thereby creating new anchoring points on the poly(styrene) particle. This may have been the reason why the electrokinetic layer thickness stability did not vary with the degree of phenoxy substitution.

The potential covalent binding of dextran chains onto the poly(styrene) nanoparticles through chain-transfer reactions is a specificity of that process which involves radical polymerization. Nevertheless, this covalent binding might be advantageous for some applications as it is an irreversible binding of macromolecules to the surface.

7.3.4.2 Poly(butylcyanoacrylate) Nanoparticles

A recent exciting application of colloidal particles was their development as a carrier for the *in vivo* delivery of drugs. The main goal of this research was to modify the surface of the nanoparticles to interfere with the distribution of drugs within the body. Indeed, it was proven that, for nanoparticles, their surface properties controlled their interactions with serum proteins and thus their biodistribution after administration [63, 64].

Polyalkylcyanoacrylates have been investigated extensively for use as drug carriers, especially in cancer therapy [65]; consequently, dextran-coated poly(alkylcyanoacrylate) nanoparticles are promising candidates for drug delivery applications. Unfortunately, cyanoacrylate esters are rated as some of the most reactive monomers in anionic or zwitterionic polymerizations [66]. Traces of bases (including water [67]) will initiate their polymerization, with chain termination occurring only in the presence of strong acids [68].

In 1979, Couvreur and colleagues were the first to develop a process for the direct generation of nanoparticles from ethyl or butyl cyanoacrylates [65]. Since that time, numerous studies have described the engineering of nanoparticle preparation but, to the best of our knowledge, only one report has been made on the miniemulsion polymerization of cyanoacrylates esters [69].

The aim of the latter study was to combine the use of polymeric surfactants prepared from polysaccharides with the miniemulsion polymerization of butylcyanoacrylate to create biocompatible nanoparticles. Very low pH values were required to avoid cyanoacrylate anionic polymerization during the emulsification step, as water initiated the polymerization. Therefore, the stability of the amphiphilic derivatives of dextran in the polymerization conditions was first checked. Surprisingly, no substitution degree variation was observed under the polymerization conditions, while polysaccharidic chain scission occurred only at high temperature and at pH 1. It is clear that, under the required pH conditions, the nature of chemical links between the dextran and hydrocarbon groups is of primary importance. With ether links, the stability under acidic conditions is enough. This conclusion would most likely be modified with ester functions, as in the case of fatty esters of dextran.

The emulsification conditions (sonication time and temperature) were optimized so as simultaneously to minimize particle size and anionic polymerization of monomer. As a result, poly(butylcyanoacrylate) nanoparticles with a polysaccharidic layer were successfully synthesized. The particle size decreased with increasing concentration of amphiphilic dextran in the aqueous phase (Figure 7.10), as noted above for the case of styrene miniemulsion polymerization. Surprisingly, the modification degree of the amphiphilic dextran has an influence on the particle size. The best results were obtained for the modified dextran $DexP_{17}$. Smaller or higher modification degrees led to a higher particle size, especially at low dextran aqueous concentrations.

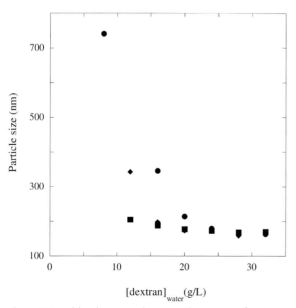

Figure 7.10 Polybutyl cyanoacrylate) particles size as a function dextran aqueous concentration for miniemulsion anionic polymerization stabilized by ● $DexP_{11}$, ■ $DexP_{16}$, and ◆ $DexP_{24}$.

The amount of adsorbed amphiphilic polysaccharide was much higher than in the case of poly(styrene) nanoparticles. Indeed, the hydroxyl groups of the polysaccharidic chain could initiate the anionic polymerization of butylcyanoacrylate, thereby increasing the affinity of the amphiphilic dextran for the nanoparticles. This higher surface coverage led to a higher electrokinetic thickness of the hydrophilic layer. In the case of DexP$_{17}$, the electrokinetic layer was even found to be close to 13 nm.

7.3.5
Colloidal Properties of the Obtained Suspensions

Here, we consider the colloidal properties of latex suspensions stabilized by amphiphilic dextrans. The discussion will be restricted to poly(styrene) nanoparticles prepared by free-radical miniemulsion polymerization.

Above 0.01 M NaCl, poly(styrene) nanoparticles covered by an anionic molecular surfactant (sodium dodecyl sulfate; SDS) were no longer stable because of the screening of the surface charges by salt ions. Therefore, flocculation due to Van der Waals attraction occurred when the salt concentration increased. In contrast, no flocculation was observed at NaCl concentrations up to 4 M for the dextran-coated poly(styrene) nanoparticles, as a result of steric stabilization by osmotic and elastic repulsion potentials (Figure 7.11).

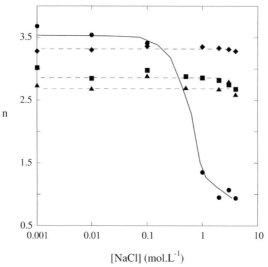

Figure 7.11 Slope n of log(optical density) versus log(wavelength) as a function of salt concentration for poly(styrene) nanoparticles synthesized by miniemulsion polymerization. Nanoparticles were obtained in the presence of ● SDS, ◆ DexP$_{23}$, ▲ DexP$_{23}$S$_7$ and ■ DexP$_{23}$S$_{25}$.

No flocculation was observed with the suspensions when heating from 25 °C up to 75 °C.

When the stability of the suspensions was checked for freeze-drying, the initial particle size was compared to the average particle size in the reconstituted suspension after freeze-drying. No cryoprotectant had been added, but the relevant parameter appeared to be the polymeric surfactant: oil weight ratio. The freeze-dried suspensions were not redispersible when the amount of polymer was too low (polymeric surfactant: oil weight ratio <0.05). However, for larger amounts of polymer only limited aggregation occurred and the average size was close to that of the initial particles [21].

Finally, for dextran-coated poly(styrene) nanoparticles prepared by emulsion polymerization, the adsorption of a protein such as bovine serum albumin (BSA) was shown to be greatly reduced when compared to uncoated ('bare') nanoparticles [18]. Similar properties could be expected for nanoparticles prepared by miniemulsion polymerization.

7.4
Conclusions

Suspensions of polymeric particles can be prepared by miniemulsion polymerization stabilized by amphiphilic derivatives of dextran. Controlling the characteristics of the initial emulsions allows defined latex particles to be obtained which are covered by a polysaccharide layer. This control can be achieved by a convenient design of polysaccharide surfactants (the nature and number of hydrocarbon moieties, introduction of limited amounts of anionic groups, nature of chemical bonds, etc.). The stability of O/W emulsions stabilized by amphiphilic dextrans has been studied and the main tendencies were used for the design of convenient conditions for miniemulsion polymerization, taking into account the characteristics of polymerization reactions.

Dextran-coated poly(styrene) and poly(butylcyanoacrylate) nanoparticles were obtained by anionic or radical miniemulsion polymerization. It has been shown that the characteristics of these materials can be controlled by those of the initial monomer miniemulsions.

The colloidal properties of the final nanoparticle suspensions were characterized and related to their surface coverage. These suspensions were shown to exhibit stability to high ionic strengths and temperatures. Moreover, is it possible to freeze-dry the suspensions and to reconstitute them, without the addition of cryoprotectant. Finally, the adsorption of other macromolecules (e.g. proteins) is efficiently prevented by a convenient surface coverage. These nanoparticles are well-suited for further applications in drug-delivery systems, as their properties with regards to drug encapsulation and release can be controlled via the molecular characteristics of the polymeric core. These considerations form the focus of current investigations.

References

1 Bechthold, N., Tiarks, F., Willert, M., Landfester, K. and Antonietti, M. (2000) Miniemulsion polymerization: applications and new materials, *Macromol. Symp.*, **151**, 549–555.

2 Schork, F.J., Poehlein, G.W., Wang, S., Reimers, J., Rodrigues, J. and Samer, C. (1999) *Colloid Surf. A*, **153**, 39.

3 Landfester, K., Bechtold, N., Förster, S. and Antonietti, M. (1999) *Macromol. Rapid Commun.*, **20**, 81.

4 Paunov, V.N., Sandler, S.I. and Kaler, E.W. (2001) *Langmuir*, **17**, 4126.

5 Landfester, K., Bechthold, N., Tiarks, F. and Antonietti, M. (1999) Formulation and, stability mechanisms of polymerizable miniemulsions, *Macromolecules*, **32**, 5222.

6 Chern, C.S. and Liou, Y.C. (1999) Kinetics of styrene miniemulsion polymerization stabilized by nonionic surfactants/alkyl methacrylate, *Polymer*, **40**, 3763–72.

7 Landfester, K. (2003) *Surfactant Science Series*, **115**, 225.

8 Wang, S. and Schork, F.J. (1994) *J. Polym. Sci. A: Polym. Chem.*, **54**, 2157.

9 Marie, E., Landfester, K. and Antonietti, M. (2002) Synthesis of chitosan-stabilized polymer dispersions, capsules, and chitosan grafting products via miniemulsion, *Biomacromolecules*, **3**, 475.

10 Yu, Z.Q., Lee, D.Y., Cheong, I.W., Shin, J.S., Park, Y.J. and Kim, J.H. (2003) *J. Appl. Polym. Sci.*, **87**, 1933.

11 Ni, P., Zhang, M., Ma, L. and Fu, S. (2006) Poly(dimethylamino)ethyl methacrylate for use as a surfactant in the miniemulsion polymerization of styrene, *Langmuir*, **22**, 6016.

12 Baskar, G., Landfester, K. and Antonietti, M. (2000) Comblike polymers with octadecyl side chain and carboxyl functional sites: scope for efficient use in miniemulsion polymerization, *Macromolecules*, **33**, 9228.

13 Lim, M.S. and Chen, H. (2000) Miniemulsion polymerization of styrene with a block copolymer surfactant, *J. Polym. Sci. Part A: Polym. Chem.*, **38**, 1818.

14 Chern, C.S. and Cheng, L. (2001) *Macromol. Chem. Phys.*, **202**, 2750.

15 Pham, B.T.T., Nguyen, D., Ferguson, C.J., Hawkett, B.S., Serelis, A.K. and Such, C.H. (2003) Miniemulsion polymerization stabilized by amphipathic macro RAFT agents, *Macromolecules*, **36**, 8907–8909.

16 Houillot, L., Nicolas, J., Save, M., Charleux, B., Li, Y. and Armes, S.P. (2005) Miniemulsion polymerization of styrene using a pH-responsive cationic diblock macromonomer and its nonreactive diblock copolymer counterpart as stabilizers, *Langmuir*, **21**, 6726.

17 Ou, J.-L., Lim, M.S. and Chen, H. (2003) A polyampholyte triblock copolymer synthesized for using as the surfactant of miniemulsion polymerization and production of highly uniform microspheres, *J. Appl. Polym. Sci.*, **87**, 2230.

18 Rouzes, C., Durand, A., Leonard, M. and Dellacherie, E. (2002) Surface activity and emulsification properties of hydrophobically modified dextrans, *J. Colloid Interface Sci.*, **253**, 217.

19 Durand, A., Marie, E., Rotureau, E., Léonard, M. and Dellacherie, E. (2004) Amphiphilic polysaccharides: Useful tools for the preparation of nanoparticles with controlled surface characteristics, *Langmuir*, **20**, 6956.

20 Nestor, J., Esquena, J., Solans, C., Levecke, B., Booten, K. and Tadros, T.F. (2005) Emulsion polymerization of styrene and methyl methacrylate using a hydrophobically modified inulin and comparison with other surfactants, *Langmuir*, **21**, 4837.

21 Rotureau, E., Léonard, M., Marie, E., Dellacherie, E., Camesano, T. and Durand, A. (2006) From polymeric surfactants to colloidal systems (2): Preparation of colloidal dispersions, *Colloids Surf. A*, **288**, 62.

22 Long, J.A., Osmond, D.W.J. and Vincent, B. (1973) The equilibrium aspects of weak flocculation, *J. Colloid Interface Sci.*, **42**, 545.

23 Tadros, T.F., Vandamme, A., Booten, K., Levecke, B. and Stevens, C.V. (2004) Stabilisation of emulsions using hydrophobically modified inulin (polyfructose). *Colloids Surf. A*, **250**, 133.

24 Durand, A. (2006) Synthesis of amphiphilic polysaccharides by micellar catalysis, *J. Mol. Catal. A*, **256**, 284.

25 Rotureau, E., Chassenieux, C., Dellacherie, E. and Durand, A. (2005) Neutral polymeric surfactants derived from dextran: a study of their aqueous solution behavior, *Macromol. Chem. Phys.*, **206**, (2038).

26 Dragan, D., Mihai, D., Mocanu, G. and Carpov, A. (1997), *React. Funct. Polym.*, **34**, 79.

27 Husemann, E. and Kafka, M. (1960) *Makromol. Chem.*, **41**, 208.

28 Rotureau, E., Léonard, M., Dellacherie, E. and Durand, A. (2004) Amphiphilic derivatives of dextran: adsorption at air/water and oil/water interfaces, *J. Colloid Interface Sci.*, **279**, 68.

29 Rotureau, E., Marie, E., Dellacherie, E. and Durand, A. (2007) From polymeric surfactants to colloidal systems (3): Neutral and anionic polymeric surfactants derived from dextran, *Colloids Surf. A*, **301**, 229–38.

30 Lorimer, J.P., Mason, T.J., Cuthbert, T.C. and Brookfield, E.A. (1995) Effect of ultrasound on the degradation of aqueous native dextran, *Ultrasonics Sonochemistry*, **2**, S55.

31 De Sousa Delgado, A., Léonard, M. and Dellacherie, E. (2001) Surface properties of polystyrene nanoparticles coated with dextrans and dextran-PEO copolymers. Effect of polymer architecture on protein adsorption, *Langmuir*, **17**, 4386–4391.

32 Lifshitz, I.M. and Slyozov, V.V. The kinetics of precipitation from supersaturated solid solutions (1961) *J. Phys. Chem. Solids*, **19**, 35.

33 Wagner, C. (1961) Theory of precipitate change by redissolution, *Z. Elektrochem.*, **35**, 581.

34 Enomoto, Y., Tokuyama, M. and Kawasaki, K. (1986) Finite volume fraction effects on Ostwald ripening, *Acta Metall.*, **34**, 2119.

35 Sadtler, V., Imbert, P. and Dellacherie, E. (2002) Ostwald ripening of oil-in-water emulsions stabilized by phenoxy-substituted dextrans, *J. Colloid Interface Sci.*, **254**, 355.

36 Demond, A.H. and Lindner, A.S. (1993) Estimation of interfacial tension between organic liquids and water, *Environ. Sci. Technol.*, **27**, 2318.

37 Meinders, M.B.J., Kloek, W. and van Vliet, T. (2001) *Langmuir*, **17**, 3923.

38 Capek, I. (2004) *Adv. Colloid Interface Sci.*, **107**, 125.

39 Kawasaki, K. and Enomoto, Y. (1988) *Physica A*, **150**, 463.

40 Manguian, M., Save, M., Chassenieux, C. and Charleux, B. (2005) Miniemulsion polymerization of styrene using well-defined cationic amphiphilic comblike copolymers as the sole stabilizer, *Colloid Polym. Sci.*, **284**, 142.

41 Taylor, P. (1998) Ostwald ripening in emulsions, *Adv. Colloid Interface Sci.*, **75**, 107.

42 Chai, X.-S., Schork, F.J., DeCinque, A. and Wilson, K. (2005) Measurement of the solubilities of vinylic monomers in water, *Ind. Eng. Chem. Res.*, **44**, 5256.

43 Fontenot, K. and Schork, F.J. (1993) Sensitivities of droplet size and stability in monomeric emulsions, *Ind. Eng. Chem. Res.*, **32**, 373.

44 Tauer, K., Imroz Ali, A.M., Yildiz, U. and Sedlak, M. (2005) On the role of hydrophilicity and hydrophobicity in aqueous heterophase polymerization, *Polymer*, **46**, 1003.

45 Alduncin, J.A., Forcada, J. and Asua, J.M. (1994) Miniemulsion polymerization using oil-soluble initiators, *Macromolecules*, **27**, 2256.

46 Sakai, T., Kamogawa, K., Nishiyama, K., Sakai, H. and Abe, M. (2002) Molecular

diffusion of oil/water emulsions in surfactant-free conditions, *Langmuir*, **18**, 1985.

47 Krause, F.P. and Lange, W. (1965), Aqueous solubilities of n-decanol, n-hexadecanol and n-octadecanol by a new method, *J. Phys. Chem.*, **69** 3171.

48 Wilke, C.R. and Chang, P. (1955) Correlation of diffusion coefficients in dilute solutions, *AIChE J.*, **1**, 264.

49 Bathfield, M., Graillat, C. and Hamaide, T. (2005) Encapsulation of high biocompatible hydrophobe contents in nonionic nanoparticles by miniemulsion polymerization of vinyl acetate or styrene: Influence of the hydrophobe component on the polymerization, *Macromol. Chem. Phys.*, **206**, 2284.

50 Chern, C.S., Chen, T.J. and Liou, Y.C. (1998) Miniemulsion polymerization of styrene in the presence of a water-insoluble blue dye, *Polymer*, **39**, 3767.

51 Kabal'nov, A.S., Pertsov, A.V., Aprosin, Y.D. and Shchukin, E.D. (1985) Influence of the nature and the composition of the disperse phase on the stability of direct emulsions to recondensation, *Colloid Journal of the USSR*, **47**, 898.

52 Reimers, J.L. and Schork, F.J. (1997) Lauroyl peroxide as a cosurfactant in miniemulsion polymerization, *Ind. Eng. Chem. Res.*, **36**, 1085.

53 Chern, C.S. and Sheu, J.-C. (2000) Effects of 2-hydroxyalkyl methacrylates on the styrene miniemulsion polymerizations stabilized by SDS and alkyl methacrylates, *J. Polym. Sci. A: Polym. Chem.*, **38**, 3188.

54 Chern, C.S., Liou, Y.C. and Chen, T.J. (1998) Particle nucleation loci in styrene miniemulsion polymerization using alkyl methacrylates as the reactive cosurfactant, *Macromol. Chem. Phys.*, **199**, 1315.

55 Mouran, D., Reimers, J.L. and Schork, F.J. (1996) Miniemulsion polymerization of methyl methacrylate with dodecyl mercaptan as cosurfactant, *J. Polym. Sci. A: Polym. Chem.*, **34**, 1073.

56 Qi, G. and Schork, F.J. (2006) On the stability of miniemulsions in the presence of RAFT agents, *Langmuir*, **22**, 9075.

57 Chern, C.S. and Chen, T.J. (1998) Effect of Ostwald ripening on styrene miniemulsion stabilized by reactive cosurfactants, *Colloids Surf. A*, **138**, 65.

58 Taylor, P. (2003) Ostwald ripening in emulsions: estimation of solution thermodynamics of the disperse phase, *Adv. Colloid Interface Sci.*, **106**, 261.

59 Scott, T.A. and Melvin, E.H. (1953) Determination of dextran with anthrone, *Anal. Chem.*, **25**, 1656.

60 Scatchard, G. (1949) The attractions of proteins for small molecules and ions, *Ann. N.Y. Acad. Sci.*, **51**, 660.

61 Rotureau, E., Léonard, M., Dellacherie, E., Camesano, T.A. and Durand, A. (2006) From polymeric surfactants to colloidal systems (1): Amphiphilic dextrans for emulsion preparation, *Colloids Surf. A*, **288**, 131.

62 Rouzes, C., Gref, R., Léonard, M. and Dellacherie, E. (2000) *J. Biomed. Mater. Res.*, **50**, 557.

63 Couvreur, P. and Vauthier, C. (2006) *Pharm. Res.*, **23**, 1417.

64 Vonarbourg, A., Passinari, C., Saulnier, P. and Benoit, J.-P. (2006) *Biomaterials*, **27**, 4356.

65 Couvreur, P., Kante, B., Roland, M., Guiot, P., Bauduin, P. and Speiser, P. (1979) *J. Pharm. Pharmacol.*, **31**, 331.

66 Ryan, B. and McGann, G. (1996) *Macromol. Rapid Commun.*, **17**, 217.

67 Eromosele, C., Pepper, D.C. and Ryan, B. (1989) *Makromol. Chem.*, **190**, 1613.

68 Pepper, D.C. and Ryan, B. (1983) *Makromol. Chem.*, **184**, 383.

69 Limouzin, C., Caviggia, A., Ganachaud, F. and Hémery, P. (2003) Anionic polymerization of n-butyl cyanoacrylate in emulsion and miniemulsion, *Macromolecules*, **36**, 667.

8
Recent Developments in Producing Monodisperse Emulsions Using Straight-Through Microchannel Array Devices

Isao Kobayashi, Kunihiko Uemura, and Mitsutoshi Nakajima

8.1
Introduction

Emulsification is an important process that is used in a variety of industrial areas including the food, cosmetic, pharmaceutical and chemical industries. *Emulsions* – the products produced by emulsification – are dispersions of two immiscible liquids in the presence of surface-active molecules and/or particles, with one of the liquids dispersed as droplets in the continuous phase of the other. Emulsification devices apply one of two possible production routes [1, 2]:

- A gradual reduction of the droplet size by rupturing a to-be-dispersed phase and droplets.
- The direct generation of droplets from a to-be-dispersed phase.

Conventional emulsification devices, such as rotor-stator systems and high-pressure homogenizers, were first used to produce emulsions and to use extensional and shear stress or impact to rupture the droplets [3, 4]. Emulsions produced using these devices usually have wide droplet size distributions. Shear-rupturing of a polydisperse emulsion in an injection couette mixer, as reported by Mason and Bibette [5], can produce emulsions with relatively narrow droplet size distributions. However, many important potential applications of emulsions require monodisperse emulsions composed of uniformly sized droplets with a coefficient of variation (CV) = [(standard deviation/average droplet size) × 100] of typically less than 5%.

The earliest major approach for producing monodisperse emulsions was membrane emulsification using shirasu porous glass (SPG) membranes, as developed by Nakashima and colleagues during the late 1980s [6, 7]. Direct SPG membrane emulsification, which applies the second production route, can produce quasi-monodisperse emulsions with average droplet diameters ($d_{av,dr}$) of 0.3 to 30 μm and CVs of approximately 10% by forcing a to-be-dispersed phase in a crossflowing continuous phase through numerous membrane pores [8, 9]. In addition to SPG membranes, microporous membranes such as ceramic, metallic and polymeric membranes and microengineered devices have been used for

Emulsion Science and Technology. Edited by Tharwat F. Tadros
Copyright © 2009 WILEY-VCH Verlag GmbH & Co. KGaA, Weinheim
ISBN: 978-3-527-32525-2

membrane emulsification, as reviewed by Vladisavljević and Williams [2]. These authors also reported direct membrane emulsification using a rotating cylindrical membrane that develops shear stress along the membrane surface [10]. Suzuki *et al.*, as well as several other research groups, have also reported (repeated) premix membrane emulsification for high-scale productions [2, 11–15]. Several authors have reviewed the current status of the membrane emulsification technique, its underlying process phenomena, and the emulsion and particulate products created using this technique [2, 16–19].

The latest major approach for producing monodisperse emulsions is to use microfluidic devices. Several types of emulsification technique for generating emulsion droplets using T-shaped microfluidic channels and flow-focusing geometries of quasi-two and three dimensions have been also proposed, originating during the early 2000s [20–29]. These microfluidic techniques apply the second production route, which enables the generation of monodisperse emulsion droplets with $d_{av,dr}$ of typically more than 10 μm and CVs of less than 5% under a forced crossflow or coflow of the continuous phase. However, as only a single channel for a to-be-dispersed phase is generally fabricated on a microfluidic device, this leads potentially to low production scales of emulsion droplets. Link *et al.* also reported the passive breakup of monodisperse larger emulsion droplets into smaller droplets using T-shaped microfluidic channels [30]. Several authors have reviewed the current status of emulsification techniques using microfluidic devices, its underlying process phenomena, and some of the emulsion and particulate products created using this technique [31–34].

During the late 1990s, the present authors' research group was the first to propose the use of microchannel (MC) emulsification using microfabricated channel arrays each having a slit-like terrace [35, 36]. MC emulsification, which utilizes the second of the production routes, enables the production of monodisperse emulsions with $d_{av,dr}$ of 1 to 100 μm and CVs of typically less than 5%. The process achieves this by forcing a to-be-dispersed phase through uniformly sized channels into a deeply etched well filled with a continuous phase (Figure 8.1) [35–38]. The MC emulsification process is unique in that it does not require a forced flow of the continuous phase [39]. Monodisperse emulsions produced by MC emulsification have been used to produce monodisperse microparticles and microcapsules [40–45]. The current status of the MC emulsification technique and its underlying process phenomena have been briefly reviewed by Kobayashi *et al.* [31], and the key principles of this technique are described in Section c08.2.

Although conventional MC emulsification devices consist of several lines of MC arrays with many channels (e.g. a few hundred channels each with a size of 10 μm), the low-density layout of the channels leads to low production scales of emulsion droplets (typically <0.1 ml h^{-1}). In order to overcome this problem, during the early 2000s Kobayashi *et al.* proposed a novel MC emulsification device which consisted of a straight-through MC array with numerous microfluidic channels arranged vertical to the device surface [46]. Straight-through MC array devices with a highly integrated layout of these uniformly sized channels are expected to achieve a high-scale production of monodisperse emulsions.

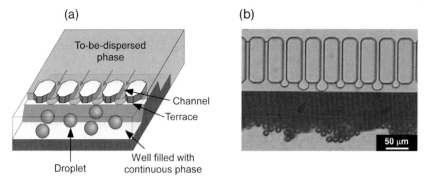

Figure 8.1 (a) Schematic illustration of the emulsification process using a grooved microchannel (MC) array. (b) Optical micrograph of production of monodisperse oil-in-water (O/W) emulsions using a grooved MC array with a pressure of 6.3 kPa applied to the to-be-dispersed phase. Refined soybean oil was used as the to-be-dispersed phase, and a 1.0 wt% sodium dodecyl sulfate (SDS) aqueous solution as the continuous phase.

This chapter provides an overview of the present authors' recent investigations into the production of monodisperse emulsions using straight-through MC array devices. These devices, together with details of the emulsification set-up are outlined in Section 8.3, while some of the important device and process factors affecting emulsification using symmetric straight-through MC arrays are discussed in Sections 8.4 and 8.5. A scaling-up study of a straight-through MC array device is described in Section 8.6, and the features of a novel asymmetric straight-through MC array and emulsification using this device in Section 8.7. The final section provides an overview of future research on emulsification techniques using straight-through MC array devices.

8.2
Principles of Microchannel Emulsification

Figure 8.1 presents a schematic illustration and optical micrograph of the MC emulsification process. In MC emulsification, droplets are generated from channels with a slit-like terrace with the applied pressure of a to-be-dispersed phase that exceeds the breakthrough pressure. The breakthrough pressure at which a to-be-dispersed phase can pass through the channels is estimated by the Young–Laplace equation taking the contact angle effect into account:

$$\Delta P_{BT} = 4\gamma \cos\theta / d_{ch} \tag{8.1}$$

where ΔP_{BT} is the Young–Laplace pressure between the two phases, γ is the interfacial tension, θ is the contact angle of the to-be-dispersed phase to the channel surface, and d_{ch} is the channel diameter [40]. Sugiura *et al.* analyzed the MC emulsification process and proposed a droplet generation mechanism for MC emulsification that exploits the interfacial tension force that is dominant on a micrometer scale [39]. The to-be-dispersed phase that has passed through the channels expands on the terrace with a distorted, disk-like shape, passes through the terrace exit and then expands in the well. When the Laplace pressure of the to-be-dispersed phase on the terrace ($\Delta P_{d,\text{terrace}}$) significantly exceeds that of the to-be-dispersed phase in the well ($\Delta P_{d,\text{well}}$) – that is, $\Delta P_{d,\text{terrace}} > \Delta P_{d,\text{well}}$ – the to-be-dispersed phase on the terrace shrinks rapidly and the neck formed on the terrace pinches off spontaneously, generating a droplet. This unique MC emulsification process enables the production of monodisperse emulsions without applying a forced flow of the continuous phase, while requiring an energy input of only 10^3 to $10^4\,\text{J\,m}^{-3}$.

The size of the droplets generated by MC emulsification is determined primarily by the channel geometry. Although the resultant droplet size is greatly affected by the channel depth and terrace length [47], it remains independent of the channel width and channel length [48]. Sugiura *et al.* proposed empirical prediction models of the droplet size for MC emulsification, using a model oil-in-water (O/W) system [47, 49]. When the aim of an MC emulsification experiment is to generate droplets of a specific size, MC arrays with narrower and longer channels can produce the monodisperse emulsion at a higher droplet generation rate, which is attributable to a greater pressure drop of the to-be-dispersed phase inside the channel [48]. The pressure drop is given by the following, known as Fanning's equation:

$$\Delta P_{d,\text{ch}} = 4f(\rho_d U_d^2/2)(l_{\text{ch}}/d_{\text{ch}}) \tag{8.2}$$

where $\Delta P_{d,\text{ch}}$ is the pressure drop of the to-be-dispersed phase inside the channel, f is the Fanning's friction factor, ρ_d is the density of the to-be-dispersed phase, U_d is the flow velocity of the to-be-dispersed phase, and l_{ch} is the channel length [48]. The character of droplet generation from channels is determined by a dimensionless number called the *capillary number* (the viscous force divided by the interfacial tension force):

$$Ca = \eta U_d/\gamma \tag{8.3}$$

where η is the viscosity of the to-be-dispersed phase [50]. Below the critical Ca where the interfacial tension force is dominant, monodisperse emulsions were produced from the channels, with a resultant droplet size independent of the Ca value. Above the critical Ca, where the viscous force becomes significant, polydisperse emulsions were produced and the droplet size increased steeply with increasing Ca. The critical Ca was independent of the channel size.

The successful production of monodisperse emulsions for MC emulsification can only be achieved when the continuous phase preferentially wets the channel surface [35, 51–53]. The surface of MC arrays must be hydrophilic to produce monodisperse O/W emulsions, and hydrophobic to produce monodisperse water-in-oil (W/O) emulsions. For example, Kawakatsu *et al.* reported that monodisperse W/O emulsions were produced using hydrophobic MC arrays with a high static contact angle (>120°) of a to-be-dispersed phase droplet to the plate surface in a continuous phase [52].

8.3
Straight-Through MC Array Device and Emulsification Set-Up

A straight-through MC array plate of standard size with a 24 × 24 mm² surface area is shown in Figure 8.2a and b. The array plate is microfabricated through repeated processes of photolithography and deep-reactive-ion etching to create the straight-through MC array in addition to a well, a thermal oxidization process and a wafer dicing process [46]. Such an array consisting of approximately 10 000 channels with a representative diameter of 10 μm is positioned within a 10 × 10 mm² area in the center of the plate. A microfabricated straight-through MC array with oblong channels which has been microfabricated vertically on a silicon plate is shown in Figure 8.2c. Straight-through MC arrays usually have channels with a very narrow size distribution of less than 1% [46]. In both MC and membrane emulsification, the channel (or pore) size distribution greatly affects the droplet size distribution of the resultant emulsions. Thus, straight-through MC arrays with uniformly sized

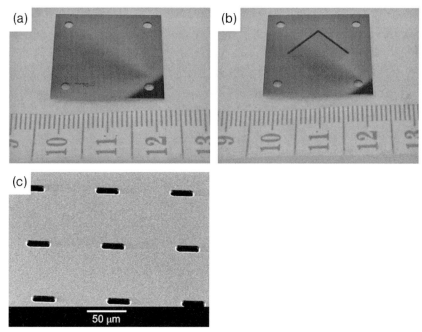

Figure 8.2 Straight-through MC array device. (a, b) Photographs of the top and bottom surfaces of a standard straight-through MC plate made from silicon. (c) Scanning electron micrograph of microfabricated oblong channels.

channels satisfy a necessary condition for producing monodisperse emulsions. The surface porosity of a straight-through MC array, defined as the ratio of the total cross-sectional area of the channels to the total area of the straight-through MC array, is generally less than 10%. An excessive increase in surface porosity may cause contact with the to-be-dispersed phase that is expanding from adjacent channels. In addition, existing straight-through MC arrays have been designed where the channel depth exceeds the diameter by more than 10-fold. These deep (and long) channels are especially advantageous for producing monodisperse emulsions at high droplet generation rates, as expressed in Equation 8.2 [48].

An experimental set-up for emulsification using a straight-through MC array device is shown in Figure 8.3. The arrangement consists of an emulsification module equipped with a straight-through MC array plate, an apparatus for supplying the two phases (e.g. syringe pumps and liquid chambers), and instruments for monitoring and recording the emulsification process [46]. A to-be-dispersed phase is pressurized using the supply apparatus and introduced into the emulsification module, filling the well under the straight-through MC array. In order to generate emulsion droplets, the to-be-dispersed phase that reaches the channel inlet is forced to pass through the channels into a continuous phase region over the channel outlet (Figure 8.3). The emulsion droplets generated can be collected by a gentle crossflow of the continuous phase along the surface of the straight-through MC array plate.

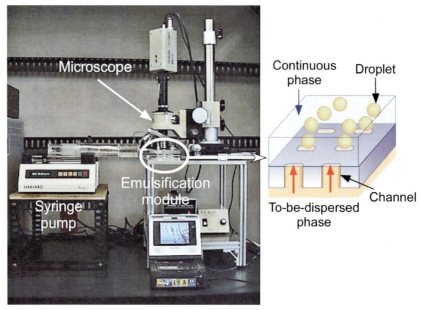

Figure 8.3 Left: A typical experimental set-up for emulsification using a straight-through MC array device. Right: A schematic illustration of the emulsification process using a straight-through MC array.

8.4
Effect of Channel Shapes on Emulsification Using Symmetric Straight-Through MC Arrays

8.4.1
Effect of Channel Cross-Sectional Shape

The first report of emulsification using symmetric straight-through MC arrays highlighted the effect of the channel cross-sectional shape on the emulsification process, using two straight-through MC array devices with circular channels and oblong channels [46]. A similar MC array with capillary-like circular channels was designed as a model device, while an array with oblong channels was designed on the basis of MC emulsification research. In the case of the straight-through MC array with circular channels (10.0 μm diameter), the to-be-dispersed phase (refined soybean oil) that passed through the circular channels was transformed into polydisperse large droplets with diameters of more than 100 μm, driven by the gentle crossflow of the continuous water phase (Figure 8.4a and b). A continuous outflow of the to-be-dispersed phase from the circular channels was also observed in the absence of any crossflow of the continuous phase. For the straight-through MC array with oblong channels (9.6 μm on the shorter line, 29.7 μm on the longer line), the to-be-dispersed phase that passed through the oblong channels was transformed into monodisperse emulsion droplets with CVs of less than 3%, independent of the applied crossflow of the continuous phase at velocities of 0.0 to 9.2 mm s^{-1} (Figure 8.4c and d). The $d_{av,dr}$ of the monodisperse emulsions produced (Figure 8.4e) was approximately 30 μm, and also independent of the applied crossflow velocity of the continuous phase. As shown in Figure 8.4d, the production of a monodisperse emulsion from the oblong channels was conducted by spontaneous-transformation-based droplet generation, as proposed by Sugiura et al. [39]. Thus, the study findings determined that the oblong channels of a simply elongated cross-sectional shape were capable of producing monodisperse emulsions, without the application of external shear stress.

8.4.2
Effect of the Aspect Ratio of Oblong Channels

As noted in Section 8.4.1, when using symmetric straight-through MC arrays the channel cross-sectional shape is the most important factor affecting the emulsification process. Based on the assumption that the aspect ratio of the oblong channels would affect the emulsification process, a series of investigations was conducted using straight-through MC arrays with oblong channels of different aspect ratios [54]. Figure 8.5a–c depicts some typical examples of the production of soybean O/W emulsions from oblong channels (about 10 μm on the shorter line; see Figure 8.5 for dimensions), with different aspect ratios (defined as the longer line divided by the shorter line). The to-be-dispersed phase that has passed through

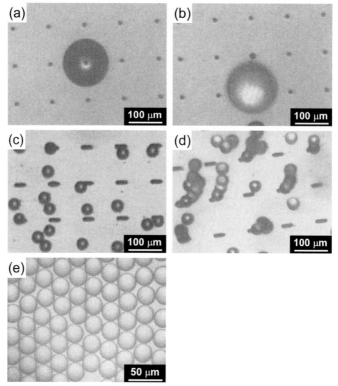

Figure 8.4 (a, b) Optical micrographs of the emulsification process using microfabricated circular channels with a 10.0 μm diameter and a 200 μm depth. The pressure applied to the to-be-dispersed phase (refined soybean oil) was 3.2 kPa and the crossflow velocity of the continuous phase (0.3 wt% SDS aqueous solution) was 0.46 mm s^{-1}. (c, d) Optical micrographs of the emulsification process using microfabricated oblong channels with a 9.6 μm shorter line, a 29.7 μm longer line, and a 200 μm depth. The pressure applied to the to-be-dispersed oil phase was 1.8 kPa, and the crossflow velocities of the continuous water phase were 0.46 mm s^{-1} (c) and 0.0 mm s^{-1} (d). (e) Optical micrograph of a monodisperse O/W emulsion produced using oblong channels [46].

oblong channels with an aspect ratio of 1.9 was transformed into large droplets with diameters of 350–400 μm under a gentle crossflow of the continuous phase (Figure 8.5a). A polydisperse emulsion composed of small droplets with diameters of about 50 μm and large droplets with diameters of 350–400 μm was produced from oblong channels with an aspect ratio of 2.7 (Figure 8.5b and d). In contrast, a monodisperse emulsion with a $d_{av,dr}$ of 42 μm and a CV of less than 2% was produced from oblong channels with an aspect ratio of 3.8 (Figure 8.5c and e), analogous to the result shown in Figure 8.4c. The results described in Sections 8.4.1 and 8.4.2 show clearly that monodisperse emulsions are produced from oblong channels when the aspect ratio of the latter exceeds a threshold value of approximately three.

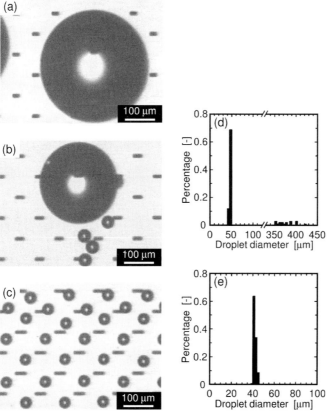

Figure 8.5 (a–c) Optical micrographs of the emulsification process using microfabricated oblong channels of different aspect ratios with a 200 μm depth. The oblong channels had a 13.3 μm shorter line and a 25.2 μm longer line (a), a 12.0 μm shorter line and a 32.8 μm longer line (b), and a 10.8 μm shorter line and a 40.8 μm longer line (c). The flux applied to the to-be-dispersed phase (refined soybean oil) was $10.0\,l\,m^{-2}\cdot h^{-1}$, and the crossflow velocity of the continuous phase (1.0 wt% SDS aqueous solution) was $1.1\,mm\,s^{-1}$. (d, e) Droplet size distributions of O/W emulsions produced using the oblong channels depicted in panels (b) and (c) [54].

8.4.3
Computational Fluid Dynamics (CFD) Simulation and Analysis

Previously, Kobayashi and colleagues showed that oblong channels with an aspect ratio that exceeded a specific value were required to generate monodisperse emulsion droplets [46, 54]. However, whilst it is difficult to observe experimentally the movement of the oil/water interface in the region near the channel outlet, it is essential that this be understood in order to analyze the effect of channel cross-sectional shape on the emulsification process. Kobayashi and coworkers used a CFD

Table 8.1 Dimensions of the modeled oblong channels and calculated results of droplet diameter and detachment time for emulsion droplet generation from the oblong channels [55].

No.	Channel dimensions[a] [μm]	Channel aspect ratio	Droplet diameter [μm][b]	Droplet detachment time [s][b]
TMC-2	20 × 10 × 200	2	[c]	[c]
TMC-2.5	25 × 10 × 200	2.5	[c]	[c]
TMC-3	30 × 10 × 200	3	[c]	[c]
TMC-3.25	32.5 × 10 × 200	3.25	35.6	0.0541
TMC-3.5	35 × 10 × 200	3.5	34.4	0.0437
TMC-4	40 × 10 × 200	4	35.4	0.0386
TMC-4.5	45 × 10 × 200	4.5	37.0	0.0395

[a] Longer line [μm] × shorter line [μm] × depth [μm].
[b] The average velocity of the to-be-dispersed phase (refined soybean oil) at the channel inlet was set at $1.0\,\mathrm{mm\,s^{-1}}$. The crossflow velocity of the continuous phase (water) was set at $0.0\,\mathrm{mm\,s^{-1}}$.
[c] Continuous outflow of the to-be-dispersed phase from the channel outlet.

method to simulate and analyze the generation process of soybean O/W emulsion droplets from a single symmetric channel (10 μm on the shorter line) of different aspect ratios [55, 56]. For oblong channels with aspect ratios ≤ 3, no droplet generation occurred during the calculations (Table 8.1). However, for oblong channels with aspect ratios ≥ 3.25, small droplets with diameters of 34–38 μm were generated from the channels, without applying any forced crossflow of the continuous phase (Table 8.1). These CFD results confirmed the existence of a threshold channel aspect ratio (of approximately 3) that was necessary for the production of monodisperse emulsions. The threshold channel aspect ratio calculated when using CFD agreed well with that obtained from experimental studies (see Sections 8.4.1 and 8.4.2). A typical example of the movements and shapes of the oil/water interface inside and outside an oblong channel below the threshold aspect ratio is shown in Figure 8.6. When the to-be-dispersed phase has begun to expand from the oblong channel, the oil/water interface blocks the entire channel outlet (Figure 8.6c), so that it becomes difficult to generate small droplets by pinching off the oil/water interface near the channel outlet. A typical example of the successful generation of an emulsion droplet from an oblong channel which exceeds the threshold aspect ratio is shown in Figure 8.7. Sufficient space for the continuous phase at the channel outlet was maintained during droplet generation, and this facilitated the rapid shrinkage and instantaneous cut-off of the pinched neck inside the channel. The generation of a small emulsion droplet from an oblong channel was completed as shown in Figure 8.7d. These visualized CFD results suggest that the maintenance of sufficient space for the continuous phase at the channel outlet during droplet generation is a prerequisite for the production of monodisperse emulsions when using symmetric straight-through MC arrays.

8.4 Effect of Channel Shapes on Emulsification Using Symmetric Straight-Through MC Arrays | 143

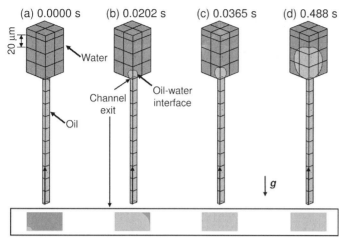

Figure 8.6 Time course of the shape of the oil/water interface inside and outside an oblong channel with a 10 μm shorter line, a 20 μm longer line, and a 200 μm depth, calculated using CFD. The average velocity of the to-be-dispersed phase (refined soybean oil) at the channel inlet was set at 1.0 mm s^{-1}. The crossflow velocity of the continuous phase (water) was set at 0.0 mm s^{-1}. The plate surface and channel walls were no-slip and no-wetting with respect to the to-be-dispersed phase [56].

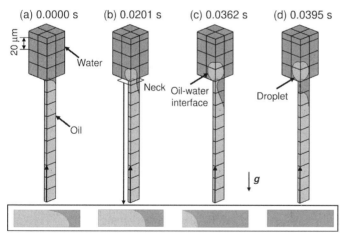

Figure 8.7 Successful generation of an O/W emulsion droplet from an oblong channel with a 10 μm shorter line, a 40 μm longer line, and a 200 μm depth, calculated using CFD. The average velocity of the to-be-dispersed phase (refined soybean oil) at the channel inlet was set at 1.0 mm s^{-1}. The crossflow velocity of the continuous phase (water) was set at 0.0 mm s^{-1} [56].

8.5
Effect of Process Factors on Emulsification Using Symmetric Straight-Through MC Arrays

8.5.1
Effect of Surfactants and Emulsifiers

Surfactants of which the molecules contain both hydrophilic and hydrophobic head groups play an important role in the emulsification process. Indeed, several reports have been made that interaction between the surfactant molecules and the channel (or membrane) surface greatly affects the droplet generation process in MC and membrane emulsification [8, 51]. Kobayashi and coworkers investigated the effect of surfactant charge on the production of soybean O/W emulsions using a straight-through MC array with oblong channels (see Figure 8.8 for the dimensions) [57]. A straight-through MC array plate, the surface of which had been treated by plasma

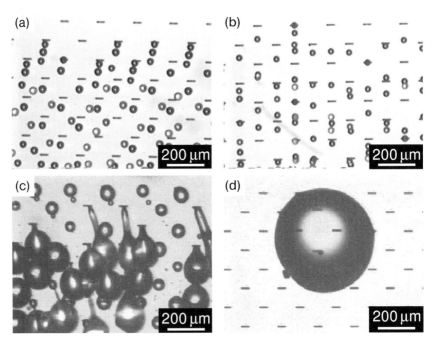

Figure 8.8 Optical micrographs of the emulsification results using microfabricated oblong channels with a 9.6 µm shorter line, a 48.7 µm longer line, and a 200 µm depth. Refined soybean O/W systems containing anionic, nonionic or cationic surfactants were used in this investigation. (a, b) Production of monodisperse O/W emulsions using SDS (a) and Tween 20 (b) as the surfactants. (c) Production of polydisperse O/W emulsions using CTAB as the surfactant. (d) Wetting of the to-be-dispersed phase on the plate surface using TOMAC as the surfactant. The fluxes applied to the to-be-dispersed phase were $10.0 \, l \, m^{-2} \cdot h^{-1}$ in (a–c) and $5.0 \, l \, m^{-2} \cdot h^{-1}$ in (d). The crossflow velocity of the continuous phase was $1.1 \, mm \, s^{-1}$ [54]. For explanations of the surfactants and experimental conditions, see Table 8.2.

Table 8.2 Preparation conditions and interfacial properties of the O/W systems with different surfactants and emulsification results using the straight-through MC array device depicted in Figure 8.8 [56].

Surfactant[a]	Type	Condition	Contact angle[d] [deg]	Interfacial tension [mN m^{-1}]	Average droplet diameter [μm]	Coefficient of variation [%]
SDS	Anionic	1.0 wt% (W)[b]	145	4.0	39.2	2.5
Tween 20	Nonionic	1.0 wt% (W)[b]	142	1.9	38.6	2.5
CTAB	Cationic	1.0 wt% (W)[b]	134	<0.1	[e]	[e]
TOMAC	Cationic	2.0 wt% (O)[c]	38	3.1	[f]	[f]

[a] SDS, sodium dodecyl sulfate; Tween 20, Polyoxyethylene (20) monolaurate; CTAB, cetyltrimethylammonium bromide; TOMAC, tri-n-octylmethlammonium chloride.
[b] Dissolved in the continuous water phase.
[c] Dissolved in the to-be-dispersed oil phase.
[d] Contact angle between an oil droplet and the surface of an oxidized straight-through MC plate in a water phase.
[e] Unstable generation of large emulsion droplets (see Figure 8.8c).
[f] Wetting of the plate surface by the to-be-dispersed phase (see Figure 8.8d).

oxidization, had a negatively charged silanol group on the plate surface that was in contact with water. For O/W systems containing an anionic surfactant (e.g. sodium dodecyl sulfate; SDS) or a nonionic surfactant (e.g. Tween 20), monodisperse emulsions with CVs of less than 3% were produced from oblong channels (Figure 8.8a and b; Table 8.2). For O/W systems containing a cationic surfactant (e.g. cetyl trimethylammonium bromide; CTAB), a polydisperse emulsion with large droplets was unstably produced from oblong channels, with some of the generated droplets sticking to the plate surface (Figure 8.8c; see also Table 8.2). In contrast, in O/W systems containing a cationic surfactant (e.g. tri-n-octylmethylammonium chloride; TOMAC), the to-be-dispersed phase that has passed through the channel outlet was then spread onto the plate surface (Figure 8.8d; see also Table 8.2). An analysis of the results obtained in these investigations indicated that a nonattractive surfactant–plate surface interaction and a high contact angle are required to produce monodisperse emulsions using a symmetric straight-through MC array.

Saito and colleagues subsequently investigated the production of soybean O/W emulsions stabilized by proteins with different isoelectric points using a straight-through MC array with oblong channels (see Table 8.3 for dimensions) [58]. When β-lactoglobulin and bovine serum albumin (BSA), each with low isoelectric points (see Table 8.3) were used as emulsifiers, monodisperse O/W emulsions with CVs of less than 3% were produced from oblong channels. These emulsifiers, which have a negative charge under experimental conditions, prevent wetting of the negatively charged plate surface by the to-be-dispersed phase. In contrast, when lysozyme with a high isoelectric point (see Table 8.3) was used as the emulsifier, the to-be-dispersed phase flowed continuously from the channel outlet. The to-be-dispersed phase also wetted the plate surface due to an attractive interaction between the positively charged emulsifier and the

Table 8.3 Isoelectric points of the proteins, preparation conditions and interfacial properties of the O/W systems with different proteins, and emulsification results using a straight-through MC device with oblong channels [58].

Protein	Isoelectric point[a] [range]	Condition	Contact angle[b] [deg]	Interfacial tension [mN m^{-1}]	Average droplet diameter[c] [μm]	Coefficient of variation[c] [%]
β-Lactoglobulin	5.14–5.41	0.45 wt% (W)[d]	18.5	11.2	43.0	1.4
BSA	4.71–4.84	0.45 wt% (W)[d]	38.5	13.5	44.1	2.1
Lysozyme	10.5–11.0	0.45 wt% (W)[d]	117.0	15.1	e	e

[a] Data from Ref. [59].
[b] Contact angle between a water droplet and the surface of an oxidized straight-through MC plate in an oil phase.
[c] The microfabricated oblong channels had a longer line of 42.8 μm, a shorter line of 13.3 μm, and a depth of 200 μm.
[d] Dissolved in the continuous phase (25 mM NaCl aqueous solution).
[e] Wetting of the plate surface by the to-be-dispersed phase.
BSA = bovine serum albumin.

negatively charged plate surface. The contact angles listed in Table 8.3 also support the preceding emulsification results. Saito and coworkers also investigated the effect of the pH of the continuous phase (BSA aqueous solutions) on the emulsification process when using the straight-through MC array. At pH 5 to 7, where BSA has a negative charge, monodisperse O/W emulsions with CVs of less than 7% were produced from oblong channels. In contrast, at pH 3 and 4, where BSA has a positive charge, the to-be-dispersed phase flowed continuously from the channel outlet, similar to the emulsification results for the O/W system containing lysozyme. In addition to the above-mentioned emulsifiers, other types of food-grade emulsifiers and several protein mixtures have been used by the present authors' research group [58, 60].

The results described in this section show clearly that interaction between the surfactant and/or emulsifier molecules and the plate surface is a key factor affecting the emulsification process when using symmetric straight-through MC arrays.

8.5.2
Effect of To-Be-Dispersed Phase Viscosity

Both Kawakatsu and coworkers [52] and Sugiura and coworkers [61] have reported that the viscosity ratio between the two phases significantly affects the resultant droplet size in MC emulsification. An MC emulsification study on the production of O/W emulsions utilized two types of chemical oil with a limited range of viscosities as the to-be-dispersed phase [62]. Kobayashi et al. investigated the generation process of O/W emulsion droplets from a straight-through MC array with oblong channels (see Figure 8.9 for dimensions), using oils with a wide range of viscosities as the to-be-dispersed phase [62]. The effect of the to-be-dispersed phase viscosity (η_d) on the

8.5 Effect of Process Factors on Emulsification Using Symmetric Straight-Through MC Arrays

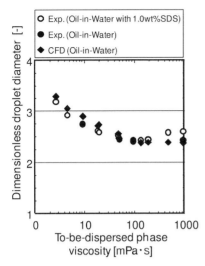

Figure 8.9 Effect of the to-be-dispersed phase viscosity on the dimensionless droplet diameter obtained from experiments and CFD calculations. The oils used were silicone (viscosity 4.6–970 mPa·s), tetradecane (viscosity 2.7 mPa·s), medium-chain triglyceride (MCT) oil (viscosity 20 mPa·s), refined soybean oil (viscosity 50 mPa·s), and liquid paraffin oil (viscosity 140 mPa·s). The microfabricated oblong channels used in the experiments had a 12.7 μm shorter line, a 43.4 μm longer line, and a 200 μm depth. The oblong channel used in the CFD calculations had a 10 μm shorter line, a 40 μm longer line, and a 200 μm depth. The pressures applied to the to-be-dispersed phase in the experiments were just above the breakthrough pressures (as defined in Section c08.2). The capillary number of the to-be-dispersed phase flowing inside the channel in the calculations was set at 2×10^{-3} for all of the O/W systems [61].

dimensionless diameter of the generated droplets ($\bar{d}_{dr}$), defined as the ratio of the droplet diameter to the channel equivalent diameter, is shown in Figure 8.9. The $\bar{d}_{dr}$ values obtained from both experiments and CFD calculations were seen to depend significantly on η_d, and decreased with increasing η_d in the range below 100 mPa·s. In this region, the $\bar{d}_{dr}$ values from experiments were independent of the concentration of a model surfactant, SDS. In contrast, the $\bar{d}_{dr}$ values obtained from experiments using the O/W system without the surfactant and from CFD calculations changed little in the η_d range above 100 mPa·s. Furthermore, $\bar{d}_{dr}$ values obtained from experiments using the O/W system with the surfactant increased significantly with increasing η_d in this region. It was assumed that this increase in $\bar{d}_{dr}$ was attributable to the significant decrease in dynamic interfacial tension during droplet generation. The results described in this section showed that there exists a threshold η_d value near which the trends in the $\bar{d}_{dr}$ values change. It should be noted that this threshold η_d value might also depend on the channel size, surfactant type and continuous-phase viscosity. In addition, the droplet generation rate for an oblong channel gradually increased with decreasing η_d over the range presented in Figure 8.9; hence, the use of a to-be-dispersed phase with low η_d values leads to the

8.5.3
Effect of To-Be-Dispersed Phase Flux

Kobayashi and coworkers also investigated the effect of the to-be-dispersed phase flux (J_d) on the production of soybean O/W emulsions using a straight-through MC array with oblong channels (see Figure 8.10 for dimensions) [57]. The micrographs presented in Figure 8.10a–c show that the droplet generation from the oblong channels may be categorized into three regions, depending on J_d. In region (a), where J_d is 60 l m^{-2}·h^{-1} or less, monodisperse emulsions with d_{av} of about 40 μm and CVs of less than 3% were generated from the oblong channels (Figure 8.10a). The d_{av} value was independent of J_d in this region (Figure 8.10d). The J_d value of 60 l m^{-2}·h^{-1} corresponds to a production rate of monodisperse emulsion droplets of 6.0 ml h^{-1} and a generation rate of 1.8×10^8 h^{-1}. In region (b), where J_d is between 70 and

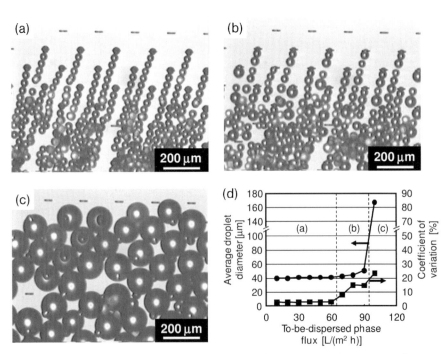

Figure 8.10 Optical micrographs of three typical examples of the emulsification process using microfabricated oblong channels with a 9.6 μm shorter line, a 48.7 μm longer line, and a 200 μm depth. (a) Production of a refined soybean O/W monodisperse emulsion from the oblong channels at J_d of 60.0 l m^{-2}·h^{-1}. (b) Production of a quasi-monodisperse O/W emulsion from the oblong channels at J_d of 70.0 l m^{-2}·h^{-1}. (c) Production of a polydisperse emulsion with large droplets from the oblong channels at J_d of 100.0 l m^{-2}·h^{-1}. The crossflow velocity of the continuous phase (1.0 wt% SDS aqueous solution) was 1.1 mm s^{-1} [57]. (d) Effect of J_d on d_{av} and CV of the generated O/W emulsion droplets.

90 l m^{-2}·h^{-1}, quasi-monodisperse emulsion droplets with CVs of 8–15% were generated from the oblong channels (Figure 8.10b). Their d_{av} value gradually increased with increasing J_d in this region (Figure 8.10d). In region (c), where J_d is 100 l m^{-2}·h^{-1}, steric hindrance among adjacent droplets restricted their further expansion and forced them to detach from the oblong channels, which resulted in the production of a polydisperse emulsion (Figure 8.10c and d). These results – and particularly those in region (a) – show a useful trend in which monodisperse emulsions precisely controlled in droplet size are produced using a symmetric straight-through MC array with J_d below the critical value.

8.6
Scaling-Up of Straight-Through MC Array Devices

Although straight-through MC array devices of standard size are capable of producing monodisperse emulsions at high to-be-dispersed phase fluxes, the emulsification device must be scaled up in order to achieve practical production scales for monodisperse emulsions. Kobayashi et al. developed a large straight-through MC plate made from silicon with a 40×40 mm^2 surface area (Figure 8.11a and b) [63].

Figure 8.11 (a, b) Photographs of top and bottom views of a large straight-through MC device made from silicon. (c) Optical micrograph of the production process for a monodisperse refined soybean O/W emulsion using the large straight-through MC device at J_d of 38.5 l m^{-2}·h^{-1}. The crossflow velocity of the continuous phase (1.0 wt% SDS aqueous solution) was 2.2 mm s^{-1} [63].

Symmetric straight-through MC arrays are positioned within four 15 × 15 mm² areas of the plate, and consisted of approximately 2.2 × 10⁵ channels (see Figure 8.11 for dimensions). The surface area of the straight-through MC arrays for the large straight-through MC plate exceeded that for the standard counterpart (see Figure 8.2a and b) by about 10-fold. A typical example of successful emulsification using a large straight-through MC plate is shown in Figure 8.11c. At a J_d of 38.5 l m^{-2}·h^{-1}, a maximum 60% of active oblong channels produced monodisperse soybean O/W emulsions with a d_{av} of 30 μm and a CV of less than 4%. This J_d value corresponds to the production rate for monodisperse emulsion droplets of 35.0 ml·h^{-1} and a droplet generation rate of 2.4 × 10⁹ ml·h^{-1}, which are rather high values for a single microfluidic emulsification device. The use of a to-be-dispersed phase with low viscosity values would achieve even higher droplet productivity from a straight-through MC array device. The straight-through MC array device can be further scaled-up to the size of a silicon wafer (to 12 cm diameter in the authors' research).

8.7
Emulsification Using an Asymmetric Straight-Through MC Array

One major drawback in emulsification with symmetric straight-through MC arrays is the difficulty in producing monodisperse emulsions when using a to-be-dispersed phase with very low viscosity (e.g. decane) (see Figure 8.12). In order to overcome this drawback, Kobayashi and coworkers developed a novel asymmetric straight-through MC array composed of numerous pairs of slits and circular channels

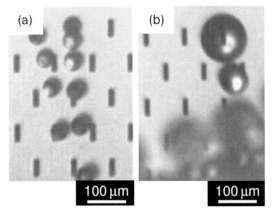

Figure 8.12 (a) Production of a monodisperse refined soybean O/W emulsion with a d_{av} of 48.1 μm and a CV of 3.8% from the microfabricated oblong channels (refer to Figure 8.9 for dimensions). (b) Unstable production of a decane-in-water emulsion composed of large droplets with a d_{av} of 89.4 μm and a CV of 7.6% from the oblong channels. The pressures applied to the to-be-dispersed phase were 0.72 kPa (a) and 1.26 kPa (b), and the crossflow velocity of the continuous phase (1.0 wt% SDS aqueous solution) was 1.1 mm s^{-1} [64].

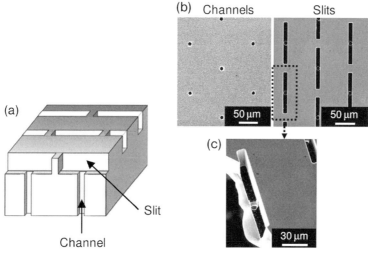

Figure 8.13 (a) Schematic illustration of an asymmetric straight-through MC array. (b, c) Scanning electron micrographs of an asymmetric straight-through MC array microfabricated on a silicon plate [64].

(Figure 8.13a) [64]. The scanning electron micrographs shown in Figure 8.13b and c demonstrate the uniformly sized slits (11 μm on the shorter line, 104 μm on the longer line, 21 μm on the depth) and circular channels (9.5 μm on the diameter, 5.4 μm on the depth) that have been precisely microfabricated in asymmetric straight-through MC array made from silicon. The to-be-dispersed phase that has passed through the slits is transformed into monodisperse droplets with d_{av} of 35–41 μm and CVs of less than 2% for O/W systems containing either refined soybean oil or decane (Figure 8.14). In particular, Figure 8.14b shows that an asymmetric straight-through MC array overcomes the unstable generation of polydisperse large droplets of very low viscosity from a symmetric straight-through MC array. The droplet generation rate from each active slit reached maximum values of approximately $50\,\text{s}^{-1}$ for decane droplets and approximately $10\,\text{s}^{-1}$ for soybean oil droplets.

The advantages of the asymmetric straight-through MC array can be summarized as follows. A circular channel with a minimum cross-sectional area more effectively controls the flow of the to-be-dispersed phase in the channel than does the oblong channel, due to the greater pressure drop in the circular channel, as expressed in Equation 8.2. Moreover, the to-be-dispersed phase expanding in the slit with a distorted, disk-like shape can function as a condenser to prevent the rapid outflow of the to-be-dispersed phase in the region over the slit outlet. Thus, both the slit and the circular channel in the asymmetric straight-through MC array are considered to contribute to the stable production of monodisperse emulsions, even when using a to-be-dispersed phase of very low viscosity. At present, we are conducting further investigation of emulsification using asymmetric straight-through MC arrays (particularly the droplet

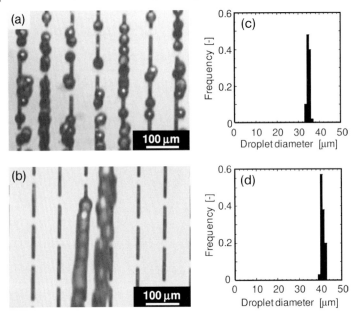

Figure 8.14 (a, b) Optical micrographs of the stable production of monodisperse O/W emulsions using an asymmetric straight-through MC array. (c, d) Droplet diameter distributions of the produced O/W emulsions. (a, c) Refined soybean oil-in-1.0 wt% SDS aqueous solution system. (b, d) Decane-in-1.0 wt% SDS aqueous solution system. The pressures applied to the to-be-dispersed phase were 1.29 kPa (a) and 1.94 kPa (b), and the crossflow velocity of the continuous water phase was 1.1 mm s^{-1} [64].

productivity for a single, asymmetric straight-through MC device) and of scaling-up the asymmetric straight-through MC array device.

8.8
Conclusions and Outlook

Straight-through MC array devices with numerous channels microfabricated vertically on a silicon plate enable the production of monodisperse emulsions with precisely controlled droplet sizes at high production scales. Cutting-edge semiconductor microfabrication techniques enable straight-through MC arrays to be created with deep, uniformly sized channels (about 10 μm diameter). The findings of our investigations have shown that an appropriate design for the channel cross-sectional shape is vital for emulsification when using straight-through MC array devices. In fact, a straight-through MC array with simple circular channels proved to be unsuitable as an emulsification device for producing monodisperse emulsions. In contrast, straight-through MC arrays with oblong channels and aspect ratios which exceeded a threshold value of about 3 were capable of producing monodisperse emulsions with d_{av} of 30–50 μm and CVs of less than 5%. The to-be-dispersed phase that has passed through the appropriate oblong channels was transformed spontaneously into uniformly sized droplets, indicating that such droplet generation is a very gentle process. The CFD method proved to be useful for simulating and analyzing the generation of emulsion droplets from channels of various shapes, and also provided useful information such as the movement of the oil/water interface near the channel outlet during droplet generation, which is difficult to observe microscopically in emulsification experiments.

The appropriate selection of surfactants and emulsifiers, and of the composition and flow conditions for two liquid phases, is also vital for emulsification when using straight-through MC devices. The plate and channel surfaces must be sufficiently wetted by a continuous phase containing surfactants and/or emulsifiers in order to produce monodisperse emulsions from straight-through MC arrays, as well as from grooved MC arrays and microporous membranes. In addition, the electrostatic surfactant–surface and emulsifier–surface interactions and the contact angle should be considered when selecting appropriate surfactants and emulsifiers. Emulsification investigations using O/W systems containing oils of a wide range of viscosity have shown that the to-be-dispersed phase viscosity affects not only the resultant droplet size but also the droplet generation rate. Below the critical value of the to-be-dispersed phase flux (several tens of $l\,m^{-2}\cdot h^{-1}$ for a soybean O/W system), monodisperse emulsions were produced from a straight-through MC array, with droplet size and size distribution independent of the flux value. This trend is especially advantageous in practical emulsification operations.

A large straight-through MC plate, developed as a first step in scaling-up of the straight-through MC device, was used successfully to produce monodisperse emulsions at droplet production rates of several tens of milliliters per hour (several billion droplets per hour). The productivity of monodisperse droplets for the large straight-through MC plate can be increased to over 100 ml per hour by using a to-be-dispersed phase with a viscosity lower than the model to-be-dispersed phase (soybean oil).

A novel asymmetric straight-through MC array composed of numerous pairs of uniformly sized slits and circular channels (about 10 μm the diameter) overcame one major drawback of symmetric straight-through MC arrays, and allowed the stable production of monodisperse emulsions, even when using a to-be-dispersed phase of very low viscosity (e.g. decane). Both, the slits and the circular channels play important roles in the stable production of monodisperse emulsions from an asymmetric straight-through MC array. Although the available data for asymmetric straight-through MC array devices are currently limited, ongoing investigations using asymmetric straight-through MC array devices are expected to verify their high performance as emulsification devices for the highly stable and efficient production of monodisperse emulsions.

Further efforts in developing straight-through MC array devices (notably of the asymmetric type) must be made in order to achieve a versatile and practicable production of monodisperse emulsions. Undoubtedly, straight-through MC arrays with channels of a wide size range will be developed in the near future, facilitating the production of monodisperse emulsions with droplet sizes of 1 to 1000 μm. Moreover, the development of straight-through MC array devices composed of nonsilicon materials should lead to more versatile applications of this emulsification technique. A further scaling-up of straight-through MC devices must be conducted in order to attain the droplet production rates of monodisperse emulsions necessary for the industrial production of high-value emulsions and particulate products, typically of at least 1 kg h^{-1}. Future experimental and CFD studies on emulsification techniques using straight-through MC array devices should enable the optimization of device design and emulsification processes. It is also expected that straight-through MC array devices will be applied to the practical production of monodisperse microparticles and microcapsules, as well as to monodisperse emulsions.

Acknowledgments

These studies were funded by the Nanotechnology Project of the Ministry of Agriculture, Forestry and Fisheries of Japan and the Program for Promotion of Basic Research Activities for Innovative Biosciences.

References

1 Williams, R.A. (2001) *Ingenia*, **7**, 1.
2 Vladisavljević, G.T. and Williams, R.A. (2005) *Adv. Colloid. Interface. Sci.*, **113**, 1.
3 Karbstein, H. and Schubert, H. (1995) *Chem. Eng. Process.*, **34**, 205.
4 McClements, D.J. (2004) *Food Emulsions: Principles Practice and Techniques*, 2nd ed. CRC Press, Boca Raton.
5 Mason, T.G. and Bibette, J. (1997) *Langmuir*, **13**, 4600.
6 Nakashima, T. and Shimizu, M. (1988) Abstracts, 21st Autumn Meeting of the Society of Chemical Engineers Japan, October Fukuoka, p. 86.
7 Nakashima, T., Shimizu, M. and Kukizaki, M. (1991) *Key Eng. Mater.*, **61**, 513.
8 Nakashima, T., Shimizu, M. and Kukizaki, M. (2000) *Adv. Drug Delivery Rev.*, **45**, 47.
9 Vladisavljević, G.T. and Schubert, H. (2002) *Desalination*, **144**, 167.
10 Vladisavljević, G.T. and Williams, R.A. (2006) *J. Colloid Interface. Sci.*, **299**, 396.
11 Suzuki, K., Shuto I. and Hagura, Y. (1996) *Food Sci. Technol. Int. Tokyo*, **2**, 43.
12 Suzuki, K. Fujiki, I. and Hagura, Y. (1998) *Food Sci. Technol. Int. Tokyo*, **4**, 164.
13 van der Zwan, E., Schröen, K., van Dijke, K. and Boom, R. (2006) *Colloids Surf. A: Physicochem. Eng. Aspects*, **277**, 223.
14 Altenbach-Rehm, J., Suzuki, K. and Schubert, H. Proceedings, 3rd World Congress on Emulsions, Lyon, 24–27 September, No. 2002 051.
15 Vladisavljević, G.T., Surh, J. and McClements, D.J. (2006) *Langmuir*, **22**, 4526.
16 Joscelyne, S.M. and Trägårdh, G. (2000) *J. Membr. Sci.*, **169**, 107.
17 Charcosset, C., LImayem, I. and Fessi, H. (2004) *Chem. Technol. Biotechnol.*, **79**, 209.
18 Gijsbertsen-Abrahamse, A.J., van der Part, A., Boom, R. and de Heiji, W.B.C. (2004) *J. Membr. Sci.*, **230**, 149.
19 Lambrich, U. and Schubert, H. (2005) *J. Membr. Sci.*, **257**, 76.
20 Thorsen, T., Roberts, R.W., Arnold, F.H. and Quake, S.R. (2001) *Phys. Rev. Lett.*, **86**, 4163.
21 Nishisako, T., Torii, T. and Higuchi, T. (2002) *Lab Chip*, **2**, 24.
22 Xu, J.H., Luo, G.S., Chen, G.G. and Wang, J.D. (2006) *Lab Chip*, **6**, 131.

23 Priest, C., Heminghaus, S. and Seemann, R. (2006) *Appl. Phys. Lett.*, **88**, 24106.
24 Umbanhowar, P.B., Prasad, V. and Weitz, D.A. (2000) *Langmuir*, **16**, 347.
25 Anna, S.L., Bontoux, N. and Stone, H.A. (2003) *Appl. Phys. Lett.*, **82**, 364.
26 Xu, Q. and Nakajima, M. (2004) *Appl. Phys. Lett.*, **85**, 3726.
27 Takeuchi, S., Garstecki, P., Weibel, D.B. and Whitesides, G.B. (2005) *Adv. Mater.*, **17**, 1067.
28 Utada, A.S., Lorenceau, E., Link, D.R., Kaplan, P.D., Stone, H.A. and Weitz, D.A. (2005) *Science*, **308**, 537.
29 Chan, E.M., Alivisatos, A.P. and Mathies, R.A. (2005) *J. Am. Chem. Soc.*, **127**, 13854.
30 Link, D.R., Anna, S.L., Weitz, D.A. and Stone, H.A. (2004) *Phys. Rev. Lett.*, **92**, 0405032.
31 Kobayashi, I. and Nakajima, M. (2005) (ed. Kockmann, N.), in *Micro Process Engineering*, Wiley-VCH, Weinheim. Chapter 5.
32 Günther, A. and Jensen, K.F. (2006) *Lab Chip*, **6**, 1487.
33 Nishisako, T., Okushima, S. and Torii, T. (2005) *Soft Matter*, **1**, 23.
34 Seo, M., Nie, Z., Xu, S., Mok, M., Lewis, P.C., Graham, R. and Kumacheva, E. (2005) *Langmuir*, **21**, 4773.
35 Kawakatsu, T., Kikuchi, Y. and Nakajima, M. (1997) *J. Am. Oil Chem. Soc.*, **74**, 317.
36 Kawakatsu, T., Komori, H., Nakajima, M., Kikuchi, Y. and Yonemoto, T. (1999) *J. Chem. Eng. Japan*, **32**, 241.
37 Kobayashi, I., Nakajima, M., Nabetani, H., Kikuchi, Y., Shono, A. and Satoh, K. (2001) *J. Am. Oil Chem. Soc.*, **78**, 797.
38 Sugiura, S., Nakajima, M. and Seki, M. (2002) *J. Am. Oil Chem. Soc.*, **79**, 515.
39 Sugiura, S., Nakajima, M., Iwamoto, S. and Seki, M. (2001) *Langmuir*, **17**, 5562.
40 Sugiura, S., Nakajima, M., Tong, J., Nabetani, H. and Seki, M. (2000) *J. Colloid Interface Sci.*, **227**, 95.
41 Sugiura, S., Nakajima, M., Itoh, H. and Seki, M. (2001) *Macromol. Rapid Commun.*, **22**, 773.
42 Kawakatsu, T., Oda, N., Yonemoto, T. and Nakajima, M. (2000) *Kagaku Kogaku Ronbunshu*, **26**, 122.
43 Iwamoto, S., Nakagawa, K., Sugiura, S. and Nakajima, M. (2002) *AAPS PharmSciTech*, **3**(3), article 25.
44 Ikkai, F., Iwamoto, S., Adachi, E. and Nakajima, M. (2005) *Colloid Polym. Sci.*, **283**, 1149.
45 Nakagawa, K., Iwamoto, S., Nakajima, M., Shono, A. and Satoh, K. (2004) *J. Colloid Interface Sci.*, **278**, 198.
46 Kobayashi, I., Nakajima, M., Chun, K., Kikuchi, Y. and Fujita, H. (2002) *AIChE J.*, **48**, 1639.
47 Sugiura, S., Nakajima, M. and Seki, M. (2002) *Langmuir*, **18**, 3854.
48 Sugiura, S., Nakajima, M. and Seki, M. (2002) *Langmuir*, **18**, 5708.
49 Sugiura, S., Nakajima, M. and Seki, M. (2004) *Ind. Eng. Chem. Res.*, **43**, 8233.
50 Sugiura, S., Nakajima, M., Kumazawa, N., Iwamoto, S. and Seki, M. (2002) *Phys. Chem. B*, **106**, 9405.
51 Tong, J., Nakajima, M., Nabetani, H. and Kikuchi, Y. (2000) *J. Surfactant Deterg.*, **3**, 285.
52 Kawakatsu, T., Trägårdh, G., Trägårdh, Ch., Nakajima, M., Oda, N. and Yonemoto, T. (2001) *Colloids Surf. A: Physicochem. Eng. Aspects*, **179**, 29.
53 Sugiura, S., Nakajima, M., Ushijima, H., Yamamoto, K. and Seki, M. (2001) *J. Chem. Eng. Japan*, **34**, 757.
54 Kobayashi, I., Mukataka, S. and Nakajima, M. (2004) *J. Colloid Interface Sci.*, **279**, 277.
55 Kobayashi, I., Mukataka, S. and Nakajima, M. (2004) *Langmuir*, **20**, 9868.
56 Kobayashi, I., Uemura, K. and Nakajima, M. *Proceedings, 4th International Congress on Emulsions*, Lyon, 3–6 October 2006, No. 610.
57 Kobayashi, I., Nakajima, M. and Mukataka, S. (2003) *Colloids Surf. A: Physicochem. Eng. Aspects*, **229**, 33.
58 Saito, M., Yin, L.J., Kobayashi, I. and Nakajima, M. (2005) *Food Hydrocolloids*, **20**, 1020.

59 Nakai, S. and Modler, H.W. (2000) *Food Proteins Processing Applications*, Wiley-VCH, New York.
60 Kobayashi, I. and Nakajima, M. (2002) *Eur. J. Lipid Sci. Technol.*, **104**, 720.
61 Sugiura, S., Kumazawa, N., Iwamoto, S., Oda, T., Satake, M. and Nakajima, M. (2004) *Kagaku Kougaku Ronbunshu*, **30**, 129.
62 Kobayashi, I., Mukataka, S. and Nakajima, M. (2005) *Langmuir*, **21**, 5722.
63 Kobayashi, I., Nakajima, M. and Mukataka, S. (2004) *Proceedings, International Congress on Engineering and Food 9*, Montpellier, 7–11 March, No. 610.
64 Kobayashi, I., Mukataka, S. and Nakajima, M. (2005) *Langmuir*, **21**, 7629.

9
Isotropic and Anisotropic Metal Nanoparticles Prepared by Inverse Microemulsion*
Ignác Capek

9.1
Introduction

During recent years, the extent of research investigations in the field of nanotechnology has exploded, bringing with it new ideas for both the processing and utilization of nanostructured materials, with applications ranging from every-day uses to advanced technologies over many scientific and commercial fields. As a consequence, a vast number of preparation methods has been established and documented for nanoscaled materials (especially particles). These processes include physical methods such as mechanical milling [1] and inert gas condensation [2], along with chemical methods such as oxidative precipitation [3], electrodeposition [4], hydrothermal [5] and sol–gel synthesis [6]. It follows that a nanotechnology would be more easily exploited and optimized when the science that allows its exploitation is understood. For this reason, much effort is presently being expended in the area of nanoscience, probing what ultimately may become nanotechnology. This vastly increased interest in nanoparticles has been driven not only by their size but also their unique physical and chemical properties, and today these properties are being exploited in commercial areas as diverse as chemical catalysis, biological sensors and electronics. Of particular interest for the application of nanoparticles is their dimensional stability on surfaces.

9.1.1
Properties of Nanoscale Particles

The preparation of nanoparticles has received considerable attention during recent decades because they possess unique physical and chemical properties [7]. Indeed, nanosized materials have shown great promise in technological applications such as microelectronics [8], photocatalysis [9], electrocatalysis [10], biomedical applications [11] and chemical processes [12]. Due to their small size (1–25 nm), nanoparticles exhibit novel materials properties that differ considerably from those of their

---

*A list of abbreviations is given at the end of this chapter.

Emulsion Science and Technology. Edited by Tharwat F. Tadros
Copyright © 2009 WILEY-VCH Verlag GmbH & Co. KGaA, Weinheim
ISBN: 978-3-527-32525-2

bulk solid state [13]. Notably, in recent years interests in nanometer-scale magnetic particles have expanded, based on their potential applications as high-density magnetic storage media [14]. Unfortunately, however, as the size of the particle decreases its reactivity increases, and the magnetic properties become increasingly influenced by surface effects [15].

The properties of nanoscale particles have attracted a great deal of attention due to the ways in which they differ from the atomic, molecular and bulk properties of those same materials [16, 17]. Nanoparticles have uses in surface-chemical, photochemical and other related fields; for example, they are required for the preparation of some catalysts, they can be used in magnetic materials, as constituents of paints, as semiconductors, as vehicles for *in vivo* drug carriers, and also in molecular devices [18, 19]. In particular, the area of magnetic materials has employed the variations in the magnetic properties of nanoparticles caused by effects such as single domains, superparamagnetism [17] and surface interactions [20]. Notably, due to their large surface-to-volume ratio, the magnetic properties of nanoparticles are dominated by surface effects and particle–support interactions. In addition, they exhibit magnetic anisotropy constants that are approximately two orders of magnitude larger than their bulk counterparts, with correspondingly enhanced coercivities [20].

Although nanomaterials are becoming increasingly important in a variety of technological applications [21], the large-scale production of tailor-made nanoparticles still represents a particular challenge. Different engineering approaches such as gas-phase synthesis or bulk-phase precipitation in liquids have highlighted several disadvantages, as the nonideal mixing in these processes often leads to broad particle size distributions and to an average particle size which is difficult to control [22].

9.1.2
Production of Nanoparticles and Microemulsions

Several techniques have been used for the production of nanoparticles, including vapor-phase techniques [23], sol–gel methods [24], sputtering [25] and coprecipitation [26]. The two main methods used to prepare metal nanoparticles are coprecipitation and *chemical reduction*. In both cases, the presence of surfactant is required to govern the growth process. Typically, the coprecipitation reactions involve the thermal decomposition of organometallic precursors [27, 28]. Chemical reduction occurring in colloidal assemblies represents another approach for the formation of size- and shape-controlled nanoparticles [29, 30].

One major benefit of the chemical methods is their relatively inexpensive investment of capital equipment. Among these chemical processes, reverse micelle (microemulsion) synthesis has been recently identified as a viable method for the production of a wide array of metals and metal oxide nanoparticles [31–33] over a relatively narrow particle size distribution.

Reverse micelle synthesis utilizes the natural phenomenon involving the formation of spheroidal aggregates in a solution when a surfactant is introduced to an organic solvent, formed either in the presence or in the absence of water [34].

Micelle formation allows for a unique encapsulated volume of controllable size through which reactions and the subsequent development of metal and metallic compounds can be produced. Aggregates containing water:surfactant molar ratios of less than 15 can be referred to as *reverse micelles* and have hydrodynamic diameters in the range of 4–10 nm [31, 35], whereas water:surfactant molar ratios in excess of 15 constitute *microemulsions*, which have a hydrodynamic diameter range between 5 and 50 nm. Microemulsions consist of a mixture of oil, water, surfactant and cosurfactant in which the amphiphilic surfactants self-assemble into spheres surrounding nanosized droplets of water [36].

The transfer of inorganic salts in reverse (water-in-oil; W/O) microemulsions has received considerable attention in the preparation of semiconductor and metal particles [37]. This is a powerful technique for obtaining ultrafine particles, and is based on the use of microreactors to control particle shape and growth [38]. When producing ultrafine particles, the W/O microemulsions used are formed by nanodroplets of water dispersed in oil. A reverse micelle (microemulsion) method, as a type of 'soft' technique, represents a suitable means of obtaining uniform and size-controlled metal nanoparticles. Both, ternary reverse microemulsions and quaternary microemulsions can be used to prepare metal nanoparticles. For example, the anionic sodium bis(2-ethylhexyl)sulfosuccinate (NaAOT) forms thermodynamic ternary reverse micelles, while the cationic cetyltrimethylammonium bromide (CTAB) forms thermodynamic quaternary reverse microdroplets. [Note: Surfactants (e.g. NaAOT, CTAB) are molecules which have a polar hydrophilic head (attracted or hydrated by water) and a hydrophobic hydrocarbon chain (soluble in oil).]

9.2
General Aspects of Microemulsions

Water-in-oil microemulsion solutions are mostly transparent, isotropic liquid media with nanosized water droplets that are dispersed in the continuous oil phase and stabilized by surfactant molecules. The microdroplets or emulsified water pools are a thermodynamically stable mixture of water, oil and surfactant, where the water and oil regions are separated by a surfactant monolayer. These surfactant-covered water pools offer a unique microenvironment for chemical reactions, including the reduction of metal salts leading to the formation of nanoparticles. Here, they not only act as microreactors for processing reactions but also exhibit the process aggregation of particles because the surfactants may adsorb to the surface of the particle when the latter's size approaches that of the water pool. As a result, the particles obtained in such a medium are generally very fine [39]. Inverse microemulsion droplets, however, can be slightly polydisperse due to a less strict transformation of monomer to assembly form. The microemulsion is a thermodynamically stable phase, and therefore the polydispersity is an equilibrium property. Reverse micelles and microemulsions that are heterogeneous on a molecular scale also have the ability to solubilize metal salts.

9.2.1
Droplet Dimensions

The droplet dimension can be modulated by a variety of parameters, notably w (w = [water]/[surfactant]) [40]. The size of the microemulsion droplets can be varied in the range 5 to 50 nm by varying the relationship between the microemulsion components. A diagram of each micellar system shows a zone where the reverse micellar phase appears. In the liquid-like phase of water/iso-octane/Na(AOT) ternary phase diagram, for example, w = [H_2O]/[Na(AOT)] mainly determines the reverse micelle size. At w-values below 15, water mobility is greatly reduced (bound water), but above w = 15 a linear increase in the water pool diameter is observed. This behavior is explained by the approach that assumes a constant area per surfactant molecule and that all surfactant molecules participate in the reverse micelle interface [41].

The size of nanoparticles prepared in the quaternary reverse micelle system is reported to be more controllable by the addition of cosurfactant [42]. For example, compared to the anionic sodium (bis(2-ethylhexyl)sulfosuccinate (NaAOT) ternary reverse micelle system, the droplet dimension of the quaternary cationic CTAB reverse micelles can be elaborately adjusted by changing w, with the additional modulation of cosurfactant at the interface of water and oil. The cosurfactant also can be used as a capping reagent to modify the surface of nanoparticles effectively and thus improve the particle growth and shape.

The reverse microdroplets are able to exchange their water contents during collision between two droplets. The intermicellar interaction was estimated to depend on the particle diameter, the attractive range, and the 'sticky parameter', which describes the attractive strength between two droplets [43]. The latter increases with the length of the bulk oil alkyl chain, is related to the decrease in percolation threshold with oil chain length, and may be explained in terms of an increase in intermicellar droplet interactions. This is due to a penetration of the solvent molecules into the interface, thereby screening the surfactant–alkyl chain interactions. In the case of long-chain oil solvents, steric hindrance does not allow solvent molecules to penetrate the interface, and this induces an increase in the attractive interactions [44]. The kinetic exchange process [45] is directly related to the sticky parameter and to the binding modulus of the film at the water/oil interface. The solvent used 'tunes' the kinetic exchange process: for short-chain solvents, the surfactant alkyl chain is well solvated and the micellar interactions are weak, thus inducing a low kinetic exchange rate constant. Conversely, large molecules are poor solvents for the alkyl chains, inducing strong interactions between micelles and, in turn, high kinetic rate constants. The kinetic exchange process is also tuned by the polar volume fraction. These two properties of size and exchange process make it possible, by mixing two micellar solutions which contain the reactants, to produce either polymer or metal nanomaterials [46].

When the droplets collide they form transient aggregates and then revert to isolated droplets. The aggregate lifetimes are typically of the orders of microseconds, and the dynamics of the exchange of solute between micelles and the continuous phase is characterized by the rate constant for entry of the solute into the micelle. This process is diffusion-controlled, as is the entry of emulsifier molecules into the micelle. Under certain critical conditions, molecules can be transported from one droplet to another, without passing through the continuous phase. A possible process here involves collisions and transient merging of the droplet cores. At low concentration of the dispersed phase, the dispersion is mostly composed of identical spherical isolated droplets, but at higher concentrations the nonspherical nanoparticles can be formed. Under such conditions the structure of the system depends on the interactions between droplets: if they are repulsive the collisions are very short and no overlapping between interfaces of colliding droplets occurs, but if the interactions are attractive then the duration of collisions increases and transient clusters of droplets are formed. Interface overlapping occurs during collisions, and this allows exchanges to be made between touching droplets. Such exchanges are achieved by of ions or molecules 'hopping' through the interfaces, or by the transient opening of these interfaces with communication between the water cores of the droplets. The electrical conductivity of the W/O microemulsion is an ideal approach to study the droplet-colliding (percolating) events. As the continuous phase of W/O systems is not conductive, the electrical conduction requires the contact of droplets to allow charge transfer between them. This transfer can be achieved either by 'charge hopping' or by the transient merger of connected droplets with communication between the water cores [47]. When this connectivity is achieved, a steep increase in conductivity is observed which has been analyzed as the 'percolation process', with the percolation threshold ϕ_{per}. Under percolation threshold conditions, water pools of inverse droplets can communicate within the microemulsion system.

9.2.2
The Use of W/O Microemulsions

The difficulties in the preparation of nanoparticles have been largely overcome by the use of W/O microemulsions [48]. The dispersed water pools that these contain act as microreactors in which a chemical reaction can be performed to generate the required product in the form of a colloidal nanodispersion. In this process, the size of the dispersant can be controlled such that the near-monodisperse state is favored and a prolonged stability is ensured. The dispersed particles that are generated can be isolated by conventional physical–chemical means. The preparation of metallic clusters [49], magnetic particles, [50], complexes [51] and metal oxides [17] have already been reported.

Microemulsions have shown their potential as an interesting alternative reaction medium for the production of nanoparticles [52]. When the nanodroplets in these microemulsions are filled with different reactants they can act as small reactors, where the fusion–fission phenomena between the droplets lead to a controlled uniform micromixing. This enables the control of both particle size and particle size distribution. In order to utilize a microemulsion system in a production technology, it is essential first to study a multitude of process parameters in order to identify suitable variables that may be controlled to adjust the properties of the particles [53].

9.2.3
Nanoparticle Preparation

When the correct microemulsions have been obtained, the particle preparation method consists of mixing the two microemulsions carrying the appropriate reactants in order to obtain the desired particles. A representation of this process is shown in Figure 9.1. Here, it can be seen that, after mixing both microemulsions containing the reactants, an interchange of the reactants (shown here as 'Metal Salt' and 'Reducing Agent') takes place during the collisions of the water droplets. This interchange of reactants occurs very rapidly [54]; indeed, for most commonly used microemulsions it occurs during the mixing process. The reaction (nucleation and growth) then takes place inside the droplets, and this controls the final size of

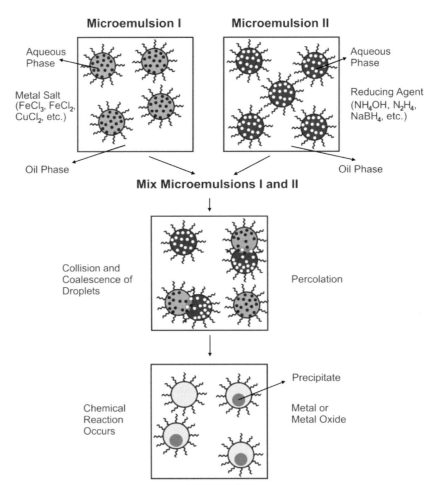

Figure 9.1 Proposed mechanism for the formation of metal particles using the microemulsion approach [58].

the particles. Any interchange of nuclei or particles between 'nucleated' droplets is hindered because this would require the formation of a large hole during the droplet collisions, which in turn would require a major change in the curvature of the surfactant film around the droplets, and this is not energetically favorable. Microemulsions should be chosen such that the curvature radius is similar to the natural radius; otherwise, the surfactant film can be opened during collisions to form 'transient droplet dimers', leading to an interchange of particles/nuclei, which is not appropriate for growth control [55]. Within the microemulsion, the droplets are in constant Brownian motion and thus collide frequently. A small fraction of these collisions result in micelle fusion, which gives rise to short-lived dimers [56]. The dimers subsequently separate to form new droplets containing a mixture of the solutions enclosed in the two original droplets. Thus, by mixing the microemulsions containing different reactants it is possible to perform chemical reactions inside the reverse micelle water pool, using it as a nanoreactor [57]. If this reaction results in a solid compound, then nanoparticles are created, the growth of which is limited by the micelle size. When chemical reactions occur in the microemulsions, with interactions between micromicelles carrying aqueous solutions of appropriate reactants, the size of the resultant product particles can be controlled via the size of the water pools of the micelles, or with the concentration of the reactants in the aqueous solutions.

The fabrication of nanoparticles within reverse microemulsions [58, 59] has been shown as a convenient route to monodisperse particles of controllable size. This method exploits two useful properties of reverse microemulsions, namely the capacity to dissolve reactants in the water core, and the constant exchange of the aqueous phase among micelles. This method has been studied for some years and is currently widely used for metal [58], semiconductor [59] and oxide [60] nanoparticle synthesis.

When the particles have attained their final size, the surfactant molecules are attached to the surface of particles, thus stabilizing and protecting them against further growth. The steps of the nanoparticle formation process are (Figure 9.1):

1. The meeting of the reducing agent with the precursor.
2. Nucleation of a nanometal particle.
3. Growth of the metal nucleus.

In this process the first step would be the most critical because the amount of reduced material present and the local environment will determine how steps 2 or step 3 proceed. The presence of a surfactant confines the reduction process that starts when the reducing agent and the precursor meet after merging of the microemulsion droplets. The size and number of the microemulsion droplets, and the probability and frequency of merging of the two types of microemulsion droplets, are important factors to consider. The nanoparticle size is reported to be limited by the nucleation process at low precursor concentrations [61], while the need for multiple collision has also been proposed [62].

The use of sonication or stirring to further enhance collisions between the micelles containing the different reactants can cause fusion between two micelles to occur, which produces a transient dimer that exchanges the water cores of the collided micelles. The dimer breaks down again into two reverse micelles with the contents from one micelle transferred into the other [34]. The mixing of the two reactants produces a precipitation reaction, from which nanoparticles can be obtained through centrifugal extraction of the solution. The process of micelle exchange, which leads to further growth, continues until the particles reach a terminal size determined by the system and the stabilization of the particles by the surfactant [31, 35]. The size of the metallic particles produced is a function of the reaction time, water content in the micelle, the concentration of reactant solutions contained within the micelle, and the solvent type [31, 35]. As reverse micelles constitute a dynamic system, they also make possible (through the interactions between droplets) the control of the nanocrystal growth. Furthermore, changing either the bulk solvent used to form the reverse micellar solution or the polar volume fraction leads to a modification of the intermicellar exchange process [63].

Controlled nucleation and the separation of nucleation from growth are keys to the synthesis of near-monodisperse metal nanoparticles in the 1–15 nm size range [64, 65]. This can be achieved either by providing a controlled number of preformed nanoparticles as nucleation centers in a growth medium where no secondary nucleation can occur (the 'seeding growth' method [65]), or by varying the ratio of strong and weak reducing agents [66]. The key goals in the synthesis of metal nanostructures are that the process provides nanostructures of a specific size and size distribution, and that the synthesis is reproducible [67]. In order to gain control of the synthesis, it was necessary to understand the mechanism by which the nanostructures are formed. A mechanism based on slow, continuous nucleation (seeding) and rapid autocatalytic growth has been used to explain the growth of metallic nanoparticles [68]. In the seeding/autocatalytic growth mechanism, for example, the slow chemical reduction of Pt ions by ascorbic acid occurs to give Pt nanoparticle growth centers (seeds) at very low concentration by the simplified redox reaction. When these seeds reach a certain size ($\sim$500 atoms) the Pt nanoparticles apparently become autocatalytic for the platinum reduction reaction [69].

9.2.4
Surfactant-Based Methods

The simplest approaches for isotropic and anisotropic nanoparticle synthesis are various surfactant-based methods [70]. Surfactant-based anisotropic micelle templates can be easily prepared [71]. For example, the $\sim$6 nm spherical micelles formed by a dilute (>1 mM) solution of CTAB surfactant converts to cylindrical micelles at higher concentrations (>20 mM), more elongated rodlike micelles in the presence of organic solubilizates [72], and wormlike micelle structures in the presence of salicylate [73]. As the self-assembled structures are dynamic aggregates of monomeric molecules and the superstructure dynamics are coupled with nanoparticle formation kinetics [74], anisotropic nanoparticles are often polydisperse in both size and shape, it is crucial to identify the ideal growth conditions to produce near-monodisperse anisotropic nanoparticles.

Surfactant molecules can also be used as 'simple' capping and stabilizing agents, as in the organometallic precursor decomposition reactions. For example, in the case of cobalt particle synthesis, when they are created in a supersaturated regime (i.e. at a high reducing agent concentration) reverse micelles are destroyed immediately after adding the reducing agent solution because of the limited water content [75]. In this way, the nucleation and growth processes take place in a poorly defined 'emulsion'. The ratio of the reducing agent (e.g. sodium tetrahydroborate; $NaBH_4$) and the metal salt or precursor (e.g. $Co(AOT)_2$) is represented by R, where R = [reducing agent]/[precursor]. At R = 0.5, the average size is 6 nm, with a rather large size distribution of the nanocrystals, σ, of 30%. When R increases from 0.5 to 8, two main behaviors are observed:

- The mean diameter of cobalt nanocrystals increases from 6 to 8 nm. This is directly related to the increase in the nuclei concentration, and subsequent particle growth induces increasingly larger particles.
- The more important feature is the drastic decrease in the size distribution, from 30 to 8%. This behavior is due to the size increase combined with the size selection that occurs at the end of the particle preparation, leading to the collection of the smaller nanocrystals [76].

Cheon and coworkers [77] have reported that, in the solution-based approach, there are four different parameters–kinetic energy barrier, temperature, time and capping molecules (the type of surfactant)–that can influence the growth pattern of nanocrystals under nonequilibrium kinetic growth conditions. For example, in the microemulsion solution, AOT used as anionic surfactant is a twin-tailed surfactant and is most commonly used to prepare microemulsions because of its bulky hydrophobic tail to the hydrophilic group [78]. AOT tends to selfassemble to form aggregates, which further leads to the formation of microemulsions with desired microstructures. This is helpful for the formation of products with some similar shapes to the microstructures of microemulsions. In the W/O microemulsions with suitable water contents, many surfactant molecules will selforganize to spherical droplets containing metal ions [79], respectively. All of these spherical microstructures which disperse in the continuous oil media are relatively stable ahead of mixing. However, the AOT microemulsions with a very low water content cannot be described in terms of spherical microstructures; rather, they usually selfassemble to form cylindrical or rod-like microstructures [80]. Such a possible growth process has been observed by Sugimoto and coworkers [81] in the synthesis of peanut-like hematite (α-Fe_2O_3) crystals with very similar shape and size. Here, these authors used a gel–sol method in the presence of sulfate ions, and disclosed that the bidentate-specific adsorption of sulfate ions to the growing surfaces parallel to the c-axis resulted in the formation of hematite nanorods [82]. The mechanism for the formation of the peanut-like shape was explained in terms of the formation of a gradual outward bending of the dense rodlike subcrystals or nanorods on both ends of ellipsoidal particles by the growth of new crystalline nanorods in the spaces between the existing subcrystals [81, 83].

9.2.5
Coprecipitation

Chemical reactions (e.g. coprecipitation, reduction) can be carried out within the aqueous cores such that the growth of the precipitates is constrained by the interior volume of the reverse micelles. Driven by the potential applications in heterogeneous catalysis, many of the earliest examples of nanoparticle syntheses in microemulsions involved the reduction of aqueous metal cations to the metallic state. Among the transition metals, Fe, Co, Ni, Cu, Rh, Pd, Ag, Ir, Pt, Au and numerous alloy nanoparticles have all been prepared in reverse micelles [84].

One drawback of the coprecipitation method is the difficulty of controlling the particle size and morphology. This is a consequence of the chemical reactions in the aqueous solutions being very rapid. The particle size of the hydroxide precursor and, as a result, of the final product is determined by the supersaturation of the precipitation agent (OH^-), which decisively influences the nucleation rate as well as the particle growth rate of the precursor hydroxides. The supersaturation is difficult to control, and it is especially difficult to maintain a homogeneous supersaturation level throughout the whole volume of the reaction vessel during coprecipitation.

One possible solution to the problems of controlling particle size and morphology during the synthesis of nanoparticles is to use a microemulsion method, which involves coprecipitation in W/O microemulsions [85]. In such a method the coprecipitation occurs in tiny droplets of aqueous phase that are embedded with a surfactant, reverse micelles or microdroplets, which are distributed in an oil phase. The chemical reactions, which occur very rapidly in aqueous solutions and are therefore difficult to control, are controlled by a much slower reaction between the microdroplets carrying different reactants in the microemulsions. The control of chemical reactions in microemulsions enables multiple-step reactions and the synthesis of composite particles [85].

9.3
Isotropic Nanoparticles

Interest in the preparation of nanoparticles using microemulsions has increased greatly since Boutonnet et al. [86] reported the first successful synthesis of platinum nanoparticles using W/O microemulsions. The microemulsion technique has been widely used to prepare isotropic nanoparticles of metal [61], metal sulfides [87] and metal halides [88]. Here, the key parameters in nanoparticle preparation are control of the particle growth and size, and also control of the reaction yield.

The increase in particle size can be attributed to a change in the water fraction in the emulsified water pool or the structure of water [89]; that is, the hydration of metallic ions. Other factors that influence nanoparticle size, such as intermicellar interactions and the intermicellar exchange rate, have also been highlighted [90]. At very low w-values, the reactants are poorly hydrated and the yield of the reduction or coprecipitation reactions is very low. Such a reaction gives rise to small nanocrystals.

Conversely, an increase in reactant hydration induces an increase in the reaction yield, so that the number of nuclei increases; then, the size of the nanocrystals is increased. By assuming that all of the system is totally hydrated at w = 15 (whatever material is used), the nanocrystal growth should evolve up to w = 15 and then stop. However, this is not the case for copper nanoparticles, the growth of which begins to decrease at w = 5 and finally stops at w = 10.

It is well known that certain surfactant molecules, whether ionic (e.g. AOT) or nonionic, when dissolved in organic solvents are capable of solubilizing water in the polar core. As the organized media, they have been widely used as spatially constrained microreactors for the growth of nanocrystallites with a desired narrow size distribution [48]. In most cases only spherical particles are prepared within the polar cores of microemulsions. Traditionally, reports have been made on the successful synthesis of microporous zincophosphate sodalite with a cubic morphology in microemulsions, which indicated that the microemulsions-based approach may not only be used as a synthetic approach but also provide the means of controlling the morphology, as well as the size of grown crystals [91]. Since then, various one-dimensional nanoscale materials have been successfully prepared in microemulsion media [92]. The results of all of these studies have suggested that the microemulsion system can be used not only as a dynamic system for the synthesis of inorganic materials through a simple and facile solution process, but also to control the size and shape of inorganic particles.

Petit *et al.* showed that, at a low water content, spherical reverse micelles of $M(AOT)_2$ were formed [93] but, in the case of $Cu(AOT)_2$, $Co(AOT)_2$ and $Cd(AOT)_2$, upon an increase in the water content, spherical water-in-oil droplets turned into cylinders. These studies were subsequently extended by others; for example, Eastoe and colleagues [94, 95] confirmed these data and showed the same outcome for other surfactants such as $Zn(AOT)_2$, $Ni(AOT)_2$ and $(C_7H_{14})_4N(AOT)_2$.

9.4
Anisotropic Nanoparticles

Anisotropic nanoparticles are of considerable current interest, due to their various shape-dependent properties [96, 97]. The synthesis of anisotropic nanoparticles on the 1–50 nm length scale represents a major challenge as they are less stable compared to spherical shapes of a similar size, and the respective symmetric bulk crystal structures often create a strong barrier. Although synthetic methods for semiconductor nanorods have been well established [98], the metallic systems have a very limited success due to their highly symmetrical cubic crystal structure. Control of the length-to-width ratio and the uniformity of the length and width distributions of metal nanoparticles, as well as synthesizing particles with a width dimension below 10 nm, represents a major challenge with large-scale bench-top methods [99]. The key parameters in nanoparticle shape control include [100]:

- The control of nucleation and growth, restricting the size to the nanometer regime [98].

- The maintenance of a high monomer concentration that induces a better stability of the anisotropic primary nanoparticles [101].
- The use of a suitable surfactant, surfactant mixture or capping molecule that selectively adsorbs onto specific planes of growing particles and induces symmetry-breaking steps [65].

It is well known that some surfactants such as CTAB form rod-like micelles in high concentrations or in the presence of certain solubilizing species [72]. This surfactant media has been widely used in the synthesis of nanomaterials [65]. Although, the micellar template mechanism is well established in the formation of mesoporous materials [74], the template mechanism in the formation of anisotropic metal nanoparticles has not been verified [102]. For example, CTAB is believed to act as a particle stabilizer [65], and its selective adsorption on certain growing faces of nanoparticles may lead to anisotropic structures [103]. Another key factor in determining particle anisotropy has been the size of the growing nanoparticle. For example, Jana and coworkers have prepared different sizes of near-monodisperse spherical gold nanoparticles, and used them as seeds for the synthesis of anisotropic nanoparticles in micellar templates [65, 104]. These authors observed that the yield and shape of the anisotropic nanoparticles depended significantly on the seed size (Figure 9.2). Typically, nanorods in 90–95% yield were obtained only when the smallest seed of 1.5 nm size was used, and the final nanorods had a 5–10 nm short axis and an aspect ratio between 1 and 5. When the seed size was increased to 3.5 nm, the nanoparticles produced were spheroids in 30–70% yield, and the final nanorods had a 10–20 nm short axis and an aspect ratio of between 1 to 5. If the seed particle size was >5 nm, the nanorod yield was very low, regardless of the seed concentration used.

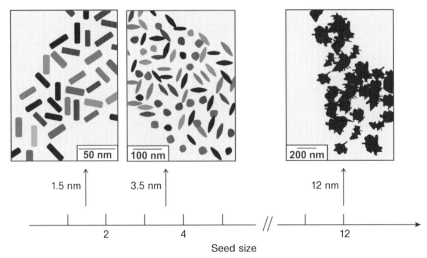

Figure 9.2 Summary of seeding-based shape control for micelle template growth, showing the effect of seed size on the final particle shape [104].

In the new method, a mixture of strong and weak reducing agents is introduced into the micellar solution of a metal salt, where the strong reducing agent initiates nucleation and the weak reducing agent helps the nanoparticles to grow. A borohydride–ascorbic acid system was reported to optimize the anisotropic nanoparticle growth condition over a wide range of CTAB concentrations [104]. Nanorods were formed when the CTAB concentration was above the second critical micelle concentration (CMC), where rod-like micelle structures were formed. Below the second CMC, the nanorod yield was low (5–10%) and most particles were spherical, and this was taken as an evidence that the micelle was acting as a template. When an attempt was made to increase the length of nanorods by lowering the nucleation rate in an optimum surfactant (above the second CMC) condition, the average aspect ratio of the nanorod was always found to be closely related to the aspect ratio of the rod-like micelle (~5). It was also shown that, if solubilizing species were used to increase the length of the rod-like micelle, the nanorod aspect ratio could be further increased up to ~10 [65]. Another observation was that the nanorod yield was temperature-dependent: if the temperature was increased to 50–100 °C, where cylindrical micelles do not exist, then the nanorod yield was decreased to <10%.

If the nucleation kinetics of nanoparticle formation are properly adjusted, the elongated rod-like micelle surface can be a useful template, and the resultant nanoparticles can be highly anisotropic and near-monodisperse. By using this micellar template approach, nanoparticles of a wide range of shapes can be prepared for metals. When using the worm-like micelle template (using CTAB–salicylic acid [73]) the yield of nanowires was found to be very low, but under mild heating conditions the nanowire yield was increased and intermediate-length nanorods were also observed. This indicated that micelle-template-based primary nanorods were converted into nanowires via oriented attachments [70, 105].

Many reports have tended to provide evidence of the surfactant-based templates. Elongated CdS particles grow in cylindrical reverse micelles, while spheres are obtained in spherical reverse micelles [92]. PbS particles are formed in a $Pb(AOT)_2$/polymer composite [106] and, according to whether this has an ordered layer structure or not, nanorods or spherical particles are obtained. On the other hand, it also appears that particle crystallinity is enhanced when there is an interconnected cylindrical phase compared to the micellar phase. Spherical particles often have defects, whereas elongated particles are mostly highly crystallized. This can most likely be attributed to the degree of confinement of the reagents, which is much higher in the interconnected microstuctures than in the micellar system. The formation of the rod-type structure in the microemulsion system under the influence of various cations and anions, which affects the rigidity of the interface between the hydrophilic polar headgroups and the aqueous core of the micelle, has been reported [107–109].

Another route to the production of nonspherical nanocrystals is to use reverse micelles in a supersaturated regime [46]. For instance, a set of syntheses of copper nanocrystals has been created in mixed AOT reverse micelles at higher reducing agent (hydrazine; N_2H_4) concentrations. Under these conditions ($R = [N_2H_4]/[Cu(AOT)_2]$), R was seen to vary from 3 to 15, whereas the water content, w, remained at 10. This meant that, whichever reducing agent concentration was used, the chemical reduction took place at the same polar volumic fraction. At a low hydrazine concentration (R = 3), the population was mainly composed of 'spherical' nanocrystals, including more or less regular spheres and pentagons, and with a mean size of 12 nm. Then, as R was progressively increased the nanocrystals become more highly facetted with the appearance of new shapes. At R = 15, copper atoms are implied in the formation of various shapes such as elongated nanocrystals, triangles, cubes and 'spheres'.

When the concentration of hydrazine (the reducing agent) is high, silver nanodisks are produced in the presence of Na(AOT)/Ag(AOT) surfactant. Under such conditions well-defined reverse micelles are no longer formed [110], while flat triangular CdS nanocrystals have been obtained by using $Cd(AOT)_2$ reverse micelles saturated by H_2S [111]. In the same way, by bubbling H_2 through an aqueous solution containing $PtCl^{4-}$, cubic platinum can be obtained [112]. For any reaction system, saturation with either the reducing agent or one of the reagents seems to be a primary factor in the formation of well-crystallized nonspherical nanoparticles. These can also be obtained in spherical reverse micelles, and this is the case for tetrahedral cadmium sulfide [111], trigonal lamellar silver [110] and copper [113] nanocrystals. In these situations, the synthesis takes place in a large excess either of reducing agent or of one of the reagents.

Anisotropic noble nanoparticle dispersions are very different in color compared to dispersions of spherical particles; this is due to the surface plasmon bands being more sensitive to particle shape than size [99]. All of the metal nanorods have two absorbance maxima that correspond to the longitudinal and transverse plasmon bands. The longitudinal plasmon band depends heavily on the aspect ratio. For example, platelets have additional quadrupole bands [114] but, upon transition from nanorods to platelets, as the aspect ratio decreases, the longitudinal band is blue-shifted and the transverse band becomes broad due to an overlap with the quadruple band. In cubes, all three plasmon bands merge into a single band. In contrast, transition from nanorods to nanowires increases the aspect ratio, and this produces a resultant red shift of the longitudinal band and a blue shift of transverse band [65].

Both, Fu et al. [115] and Jana [104] have proposed a micelle template mechanism that can explain the origin of various shapes of nanoparticles (Figures 9.3 and 9.4). The new mechanism introduces nanoparticle size-dependence for the effective template mechanism, and thus is applicable to a wide range of systems and experimental conditions. There are three important parts of this mechanism:

- The cylindrical micelle induces breaking of symmetry in the growing nanoparticle.
- Symmetry-breaking performance is dependent on nanoparticle size.
- Beyond a size limit (> 5–10 nm) the nanoparticle detaches from the micelle surface because of its comparatively larger mass, and the micelle template mechanism does not work.

Nanorods were obtained in high yield only when the nonseeding method was used or when the smallest seed of 1.5 nm was used in the seeding growth method.

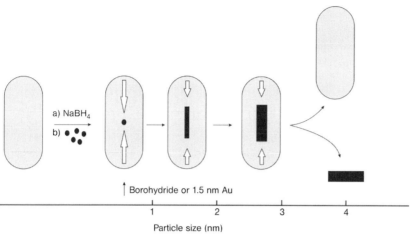

Figure 9.3 Proposed rod-like micelle template mechanism for (a) borohydrate or (b) 1.5 nm Au seed nanoparticles [104].

However, lower symmetric spheroids, platelets, cubes and stars were formed if the seed size was between 3–5 nm or the nanorods were allowed to grow further.

If the seed particle size was >5 nm, the anisotropic nanoparticle yield was very low, regardless of the seed concentration used.

According to earlier reported mechanisms [74, 115, 116], monomers organize on the micelle surface and then produce a covalent organic shell surrounding the micelle, after which these primary core–shell particles are connected to each other via interparticle bonding. Primary rod-like particles are formed while occupying only a part of each micelle surface [104]. When the particle–particle interaction is weak, the primary particles remain isolated during the entire growth processes.

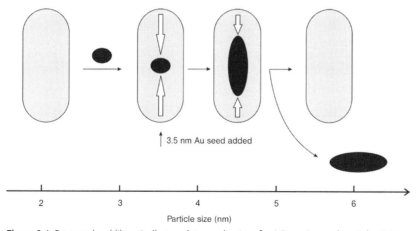

Figure 9.4 Proposed rod-like micelle template mechanism for 3.5 nm Au seed particles [104].

The fine control of nucleation-growth kinetics is extremely important for the effective template mechanism, and the final particle size is dictated by the size of the micelle template and the relative rates of nucleation and growth. Hence, the kinetic parameters of nanoparticle growth have significant roles in controlling the particle shape. It has been observed that the key factors in determining particle anisotropy are the anisotropic micelle-induced breaking of the symmetry of the growing nanoparticle (typically in the size range <3 nm), and the relative rates of nucleation and growth, which determine the aspect ratio of the final nanoparticle [104].

In particular, the growth of gold nanorods induced by the presence of the cationic surfactant CTAB, initially devised by Murphy and coworkers, [65, 117], has become very popular and a variety of mechanisms has been proposed for its explanation. Such mechanisms focus primarily on the selective blocking of certain crystallographic faces or anisotropic surface properties that control the approach of metal ions from solution. Variations of noble metal nanorods to spheres were initiated by using cyanide as oxidant in the presence of CTAB micelles. In the presence of cyanide, short spheroids with sharp tips were seen to dissolve preferentially from the tips, leading to lower-aspect ratio nanorods and eventually to spheres. However, for longer nanorods, cyanide oxidation occurred at various spots along the side edges, with no intermediate shorter rods being formed.

Rodrıguez-Fernandez et al. have recently shown that the $AuCl_4^-$ ions are quantitatively bound to CTAB micelles and can be used to vary the particle size and shape [118]. The crystal growth can be manipulated and spatially controlled by CTAB if the active solution species that deposits onto the growing particle surface is adsorbed to the CTAB micelles. Conversely, in the absence of CTAB corrosion should be homogeneous over the particle surface. The dissolution of gold rods by $AuCl_4^-$ is as follows:

$$AuCl_4^- + 2 Au^0 + 2 Cl^- \leftrightarrows 3 AuCl_2^- \tag{9.1}$$

This reaction only takes place in the presence of CTAB because of a change in the reduction potential of $AuCl_4^-$ upon complexation, and leads to a gradual oxidation of the Au nanoparticles. A careful analysis of this reaction shows that it occurs preferentially at nanoparticle surface sites with a higher radius of curvature, and provides an elegant procedure for the narrowing of the nanoparticle size and shape distribution.

If $AuCl_4^-$ is added to the seed Au nanoparticles in the absence of CTAB, no changes in either the optical response (plasmon band, position, shape, or intensity) or morphology of the particles are observed. However, if both the seed Au nanoparticles and the Au salt are premixed with CTAB, then upon mixing there is a gradual change in the spectrum [118]. There is a systematic blue-shift of the absorption peak, a clear decrease in the peak intensity, a narrowing of the absorption band, and an increase in the symmetry of the band profile. Such spectral changes are consistent with the gradual oxidation of Au^0, since a decrease in particle size will lower the absorbance (A) and cause a blue-shift. In the time evolution of the ultraviolet (UV)-visible spectra during the oxidation of Au nanoparticles in water before (0) and after the addition of

HAuCl$_4$ in the presence of CTAB (0.05 M), the Au: Au^{III} molar ratios were (a) 1 : 0.25 and (b) 1 : 0.5 [118]:

(a) λ_{max} (nm)/time (min): 550/0, 545/36, 540/131, 535/300, 530/1036
    A (a.u)/time (min): 1.0/0, 0.9/36, 0.8/131, 0.7/300, 0.6/1036

(b) λ_{max} (nm)/time (min): 500/0, 490/36, 478/131, 465/300, 450/1036
    A (a.u)/time (min): 1.05/0, 0.85/36, 0.6/131, 0.3/300, 0.13/1036

Furthermore, the narrowing and increased symmetry of the surface plasmon (SP) absorption band also suggested that there was a progressive increase in monodispersity [119].

This reaction only takes place in the presence of CTAB, above the CMC, when CTAB forms micelles which carry the complexed AuCl$_4^-$ ions. If the concentration of CTAB is lower than the CMC, then precipitation of the AuCl$_4^-$ ions complexed with CTAB monomers is observed [120]. Importantly, the differently shaped gold nanoparticles expose different crystallographic facets to the solution at the protuberances. As they all display the same surface smoothing during oxidation, it is concluded that the mechanism that leads to spatially directed oxidation of the particles is independent of the exact crystallographic facets present at the sharply curved points on the particle surface. Hence, the mechanism is not determined by the specific binding of CTAB to particular facets, but rather is a consequence of the high local curvature. This suggests that both reduction of AuCl$_4^-$ to form gold rods and dissolution of the rods both take place at the rod termini. Micellization of the gold salt leads to preferred electron transfer at the tips, and implies that the collision frequency of micelles with the gold rods is higher at the tips.

The influence of particle size on the oxidation rate was studied by addition of AuCl$_4^-$/CTAB solutions to both Au colloids with different particle size (15 and 60 nm) and binary mixtures of these. The time evolution of the absorbance at 400 nm upon AuCl$_4^-$/CTAB addition indicated that, for smaller particle sizes, the decay rate is indeed faster. During the oxidation of binary mixtures, it was observed that the smaller nanoparticles dissolved first, and when these had been totally consumed the larger nanoparticles were still almost unaffected. Although the SP bands of the two different populations within the sample are close to each other, the spectral evolution is quite conclusive. Initially, the absorption band results from the sum of the contributions from both sizes. However, once the oxidation starts there is a red-shift due to consumption of the smaller particles (smaller contribution at lower wavelengths), and thereafter the band slowly blue-shifts because of the decrease in ellipticity, as described above.

In order to provide a clear demonstration of this effect, Rodrıguez-Fernandez et al. have chosen samples with relatively high aspect ratio (around 5) [118]. These authors used a mixture of mostly rods (68%) and cubes (32%), as shown in Figure 9.5a. When HAuCl$_4$ was added in the presence of CTAB, absorption spectroscopy [the absorption band shifted from 900 nm before oxidation to 700 nm after oxidation (120 min) and simultaneously the absorbance decreased from 2.2 to 1.0 (a.u.)] revealed a gradual blue shift and decrease in the longitudinal surface plasmon band, indicating the expected decrease in aspect ratio, which was also confirmed with transmission

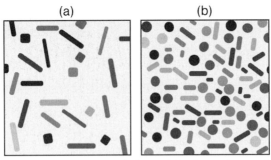

Figure 9.5 Au nanorods (a) before and (b) after oxidation with HAuCl$_4$ ([Au]:[Au^{III}] = 1:0.25) in the presence of CTAB (0.05 M) [118].

electron microscopy (TEM). The TEM images also showed that nanocubes present in the sample became more spherical after the oxidation (before: L = 48 nm, W = 10, L/W = 5; after: L = 29 nm, W = 10.4, L/W = 2.8, [Au]:[Au^{III}] = 1:0.25, 0.05 M CTAB) (see Table 9.1). This situation is most likely related to the extreme curvature associated with such acute tips. This cube-to-sphere conversion is also reflected in the UV-visible spectra, as the band at 545 nm (due to nanocubes) in the spectrum at time zero quickly merged with the absorption band (around 520 nm) due to excitation of the surface plasmons in spherical nanoparticles.

By controlling the experimental parameters such as concentration of the anionic surfactant, water content and reaction temperature, inorganic nanostructures with morphologies of bundles of rods, ellipsoids, spheres, dipyramids and nanoparticles can be efficiently achieved by using the microemulsion approach, respectively [121]. It was shown that uniform bundles of PbWO$_4$ nanorods formed when the concentration of AOT was 0.06 M (Figure 9.6b). However, when the concentration of AOT was decreased to 0.03 M, magnificent dipyramidal PbWO$_4$ nanostructures were obtained in the AOT-containing microemulsion system (Figure 9.6a). From this image, it is can be seen clearly that the samples are composed of a dipyramid shape with a typical side length of about 2 μm. The reasons for this outcome may lie in fewer surfactant molecules being adsorbed on the particle surface, a low growth rate, and the unusual crystal habits of PbWO$_4$ playing an important role, with the result being

Table 9.1 Average values of the dimensions [width (W) and length (L)] of Au nanoparticles oxidized with varying amounts of HAuCl$_4$ [118].

[Au]:[Au^{III}]	L (nm)	W (nm)	L/W
1:0.0000	96.3	68.9	1.40
1:0.125	76.0	57.5	1.35
1:0.300	57.3	50.0	1.15
1:0.340	51.5	45.5	1.13

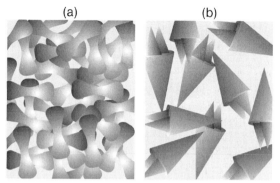

Figure 9.6 Variation of particle size and shape with the AOT concentrations. (a) 0.03 M; (b) 0.06 M [121].

formation of the final morphology. Continuing to increase the AOT concentration from 0.03 to 0.16 M resulted in the formation of short PbWO$_4$ nanorods of about 1∼3 μm in length and several hundred nanometers in diameter, with some similar shapes to those of rod bundles.

When trace water [ca. 4 vol%, derived from ethylene glycol (EG) and hydrated metal salts] was present in the microemulsion system, only bundles of PbWO$_4$ nanorods (just as shown in Figure 9.7a) were obtained. However, if a larger amount (15 vol%) of water was added into the microemulsion system, then spherical PbWO$_4$ nanostructures with uniform diameters were synthesized, as described above (Figure 9.7c). By controlling the water content of the AOT-containing microemulsion solution at 4∼15 vol%, PbWO$_4$ nanostructures with different proportions of nanorod bundles and spheres (Figure 9.7b) were formed at room temperature. It can be estimated that, at different percentages of water content, microemulsions with different microstructures are formed, which not only exhibit a considerable inhibition effect on the PbWO$_4$ crystallization but also show a considerable influence on the crystal growth of PbWO$_4$. When the water content was increased to 50 vol% or above, only shapeless PbWO$_4$ nanoparticles with a mean size of ca. 20 nm were obtained; this may be caused by the collapse of microstructures of the formed microemulsions at a high water percentage.

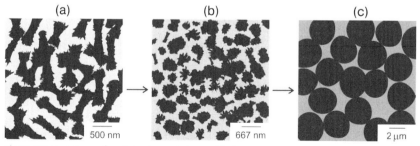

Figure 9.7 Variations of particle size and shape with the water content: (a) 4 vol%; (b) 10 vol%; (c) 15 vol.% [121].

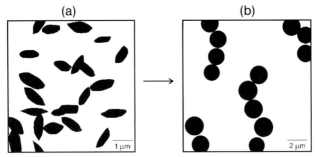

Figure 9.8 Variation of particle shape with temperature: (a) 40 °C; (b) 80 °C [121].

At 40 °C, the as-prepared particles are of an ellipsoid morphology, as shown in Figure 9.8a. Increasing the reaction temperature resulted in the formation of spherical PbWO$_4$ samples. When the reaction temperature was raised to 80 °C, all of the samples were of a spherical morphology, but by keeping the reaction temperature between 40 and 80 °C the products were PbWO$_4$ with different proportions of ellipsoids and spheres. Compared with the spherical PbWO$_4$ obtained in the AOT/EG/H$_2$O microemulsion system, the diameter of the spheres at 80 °C was much smaller, at about 1 μm (Figure 9.8b). Namely, spheres obtained at higher temperature have a smaller diameter compared to those synthesized at room temperature. Therefore, the products with different morphologies can be synthesized in the microemulsions by controlling the reaction temperature.

When the solutions dissolving equivalent [Pb^{2+}] or [WO$_4^{2-}$] are mixed, microemulsions with sphere-like microstructures would collide and fuse, simultaneously exchanging ions of lead or tungstate that are surrounded in the spherical microemulsion droplets but still free to move. The nucleation takes place very rapidly, as can be verified by the appearance of a white precipitate at only 2~3 s after the microemulsions are mixed. Further self-organization of the nucleus would lead to the formation of sphere-like PbWO$_4$ samples through the Ostwald ripening process. However, in this case the dimensions of samples produced from the reaction are many times larger than the typical dimensions for microemulsion droplets of around 5~100 nm [122]. The sizes of the spheres are not confined strictly by those of the AOT spherical microstructures. Due to the flexibility of the microstructure interface, PbWO$_4$ spheres will grow continually at the base of the initially formed spherical microstructures. In this case, it was assumed that the sulfonic acid groups of AOT would be adsorbed preferentially onto the growing surfaces, parallel to a certain crystallographic direction of PbWO$_4$, and result in the formation of rod-like nanostructures. Therefore, rod bundles of PbWO$_4$ nanostructures could be formed in a similar manner to that of peanut-like hematite particles – that is, through an outward bending of adjoining subcrystals by nucleation and growth of a new subcrystal in each space between them. It should be noted that all of the PbWO$_4$ samples obtained in the microemulsion solution exhibit nanocrystalline structure, and are very likely composed of rod-like nanocrystals that act as building blocks or subcrystals. Meanwhile, the presence of a double-tailed hydrophobic group in the AOT would play the role of a spacer; this would be present between the growing subcrystals which were jostling one another, and enhance the outward bending of the subcrystals, favoring the formation of bundles of rods. Following prolonged ultrasonic treatment of the sample, the rod-like bundle morphology was slightly damaged and many single, shorter nanorods had appeared. With increasing temperature, the microemulsion system – under the delicate balance between kinetic growth and thermodynamic growth regimes – might be of low viscosity and instability.

It is clear that the temperature and capping molecules (AOT) are key parameters here. Spherical microstructures were formed due to a low surface energy and high stability, and further aggregated into spherical products at higher temperature. From these results, it is clear that anionic surfactant AOT in the microemulsion solutions, which freely self-organize into rod-like, cylindrical or spherical microstructures under appropriate reaction conditions, could be used as a soft template to synthesize products with unique morphologies. As a consequence of this soft template effect, the products could duplicate the microstructures of the AOT microemulsions and grow sequentially. This can be identified as the AOT-microemulsion-based formation and evolution route of $PbWO_4$ nanostructures.

Kuiry et al. have focused on the formation of nanorods by selfassembling crystalline cerium oxide nanoparticles into nanorods, using the microemulsion synthesis technique [123]. In the present case, micelles in microemulsion I contain the precursor solution bearing Ce ions. After the addition of microemulsion II (which contains hydrogen peroxide solution) to microemulsion I, the mutual interaction between the two types of micelle took place by random collision. During such interactions a transfer of reactants occurred and the nanoparticles were formed by a process of nucleation and growth inside the micelle cores. The ceria nanoparticles were formed as per the following scheme of reactions:

$$2\,Ce^{3+} + 2\,OH^- + H_2O_2 \rightarrow 2\,Ce(OH)_2^{2+} \tag{9.2}$$

$$Ce(OH)_2^{2+} + 2\,OH^- \rightarrow CeO_2 + 2\,H_2O \tag{9.3}$$

Hydroxyl ions are formed locally inside each micelle core as an intermittent product of the dissociation of hydrogen peroxide [124] during the interaction. The nanorods formed during such a microemulsion synthesis are composed of a number of ceria nanocrystals, as shown by the presence of diffraction rings (see Figure 9.9). The selected area electron diffraction (SAED) pattern of the ceria particles shows the presence of three distinct diffraction rings corresponding to (111), (220) and (311) lattice planes. The self-assembly process was observed during aging of the sol, a few days after the formation of such cerium oxide nanoparticles, leading to the gradual evolution of ceria nanorods (Figure 9.10). The aspect ratio of the nanorods was seen be ~6, while their diameter and length were approximately 40 and 250 nm, respectively.

The nanocrystals of cerium oxide were initially formed in the microemulsion-mediated synthesis process, followed by self-assembly of the particles into nanorods during the aging process. This self-assembly resembles a controlled agglomeration process with the help of random collisions among the micelles that contain the nanocrystals. The formation of such 'supra-aggregates' was also observed [108]

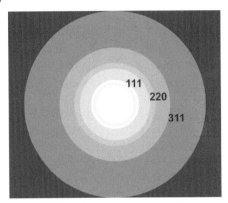

Figure 9.9 The selected area electron diffraction (SAED) pattern of nanorods of cerium oxide [123].

during the synthesis of Cu nanoparticles in the AOT/isooctane microemulsion system. In the cerium oxide nanocrystals, the alkyl chains of the surfactant were highly solvated and most likely adjusted by rotation to form the cylindrical shape. The presence of NO^{3-} ions from the precursor solution may also have influenced the interfacial properties of the micelles in such a way that the distance of separation between the two anionic polar heads of the AOT was increased. This in turn facilitated formation of the cylindrical shape by decreasing the local curvature of the micelle surface, as suggested by others [125]. The probable mechanism by which the ceria nanorods were self-assembled is shown schematically in Figure 9.11. A ceria nanoparticle inside the micelle (a supra-aggregate) and the nanorod formed from the supra-aggregate are shown in Figure 9.11a–c, respectively. The cone-shaped portions at both the ends of the nanorods were possibly formed to accommodate the abrupt change in surface free energy that would otherwise exist due to the presence of a sharp curvature. The growth of the nanorods took place by the self-assembly of ceria nanoparticles at both the ends.

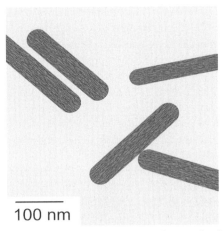

Figure 9.10 Diagrammatic images of nanorods of cerium oxide at high magnification [123].

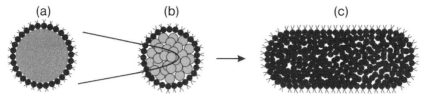

Figure 9.11 Schematic representation of the self-assembly of (a) ceria nanoparticles to form (b) a supra-aggregate and finally to form (c) nanorods [123].

9.5
Conclusions and Outlook

The microemulsion technique has been used widely to prepare isotropic nanoparticles of metal, metal oxides and metal halides. The key parameters in nanoparticle preparation are the control of growth and size, and of reaction yield. Increases in size can be attributed to a change in the water fraction in the emulsified water pool or in the structure of water–that is, the hydration of metallic ions. Other factors that influence nanoparticle size, such as stabilizer concentration, intermicellar interactions and the intermicellar exchange rate, have also been identified. In most cases only spherical particles are prepared within the polar cores of microemulsions. The microemulsions-based approach can be used not only as a synthetic approach but also to provide a means of controlling the morphology and size of the grown crystals.

Anisotropic nanoparticles are of considerable current interest due to their shape-dependent properties. One key factor in determining particle anisotropy is the size of the growing nanoparticle. Typically, the yield and shape of anisotropic nanoparticles depend heavily on the seed size. For example, with a size of 1.5 nm the final nanorods had a 5–10 nm short axis and an aspect ratio between 1 and 5, but when the size was increased to 3.5 nm the nanoparticles produced were spheroidal with a 30–70% yield, while the nanorods had a 10–20 nm short axis and an aspect ratio of 1 to 5. If the seed particle size was raised further to >5 nm the nanorod yield was very low, regardless of the seed concentration used.

When a mixture of strong and weak reducing agents is introduced into the micellar solution of a metal salt, the strong agent initiates nucleation while the weak agent helps the nanoparticles to grow. Nanorods were formed when the CTAB concentration was above the second CMC, when rod-like micelle structures were formed, but below the second CMC the nanorod yield was low (5–10%) and most particles were spherical. By using this micellar template approach, nanoparticles of a wide range of shapes can be prepared for metals. When using the worm-like micelle template the yield of nanowires was found to be very low; however, under mild heating conditions the yield was increased and intermediate-length nanorods were also observed. The micelle–template-based primary nanorods were converted into nanowires via oriented attachments. Another possibility to produce nonspherical nanocrystals is to use reverse micelles in a supersaturated regime. At low reducing agent concentrations, the population is composed mainly of 'spherical' nanocrystals, including more or less regular spheres and pentagons. Yet, by progressively increasing R, the nanocrystals become more highly facetted with the appearance of new shapes. At R = 15, metal atoms are implied in the formation of various shapes such as elongated nanocrystals, triangles, cubes and 'spheres'.

The crystal growth can be manipulated and spatially controlled by using an ionic stabilizer (e.g. CTAB) if the active solution species that deposits on the growing particle surface is adsorbed onto the CTAB micelles. Conversely, in the absence of CTAB corrosion should occur homogeneously over the particle surface. The dissolution of gold rods by $AuCl_4$ takes place only in the presence of CTAB due to a change in the reduction potential of $AuCl_4^-$ upon complexation, and this leads to a gradual oxidation of the Au nanoparticles. This reaction has shown to occur preferentially at those nanoparticle surface sites with a higher radius of curvature, and provides an elegant procedure for the narrowing of nanoparticle size and shape distribution.

By controlling experimental parameters such as anionic surfactant concentration, water content and reaction temperature, inorganic nanostructures with morphologies of bundles of rods, ellipsoids, spheres, dipyramids and nanoparticles, respectively, can be efficiently achieved by using the microemulsion approach. Uniform bundles of $PbWO_4$ nanorods were shown to be formed when the concentration of AOT was 0.06 M, but when this was reduced to 0.03 M magnificent dipyramidal $PbWO_4$ nanostructures were obtained in the AOT-containing microemulsion system. By controlling the water content of the AOT-containing microemulsion solution at between 4 and 15 vol%, $PbWO_4$ nanostructures with different proportions of nanorod bundles and spheres were formed at room temperature. However, when the water content was increased to 50 vol% or above, $PbWO_4$ nanoparticles with a mean size of only ca. 20 nm were obtained. At 40 °C, the as-prepared particles were of an ellipsoid morphology, but increasing the reaction temperature led to the formation of spherical $PbWO_4$ samples. At a reaction temperature of 80 °C all of the samples were of a spherical morphology.

Nanocrystals of cerium oxide were initially formed in the microemulsion-mediated synthesis process, and these subsequently self-assembled into nanorods during the aging process. The self-assembly resembled a controlled agglomeration process, aided by random collisions among the micelles containing the nanocrystals. The formation of such 'supra-aggregates' was observed in other microemulsion systems during the synthesis of Cu nanoparticles in the AOT/isooctane microemulsion system.

In general, dispersions of anisotropic noble nanoparticles demonstrate very different colors compared to those of spherical particles; this is due to the surface plasmon bands being more sensitive to particle shape than to size. All of the metal nanorods have two absorbance maxima that correspond to the longitudinal and transverse plasmon bands. The longitudinal plasmon band depends heavily on the aspect ratio. For example, platelets have additional quadrupole bands, but on transition from nanorods to platelets (as the aspect ratio decreases) the longitudinal band is blue-shifted while the transverse band is broadened due to overlap with the quadruple band. In cubes, all three plasmon bands merge into a single band. In contrast, the transition from nanorods to nanowires causes an increase in the aspect ratio and this produces a resultant red-shift of the longitudinal band and a blue-shift of the transverse band.

Metal nanoparticles, with their small size (1–100 nm), exhibit novel materials' properties that differ considerably from those of the bulk solid state. Metal nanostructures have been synthesized in many forms, ranging from conventional metal colloids to modern, near-monodispersed nanoclusters, shape-controlled nanocrystals and other nanostructures such as wires. Nanoparticles are used in surface-chemical, photochemical and other related fields, and can also be employed in magnetic materials, as constituents of paints, as vehicles for in-vivo drug carriers, and in molecular devices.

When the reaction takes place in spherical reverse micelles, most of the nanoparticles are spherical, their size being controlled by the use of functionalized reverse micelles, where the counterion of the surfactant is a reactive agent. Under these conditions, one of the key parameters involved in the nucleation process is the degree of hydration of the reactant; that is, the micellar size that can be well controlled.

In order to provide reproducibility and control of the properties required for advanced technological applications of metal nanostructures, new methods for their synthesis are vitally important. Recently, several techniques have been developed for the production of nanoparticles, including vapor-phase and sol–gel methods, sputtering and coprecipitation. One recognized goal of these new synthetic approaches is to achieve control over the composition, size, surface species, solubility, stability, isolability and other functional properties of the nanostructures.

Several recent reports have described a growing interest in new approaches to nanoparticle synthesis, driven primarily by an ever-increasing awareness of the unique properties and technological importance of these materials. The major issues associated with nanoparticle preparation include the control of particle size and internal structure. The fabrication of nanoparticles within reverse microemulsions has been shown to be a convenient method for creating monodisperse particles of controllable size. This method exploits two useful properties of reverse microemulsions: (i) the capacity to dissolve reactants in the water core; and (ii) a constant exchange of the aqueous phase among micelles. Thus, by mixing microemulsions containing different reactants it is possible to perform chemical reactions inside the reverse micelle water pool, using it as a nanoreactor. If this reaction results in a solid compound, the nanoparticle growth is limited by the micelle size. This method has been studied for some years and is used widely for metal, semiconductor and oxide nanoparticle syntheses.

A controlled method for the preparation of metal oxide nanoparticles via the sol–gel route is to use a W/O microemulsion in which the stabilized aqueous micelle acts as a type of nanoreactor. The microemulsion–microwave method has been used to control zeolite crystal morphologies, with uniform and smaller template-free zeolite nanocrystals having been synthesized in reverse microemulsion by microwave heating. Microwave heating has found a number of applications in synthetic chemistry since, compared to conventional heating, it has the advantage of a short reaction time and produces small particles with a narrow size distribution and high purity. The combination of reverse microemulsion and microwave heating has the added advantage that the oil phase in the reverse microemulsion system is transparent to microwaves, so that the aqueous domains are heated directly, selectively and rapidly.

Recently, reverse micelle and/or microemulsion synthesis has been demonstrated as a viable method for producing a wide array of metals and metal oxide compounds over a relatively narrow particle size distribution. Reverse micelle synthesis utilizes a natural phenomenon which involves the formation of spheroidal aggregates in a solution when a surfactant is introduced to a polar organic solvent, formed either in the presence or absence of water. Micelle formation allows for a unique encapsulated volume of controllable size through which reactions can be conducted and metal and metallic compounds subsequently produced. Moreover, the reactants for the chemical processes can be contained within the micelles. The main attractions of the two-step microemulsion reduction technique for preparing mixed metal nanoparticles are the ease and accuracy of composition control. The reduction reaction not only occurs in a confined reaction zone within the microemulsions, but its extent and uniformity can also be controlled. The final nanoparticles should follow the metal composition in the precursor solution, without any loss in control of the particle size.

Magnetic nanoparticles are of interest for a wide variety of applications: technologically they can be used as magnetic seals and in printing and recording, whilst in biology they can be used as magnetic resonance imaging (MRI) agents and for cell tagging and sorting. Such nanoparticles also have exciting potential applications in biological diagnosis. In all of these areas of research, the particle size, shape and surface properties are of the utmost importance, and in recent years excellent progress has been made in the control of these parameters. Iron oxides such as Fe_2O_3 and Fe_3O_4 can be prepared as monodispersed surface-derivatized nanoparticles, while Co and Fe can be created as both nanoparticles and nanorods by using solution methods.

Due to their small dimensions, magnetic nanoparticles exhibit a rich variety of interesting phenomena, such as a single-domain state, coercivity enhancement and quantization of spin waves. The field of magnetic materials has made use of variations in the magnetic properties of nanoparticles caused by effects such as single domains, superparamagnetism and surface interactions. In particular, due to their large surface-to-volume ratio, the magnetic properties of nanoparticles are dominated by surface effects and particle–support interactions. Typically, they exhibit magnetic anisotropy constants that are one or even two orders of magnitude larger than that of their bulk counterparts, with correspondingly enhanced coercivities. The properties of nanometer-sized magnetic particles of iron, cobalt or nickel have been studied extensively both experimentally and theoretically. Although these particles show promise for practical applications such as catalysis, magnetic recording, magnetic fluids and biomedical applications, their use has been limited due to uncontrolled oxidation.

9.5 Conclusions and Outlook

The unique physical properties of nanoscale magnetic materials, such as superparamagnetism, have generated considerable interest for their use in a wide range of diverse applications, from data information storage to the *in vivo* magnetic manipulation in biomedical systems. Many technological applications require the magnetic nanoparticles to be embedded in a nonmagnetic matrix; hence, over the past few years increasing attention has been paid to preparing nanostructures with magnetic nanoparticulate components, and on understanding the magnetic behavior of nanoparticles due to new possible surface, interparticle and exchange interactions in a magnetic/nonmagnetic matrix.

Although the viability of synthesizing and assembling metal nanoparticles has been well demonstrated in many studies, major challenges persist with regards to controlling the continuous nature of the metal coating, the coating thickness, the size monodispersity and thin-film assembly.

The synthesis of anisotropic nanoparticles on the 1–100 nm length scale represents a major challenge as they are less stable than spherical shapes of similar size, and the respective symmetric bulk crystal structures often create a strong barrier. The key factors in determining particle anisotropy have been recognized to be the anisotropic micelle-induced breaking of the symmetry of the growing nanoparticle, and the relative rates of nucleation and growth, as both factors determine the aspect ratio of the final nanoparticle. Nanorods are obtained in high yield only when a nonseeding method is used, or when small-seed nanoparticles are used in the seeding growth method. Lower-symmetry spheroids, platelets, cubes and stars were formed if the seed size was between 3–5 nm or if the nanorods were allowed to grow further. Anisotropic nanoparticles are of considerable current interest, due to a variety of shape-dependent properties. Although, synthetic methods for nanorods have been well established, the metallic systems have achieved very limited success due to their highly symmetric cubic crystal structure. Controlling the length-to-width ratio and uniformity of the length and width distributions of metal nanoparticles, as well as synthesizing particles with a width dimension below 10 nm, represents a major challenge for large-scale, bench-top methods.

A new, nonseeding method has been developed which is simple and useful for the gram-scale synthesis of nanorods. Here, a mixture of strong and weak reducing agents is introduced into the micellar solution of a metal salt, at which point the strong agent initiates nucleation while the weak agent helps the nanoparticles to grow. By using this micellar template approach, nanoparticles of a wide range of shapes can be prepared for metals.

Although, the intermicellar interactions, surfactant capping and the reducing agent concentration all govern the growth process, the shape is controlled by using nonspherical templates. In this way, elongated copper nanocrystals can be obtained when the synthesis occurs in cylindrical reverse micelles or in an interconnected cylinder phase. The mimicry related to shape is not optimal, however, and the particle shape may be more efficiently controlled by combining the strategy of the surfactant-based template with the capping involving salts or specific molecules. Nonspherical shapes can also be obtained in spherical reverse micelles, when the synthesis occurs in the presence of a large excess (the use reverse micelles in a supersaturated regime) of either the reducing agent or one of the reagents. For example, in the presence of a Na(AOT)/Ag(AOT) surfactant silver nanodisks are produced that no longer form well-defined reverse micelles when the concentration of the hydrazine reducing agent is high. Flat, triangular CdS nanocrystals are obtained by using $Cd(AOT)_2$ reverse micelles saturated with H_2S. For any reaction system, saturation with either the reducing agent or one of the reagents seems to be a primary factor in the formation of well-crystallized, nonspherical nanoparticles such as elongated nanocrystals, triangles, cubes and nanorods. In some studies, the fabrication of ZnO, $BaCO_3$ nanowires, $BaSO_4$ nanofilaments and CdS, $BaWO_4$ and $K_3[PMo_{12}O_{40}]\cdot xH_2O$ nanorods have shown that, under certain conditions, the microemulsion method can also be used to prepare some one-dimensional (1-D) nanostructures.

Although significant progress has been made recently in the development of nanowire-based sensors, field-effect transistors and lasers, the synthesis (via W/O microemulsions) and assembly of nanowires for the reproducible mass fabrication of these devices remain a major challenge.

In 1991, shortly after the discovery of carbon nanotubes [126], 1-D nanostructured materials attracted worldwide interest because of their unique electronic, optical and mechanical properties. Nanotubes, in particular, were shown to act as macroporous materials for macromolecular catalysis and, during the past few years, many methods have been developed–including high-temperature and soft-chemistry techniques– for the preparation of 1-D inorganic nanotubes of metals, metal chalcogenides, metal oxides, BN and $NiCl_2$.

Today, the use of magnetic nanoparticles in biology shows great promise, with one particular area of interest being the ability to increase the circulation time of magnetic nanoparticles in the blood. For this, the integration of magnetic nanoparticles into stealth liposomes [127] or artificial hollow capsules have shown much promise. Another recent goal has been to develop nanoattractors capable of concentrating magnetic nanoparticles into a desired region of the body.

Although, genetic engineering has provided new challenges and opportunities to medicine and biomedical research and development [128], the manipulations are often limited when the DNA strands are cleaved in a cellular environment [129]. Although this can be prevented to a large degree (e.g. by adding inhibitors or entrapping DNA in liposomes [130]), it may well hinder any further manipulation of DNA. Nanomaterials with unique properties such as large surface areas, pore structures, embedding effects and size effects, have been used effectively in both bioanalysis and drug delivery [131] as they appear capable of shielding the embedded DNA sequences.

Among applications in biology and medicine, both *in vivo* and *in vitro*, magnetic nanoparticles and magnetic fields may be used to remotely position or selectively filter biological materials [132]. In MRI, the presence of the particles at a given site can increase the contrast of certain cell types by several orders of magnitude, while the use of magnetic nanoparticles to improve cell manipulation and DNA sequencing may also help in the development of pharmaceuticals and drug delivery systems. Each of these applications should benefit greatly from an ability to control the surface and interparticle spatial properties of magnetic nanoparticles.

Today, the scientific and technological issues of nanostructured particles and materials continue to attract considerable attention, mainly because nanostructured particles and materials – as well as a physical and/or chemical combination of these materials at the nanometer or subnanometer scale – may lead to the development of innovative materials. Undoubtedly, future applications will impact greatly on areas such as catalysis, technical ceramics, membrane technology, optoelectronics and solid-state ionics, as well as on systems for clean energy conversion and storage. Such progress will depend very heavily, however, on the pace of fundamental research into nanostructured particles and materials in solid-state chemistry, solid-state physics, materials science and colloid chemistry.

Nomenclature/Abbreviations

$\alpha\text{-Fe}_2\text{O}_3$	hematite
ϕ_{per}	percolation threshold
CMC	critical micelle concentration
CTAB	cetyltrimethylammonium bromide
NaAOT	sodium bis(2-ethylhexyl)sulfosuccinate
$NaBH_4$	sodium tetrahydroborate
R	reducing agent to precursor mole ratio ([reducing agent]/[precursor])
SAED	selected area electron diffraction
w	water to surfactant mol ratio ([water]/[surfactant])

Acknowledgments

This research is supported by the Slovak Grand Agency (VEGA) through the grant number 2/7013/27 and 2/7083/27, SAV-FM-EHP-2008-01-01 project, and Science and Technology Assistance Agency through the APVT projects (0173-06, 0362-07, 0030-07, 0562-07, 0592-07).

References

1 Gotic, M., Czako-Nagy, I., Popovic, S. and Music, S. (1998) *Philos Mag Lett*, **78**, 193.
2 Schaefer, J.E., Kisker, H., Kronmüller, H. and Würschum, R. (1992) *Nanostructured Mater*, **1** (6), 523.
3 Sartale, S.D. and Lokhande, C.D. (2000) *Indian J Eng Mater Sci*, **7**, 404.
4 Erb, U. (1995) *Nanostructured Mater*, **6** (5–8), 533.
5 Athawale, A.A. and Bapat, M.J. (2005) *Metastable Nanocrystalline Mater*, **23**, 3.
6 Moschini, F., Guillaume, B., Belmont, A. and Clots, R. (2004) *Key Eng Mater*, **264–268**, 2335.
7 Service, R.F. (1996) *Science*, **271**, 920.
8 Sugimoto, M. (1999) *J. Am Ceram Soc*, **82**, 269.

9 Hoffman, A.J., Mills, G., Yee, H. and Hoffmann, M.R. (1992) *J Phys Chem*, **96**, 5546.

10 Joo, S.H., Choi, S.J., Oh, I., Kwak, J., Liu, Z., Terasaki, O. and Ryoo, R. (2001) *Nature*, **412**, 169.

11 Santra, S., Wang, K. and Tapec, R.J. (2001) *Biomed Opt*, **6** (2), 160.

12 Bach, U., Lupo, D., Comte, P. and Moser, J.E. (1998) *Nature*, **395**, 583.

13 Alivisatos, A.P. (1996) *Science*, **271**, 933.

14 Sun, S.C., Murray, B., Weller, D., Folks, L. and Moser, A. (2000) *Science*, **287**, 1989.

15 Bodker, F., Morup, S. and Linderoth, S. (1994) *Phys Rev Lett*, **72**, 282.

16 Nickolov, Z.S., Paruchuri, V., Shah, D.O. and Miller, J.D. (2004) *Colloids Surf A*, **232**, 93.

17 Han, M., Vestal, C.R. and Zhang, Z.J. (2004) *J Phys Chem B*, **108**, 583.

18 Yamada, M., Arai, M., Kurihara, M., Sakamoto, M. and Miyake, M. (2004) *J Am Chem Soc*, **126**, 9482.

19 Hammond, P.T. (2004) *Adv Mater*, **16**, 1271.

20 Feldmann, C. (2003) *Adv Funct Mater*, **13**, 101.

21 Schmid, G. (2004) *Nanoparticles: From Theory to Application*, Weinheim, Wiley VCH.

22 Chen, G., Luo, G.T., Xu, J.H. and Van, J.D. (2004) *Powder Technol.*, **180** (5), 139.

23 Siegel, R.W. (1998) *J Mater Rev*, **3**, 1367.

24 Fegley, B.J., White, P. and Bowen, H.K. (1985) *Am Ceram Soc Bull*, **64**, 1115.

25 Fayet, P. and Woste, L.Z. (1986) *Phys D*, **3**, 177.

26 Jang, Z.X., Sorensen, C.M., Klabunde, K.J. and Hadjipanayis, G.C. (1991) *J Colloid Interface Sci*, **146**, 38.

27 Cordente, N., Respaud, M., Senocq Casanove, M.J., Amiens, C. and Chaudret, B. (2001) *Nano Lett.*, **1**, 565.

28 Dumestre, F., Chaudret, B., Amiens, C., Fromen, M.C., Casanove, M.J., Renaud, P. and Zurcher, P. (2002) *Angew Chem*, **114**, 4462.

29 Pileni, M.P. (2003) *Nat Mater*, **2**, 145.

30 Lisiecki, I. (2004) *Colloids Surf A*, **250**, 499.

31 McLeod, M.C., McHenry, R.S., Beckman, E.J. and Roberts, C.B. (2003) *J Phys Chem B*, **107**, 2693.

32 Sun, W., Xu, L., Chu, Y. and Shi, W. (2003) *J Colloid Interface Sci*, **266**, 99.

33 Yuasa, M., Sakai, G., Shimano, K., Teraoka, Y. and Yamazoe, N. (2004) *J Electrochem Soc*, **151** (A1477), 9.

34 Misra, D.K., Gubala, S., Kale, A. and Tegelhoff, W.F.Jr. (2004) *Mater Sci Eng*, **B111**, 164.

35 Fendler, J.H. (1987) *Chem Rev*, **87**, 877.

36 Zana, R. (1994) *Heterogen Chem Rev*, **1**, 145.

37 Chan, J.P., Lee, K.M., Sorensen, C.M., Klabunde, K.J. and Hadjipanyis, G.C. (1994) *J Appl Phys*, **75**, 5876.

38 Gobe, M., Kon-no, K., Kandori, K. and Kitahara, A. (1983) *J Colloid Interface Sci*, **93**, 293.

39 Paul, B.K. and Moulik, S.P. (1997) *J Dispersion Sci Technol*, **18**, 301.

40 Zhang, J., Sun, L.D., Qian, C., Liao, C.S. and Yan, C.H. (2001) *Chin Sci Bull*, **46**, 1873.

41 Pileni, M.P., Zemb, T. and Petit, C. (1985) *Chem Phys Lett*, **118**, 414.

42 Maria, M.L., Agostiano, A., Manna, L., Monica, M.D., Catalano, M., Chiavarone, L., Spagnolo, V. and Lugara, M. (2000) *J Phys Chem B*, **104**, 8391.

43 Cassin, G., Badiali, J.P. and Pileni, M.P. (1995) *J Phys Chem*, **99**, 12941.

44 Van Dijk, M.A. (1985) *Phys Rev Lett*, **9**, 1003.

45 Robinson, B.H., Toprakcioglu, C., Dore, J.C. and Chieux, P.J. (1984) *Chem Soc Faraday Trans*, **1**, 80.

46 Lisiecki, I. (2005) *J Phys Chem B*, **109**, 12231.

47 Cazabat, A.M. and Langevin, D. (1981) *J Chem Phys*, **74**, 3148.

48 Pileni, M.P. (1997) *Langmuir*, **13**, 3266.

49 Martin, J.E., Wilcoxon, J.P., Odinek, J. and Provencio, P. (2002) *J Phys Chem B*, **106**, 971.

50 Sun, S., Zeng, H., Robinson, D.B., Raoul, S., Rice, P.M., Wang, S.X. and Li, G.J. (2004) *Am Chem Soc*, **126**, 273.
51 DeLongchamp, D.M. and Hammond, P.T. (2004) *Adv Funct Mater*, **14**, 224.
52 Lisiecki, I., Billoudet, F. and Pileni, M.P. (1997) *Journal of Molecular Liquids*, **72**, 251.
53 Rauscher, F., Veit, P. and Sundmacher, K. (2004) *Colloids Surf A*, **254**, 183.
54 Eicke, H.F. and Zinsli, P.E. (1978) *J Colloid Interface Sci*, **65**, 131.
55 Zana, R. and Lang, J. (1987) in *Microemulsions: Structure and dynamics*, (eds Friberg, S.E. and Bothorel, P.), CRC Press, Boca Raton, Florida, p. 153.
56 Jain, T.K., Cassin, G., Badiali, J.P. and Pileni, M.P. (1996) *Langmuir*, **12**, 2408.
57 Lopez-Quintela, M.A. (2003) *Curr Opin Colloid Interface Sci*, **8**, 137.
58 Capek, I. (2004) *Adv Colloid Interface Sci*, **110**, 49–74.
59 Pinna, N., Weiss, K., Sack-Kongehl, H., Vogel, W., Urban, J. and Pileni, M.P. (2001) *Langmuir*, **17**, 7982.
60 Wu, M., Long, J., Haung, A. and Luo, Y. (1999) *Langmuir*, **15**, 8822.
61 Chen, D.H. and Wu, S.H. (2000) *Chem Mater*, **12**, 1354.
62 Hanna, H.I., Rahul, B., Anders, P., Magnus, S., Christer, S., Krister, H. and Dinesh, O.S. (2001) *J Colloid Interface Sci*, **241**, 104.
63 Lisiecki, I. and Pileni, M.P. (1995) *J Phys Chem*, **99**, 5077.
64 Jana, N.R., Gearheart, L. and Murphy, C.J. (2001) *Chem Mater*, **13**, 2313.
65 Murphy, C.J. and Jana, N.R. (2002) *Adv Mater*, **14**, 80.
66 Jana, N.R. and Peng, X. (2003) *J Am Chem Soc*, **125**, 14280.
67 Roucoux, A., Schultz, J. and Patin, H. (2002) *Chem Rev*, **102**, 3757.
68 Watzky, M.A. and Finke, R.G. (1997) *J Am Chem Soc*, **119**, 10382.
69 Greenbaum, E. (1988) *J Phys Chem*, **92**, 4571.
70 Jana, N.R. (2004) *Angew Chem*, **116**, 1562.
71 Jana, N.R. and Pal, T. (2001) *J Surf Sci Technol*, **17**, 191.
72 Toernblom, M. and Henriksson, U. (1997) *J Phys Chem B*, **101**, 6028.
73 Lin, Z., Cai, J.J., Scriven, L.E. and Davis, H.T. (1994) *J Phys Chem*, **98**, 5984–5993.
74 John, V.T., Simmons, B., McPherson, G.L. and Bose, A. (2002) *Curr Opin Colloid Interface Sci*, **7**, 288.
75 Lisiecki, I. and Pileni, M.P. (2003) *Langmuir*, **19**, 9486.
76 Legrand, J., Petit, C. and Pileni, M.P. (2001) *J Phys Chem B*, **105**, 5643.
77 Lee, S.M., Cho, S.N. and Cheon, J.W. (2003) *Adv Mater*, **15** (5), 441.
78 Tanori, J. and Pileni, M.P. (1997) *Langmuir*, **13**, 639.
79 Dixit, S.G., Mahadeshwar, A.R. and Haram, S.K. (1998) *Colloids Surf A.*, **133**, 69.
80 Ravey, J.C. and Muziker, M.J. (1987) *Colloid Interface Sci*, **116**, 30.
81 Shindo, D., Park, G.S., Waseda, W. and Sugimoto, T. (1994) *J Colloid Interface Sci*, **168**, 478.
82 Sugimoto, T. and Hwang, Y.J. (1998) *Colloid Interface Sci*, **207**, 137.
83 Sasaki, N., Murakami, Y., Shindo, D. and Sugimoto, T. (1999) *J Colloid Interface Sci*, **213**, 121.
84 Cushing, B.L., Kolesnichenko, V.L. and O'Connor, C.J. (2004) *Chem Rev*, **104**, 3893.
85 Pileni, M.P. (1993) *J Phys Chem*, **97**, 6961.
86 Boutonnet, M., Kizting, J. and Stenius, P. (1982) *Colloids Surf*, **5** (3), 209.
87 Kortan, A.R., Hull, R., Opila, R.L., Bawendi, M.G., Steigerwald, M.L., Carroll, P.J. and Brus, L.E. (1990) *J Am Chem Soc*, **112**, 1327.
88 Pillai, V., Kumar, P., Multani, M.S. and Shah, D.O. (1993) *Colloids Surf A*, **80**, 69.
89 Motte, L., Lisiecki, I. and Pileni, M.P. (1994) in *Hydrogen Bond Networks*,

(eds. Dore, J.C. and Bellissent-Funel, M.C.), NATO Publisher, p. 447.
90 Vincenzo T.L., (2001) in *Nano-Surface Chemistry*, (ed. Rosoff, M.), Marcel Dekker, New York, p. 492.
91 Reddy K.S.N., Salvati, L.M., Dutta, P.K., Abel, P.B., Suh, K.I. and Ansari, R.R. (1996) *J Phys Chem*, **100**, 9870.
92 Simmons, B.A., Li, S., John, V.T., McPherson, G.L., Bose, A., Zhou, W. and He, J. (2002) *Nano Lett*, **2**, 263.
93 Petit, C., Lixon, P. and Pileni, M.P. (1991) *Langmuir*, **7**, 2620.
94 Eastoe, J., Steytler, D.C., Robinson, B.H., Heenan, R.K., North, A.N. and Dore, J.C. (1994) *J Chem Soc Faraday Trans.*, **90**, 2479.
95 Eastoe, J., Robinson, B.H. and Heenan, R.K. (1993) *Langmuir*, **9**, 2820.
96 Sonnichsen, C. and Alivisatos, A.P. (2005) *Nano Lett*, **5**, 301–304.
97 Sun, Y. and Xia, Y. (2002) *Science*, **298**, 2176–2179.
98 Peng, X., Manna, L., Yang, W., Wickham, J., Scher, E., Kadavanich, A. and Alivisatos, A.P. (2000) *Nature*, **404**, 59.
99 Jana, N.R. (2003) *Chem Commun*, 1950.
100 Sun, Y., Mayers, B., Herricks, T. and Xia, Y. (2003) *Nano Lett*, **3**, 955.
101 Peng, X. (2003) *Adv Mater*, **15**, 459.
102 Pileni, M.P. (2003) *Nat Mater*, **2**, 145.
103 Johnson, J.C., Dujardin, E., Davis, S.A., Murphy, C.J. and Mann, S. (2002) *J Mater Chem*, **12**, 1765.
104 Jana, N.R. (2005) *Small*, **1** (8–9), 875.
105 Jana, N.R. (2004) *Angew Chem Int Ed*, **43**, 1536.
106 Wang, S. and Yang, S. (2000) *Langmuir*, **16**, 389.
107 Lemaire, B.J., Davidson, P., Ferré, J., Jamet, J.P., Panine, P., Dozov, I. and Jolivet, J.P. (2002) *Phys Rev Lett*, **88**, 1255071.
108 Filankembo, A., Giorgio, S., Lisiecki, I. and Pileni, M.P. (2003) *J Phys Chem B*, **107**, 7492.
109 Jana, N.R., Gearheart, L. and Murphy, C.J. (2001) *Adv Mater*, **13**, 1389.
110 Germain, V., Li, J., Ingert, D., Wang, Z.L. and Pileni, M.P. (2003) *J Phys Chem B*, **107**, 34.
111 Pinna, N., Weiss, K., Urban, J. and Pileni, M.P. (2001) *Adv Mater*, **13**, 261.
112 Heiglein, A. and Giersig, M. (2000) *J Phys Chem B*, **104**, 6767.
113 Salzemann, C., Lisiecki, I., Urban, J. and Pileni, M.P. (2004) *Langmuir*, **20**, 11772.
114 Jin, R., Cao, Y.W., Mirkin, C.A., Kelly, K.L., Schatz, G.C. and Zheng, J.G. (2001) *Science*, **294**, 1901.
115 Fu, H., Xiao, D., Yao, J. and Yang, G. (2003) *Angew Chem*, **115**, 2989.
116 Fu, H., Xiao, D., Yao, J. and Yang, G. (2003) *Angew Chem Int Ed*, **42**, 2883.
117 Jana, N.R., Gearheart, L., Obare, S.O. and Murphy, C.J. (2002) *Langmuir*, **18**, 922.
118 Rodrıguez-Fernandez, J., Perez-Juste, J., Mulvaney, P. and Liz-Marzan, L.M. (2005) *J Phys Chem B*, **109**, 14257–14261.
119 Link, S. and El-Sayed, M.A. (1999) *J Phys Chem B*, **103**, 8140.
120 Esumi, K., Matsuhisa, K. and Torigoe, K. (1995) *Langmuir*, **11**, 3285.
121 Chen, D., Shen, G., Tang, K., Liang, Z. and Zheng, H. (2004) *J Phys Chem B*, **108**, 11280.
122 Moulik, S.P. and Paul, B.K. (1998) *Adv Colloid Interface Sci.*, **78**, 99.
123 Kurzy, S.C., Patil, S.D., Deshpande, S. and Seals, S. (2005) *J Phys Chem B Condens Matter Mater Surf Interfaces, Biophys* **109**, 6936.
124 Kurzy, S.C., Seals, S., Fei, W., Ramsdell, J., Desai, V.H., Li, Y., Babu, S.V. and Wood, B.J. (2003) *Electrochem Soc*, **150**, C36–C43.
125 Alargova, R.G., Petkov, J.T. and Petsev, D.N. (2003) *J Colloid Interface Sci*, **261**, 1–11.
126 Iijima, S. (1991) *Nature*, **354**, 56–58.
127 Kuznetsov, A.A., Filippov, V.I., Alyautdin, R.N., Torshina, N.L. and Kuznetsov, O.A. (2001) *J Magn Magn Mater*, **225**, 95.
128 Anderson, L. (1999) *Genetic Engineering*, Chelsea Green Publishing Co. White River Junction, VT.

129 Dong, S.M., Fu, P.P., Shirsat, R.N., Hwang, H.M., Leszczynski, J. and Yu, H.T. (2002) *Chem Res Toxicol*, **15**, 400.

130 Wang, J., Zhang, P.C., Mao, H.Q. and Leong, K.W. (2002) *Gene Therapy*, **9**, 1254–1261.

131 Cui, Z. and Jumper, R.J. (2002) *Bioconjugate Chem*, **13**, 1319.

132 Cui, Y.L., Wang, Y.N., Hui, W.L., Zhang, Z.F., Xin, X.F. and Chen, C. (2005) *Biomed Microdevices*, **7**, 153.

10
Preparation of Nanoemulsions by Spontaneous Emulsification and Stabilization with Poly(caprolactone)-*b*-poly(ethylene oxide) Block Copolymers

Emmanuel Landreau, Youssef Aguni, Thierry Hamaide, and Yves Chevalier

10.1
Introduction

Today, the use of amphiphilic polymers in place of classical surfactants represents a trend in the field of emulsion formulation and stabilization of suspensions [1, 2]. These polymers are referred to as polymeric surfactants, dispersants or stabilizers, according to the terminology of the specific application. The dispersion of organic particles in water, and stabilization of the resultant suspensions, require amphiphilic polymers which have well-separated hydrophilic and hydrophobic parts, the separation of which as blocks in the macromolecule is essential. Various oil-in-water (O/W) dispersions such as emulsions, latexes and suspensions of polymer particles can be fabricated and stabilized with the help of polymeric surfactants. In most cases, the technology of emulsification/dispersion requires that the polymeric surfactant is soluble in the aqueous phase; consequently, water-soluble polymers have undergone the greatest degree of investigation.

Essentially two types of amphiphilic polymer are encountered, namely grafted polymers and block copolymers:

- *Grafted polymers* are generally hydrophilic polymers grafted with hydrophobic substituents (fatty acids, alkyl chains); alkyl-modified polysaccharides are typical examples. An alternative is a hydrophobic polymer grafted with hydrophilic substituents such as polyethylene glycol (PEG)-grafted silicones. Owing to the constraints of synthetic chemistry, the architecture of such amphiphilic polymers is generally of the grafted type, consisting of one block as the polymer backbone and the counterpart as side chains.

- *Block copolymers* are prepared by the copolymerization of hydrophilic and hydrophobic monomers. The degree of 'blockiness' can be adjusted by the correct choice of polymerization process. Living polymerization reactions have most often been used for the synthesis of such block copolymers, either diblock or triblock. The well-known ethylene oxide/propylene oxide copolymers are prepared by a two-step anionic polymerization of these monomers. The synthesis of block copolymers is

Emulsion Science and Technology. Edited by Tharwat F. Tadros
Copyright © 2009 WILEY-VCH Verlag GmbH & Co. KGaA, Weinheim
ISBN: 978-3-527-32525-2

not restricted to living anionic polymerization; rather, cationic, coordinated and various ring-opening polymerization reactions also afford block copolymers. Controlled polymerization can also be used instead of true living polymerization; the first block copolymers to be prepared by means of controlled radical polymerization appeared some ten years ago, increasing the choice of the monomers that can be used.

Among emulsifiers, *polymeric surfactants* [1, 2] show the clear advantages that have motivated the studies reported in this chapter. In particular, the structure of diblock copolymers having hydrophilic and hydrophobic (nonpolar) blocks is similar to that of the classical emulsifiers, although the properties of each block are enhanced. The hydrophilic portion is much more polar that that of a classical surfactant, and the same is true for the hydrophobic portion. The consequence is excellent surface properties at low concentrations [3], and this allows lower concentrations of emulsifier to be used to stabilize the emulsions. In addition to the obvious economical benefits of this, the residual concentration of emulsifier in the aqueous phase is low, and this limits secondary effects such as air entrapment and foaming, as well as adsorption at various interfaces, notably at the cell membranes in case of their application is the pharmaceutical domain.

10.1.1
Block Copolymers

Although block copolymers are potentially useful emulsifiers, their use in place of the traditional nonionic emulsifiers requires particular attention. Difficulties arise from some of the specific properties of polymeric surfactants. For example, their adsorption at interfaces is so strong that it appears quasi-irreversible, and therefore the rapid interfacial rearrangements that are possible with classical surfactants can no longer take place. Once a polymeric surfactant has been adsorbed at the surface of one droplet, it is no longer available for another droplet, even if its surface is bare (i.e. free of emulsifier). The kinetics of adsorption is also very slow, so that several emulsification processes cannot function when polymeric surfactants are used alone, because the emulsion droplets produced are not stabilized sufficiently quickly by polymer adsorption and undergo rapid coagulation during the fabrication process. Classical surfactants made from small molecules that are adsorbed faster must be mixed with polymeric surfactants in order to ensure the successful emulsification, with the polymeric surfactant allowing good long-term stabilization after it has been adsorbed. When the polymeric surfactant is dissolved in the aqueous phase, it must be chosen as being sufficiently hydrophilic, and its extensive swelling by water causes a strong thickening effect, especially if the polymeric surfactant has a high molar mass and is not adsorbed at the surface of particles in suspension. Thickening by the polymeric emulsifier alters the emulsification process with respect to a classical emulsifier; this effect is especially strong with poly(vinyl alcohol) [4, 5].

Therefore, a polymeric emulsifier is chosen with regards to both the long-term stability of the final emulsion and the possibility of preparing the emulsion. Both of these aspects were assessed in the present investigations of the emulsification of polycaprolactone (PCL) or various liquid oils by the 'spontaneous emulsification' process with the help of the poly(caprolactone)-b-poly(ethylene oxide) block copolymers (PCL-b-PEO) as emulsifiers.

10.1.1.1 Spontaneous Emulsification

Spontaneous emulsification is a mild emulsification method that does not require a high energy input, for example by means of high shear mixing (using a propeller, homogenizer, etc.). The emulsification is said to be spontaneous because it does not require an external energy input. An O/W emulsion is formed when a solution of oil in a water-soluble organic solvent (acetone) is mixed with water; the polar organic solvent then dissolves into the water to leave a supersaturated solution of oil in water. The subsequent liquid–liquid phase separation results in the formation of oil droplets in an aqueous continuous phase. Previously, this process was described as the 'diffusion and stranding' mechanism by Davies and colleagues [6, 7]; alternatively, Miller and coworkers [8, 9] and Vitale and Katz [10] proposed the classical 'nucleation and growth' mechanism as the possible route for oil droplet formation. The latter mechanism explains the main features of the process [11], in particular the failure of emulsification at high oil contents; although a definite proof of the actual mechanism has not yet been proposed. The 'spontaneous emulsification' is referred to as the 'nanoprecipitation' process [12–14] when the oil is an organic polymer (i.e. insoluble in water), and this leads to the preparation of aqueous suspensions of polymer nanoparticles. The same type of precipitation method is used in the production of inorganic nanocolloids [15].

The studies described here deal with an evaluation of the emulsifying properties of block copolymers of the PCL-b-PEO type, the general chemical structure of which is as follows:

$$CH_3-(OCH_2CH_2)_n-(\underset{\underset{O}{\|}}{O}CCH_2CH_2CH_2CH_2CH_2)_m-OH$$

The hydrophobic PCL block is a well-known biocompatible and biodegradable polyester [16–18] that is widely used as a matrix for the encapsulation of active substances inside polymeric microspheres or nanospheres [19–21]. The hydrophilic block consists of poly(ethylene oxide) (PEO), which is biocompatible but barely biodegradable. The use of PEO in many nonionic emulsifiers such as polysorbates (Tween®) and monoalkyl ethers of PEO (Brij®) is, however, both common and well accepted. It is difficult to determine which hydrophilic materials may be substituted satisfactorily, as the two parameters that control the properties of these block copolymers are the degrees of polymerization (DP) of the PCL and PEO blocks. The synthesis process that has been used to date allows the DPs of both blocks to be tuned in order to allow a systematic investigation of the relationships between chemical structures and properties.

10.1.1.2 Biodegradability and Biocompatibility

Biodegradability and biocompatibility are claimed to be especially important for applications in the areas of pharmaceuticals and cosmetics. The requirements for such applications [19–21] are quite severe, as a formulation can only be proposed if the safety of all excipients (including emulsifiers) has been demonstrated via toxicology tests. As a consequence, the common practice in pharmacy is to use compounds that have been described and validated in the European Pharmacopoeia [22], because the necessary toxicology studies have already been conducted. The choice of emulsifiers in a positive list is a difficult task, and a suitable emulsifier often cannot be found because of the too-small number of described compounds. An application for the agreement of a new compound requires long and expensive toxicology studies to be conducted, with the decision being taken only when the optimal properties of the compound have been demonstrated. This in turn provides the motivation to seek new compounds with improved properties and which meet the constraints of the proposed pharmaceutical and/or cosmetic applications.

Emulsifiers used in pharmaceutical applications allow the stabilization of emulsions or suspensions, and are both biocompatible and biodegradable. In general, they also contribute to the stabilization of the active substance encapsulated within the particles. It is imperative that the active substance should reach the therapeutic target, and also circulate within the body over a long time period. Coating the surface with a protective layer made from a hydrophilic polymer is an efficient means of preventing the detection and elimination of particles by the immune system [23–25]. In this respect, the hydrophilic portion of a block copolymer emulsifier provides stabilization of the active substance, whereas a too-small and mobile emulsifier does not.

10.1.2
Block Copolymer Micelles

The majority of studies reported to date have dealt with 'block copolymer micelles'; these are nanoparticles made from pure block copolymer, and are quite different from the micelles of classical surfactants. As block copolymer micelles do not dissolve spontaneously in water, their preparation requires an emulsification process that strongly resembles the nanoprecipitation process or spontaneous emulsification. Although the structure of block copolymer micelles is almost identical to that of classical micelles, their size is larger and the dynamic phenomena are very slow. The structural and dynamic properties of block copolymer micelles, together with their encapsulation properties, have been extensively studied [3, 25–42]. The overall structure of the nanoparticles in the present investigation is an organic core (either oil or polymer) surrounded by a shell made from block copolymers that ensure colloidal stability of the suspension. While this is similar to block copolymer micelles, the particles described here differ because they contain a large amount of organic material inside the particles, and their size is much larger. In fact, they are true emulsions. The block copolymers ensure colloidal stability of the suspension and prevent recognition by the immune system, owing to the PEO block.

10.1.3
Diblock Copolymers

Diblock copolymers of the present type have been widely studied for their applications in the encapsulation and vectorization of drugs by means of block copolymer micelles [25, 30–42]. Surprisingly, the use of these block copolymers as emulsifiers has not received a great deal of attention, although the main applications of surfactants and polymeric surfactants in pharmaceutical formulations are stabilization of emulsions and suspensions of polymer particles. The choice of emulsifier is of paramount importance in the field of nanoemulsions (i.e. emulsions of submicronic size) because of their large specific area. However, technical data covering the use of block copolymers as emulsifiers for the stabilization of nanoparticle suspensions is required in order to evaluate the benefits and possible drawbacks of these polymeric surfactants.

Two main parameters have been studied, namely the feasibility of polymer emulsions with the help of these block copolymer surfactants, and the long-term stability of the emulsions. Hence, the chapter comprises two main parts:

- The emulsification of PCL in water by using the 'nanoprecipitation' process with the block copolymers as emulsifiers is first reported, with special emphasis placed on the control by the formulation parameters, namely the chemical compositions of the aqueous and organic phases and the choice of block copolymer structure. No active substance has been incorporated in this model study. Loading with active substances that are known to enter PCL nanoparticles would be more straightforward, as the block copolymer emulsifiers adsorbed at the surface of the particles do not alter the properties of PCL inside the particles.

- The extension of this emulsification process to a variety of common oils used in cosmetic and pharmaceutical formulations is also reported. These block copolymers allow the efficient and sustainable fabrication of O/W emulsions with several types of oil, although failure to emulsify has been reported with some oils.

10.2
Materials and Methods

10.2.1
Materials

The PEO block of copolymers was obtained from PEG mono methyl ether of different molar masses (purchased from Fluka and used as received). Reagent-grade caprolactone (Aldrich) was used as received, and triethyl aluminum (TEA; Aldrich) was used as a 25 wt% solution in toluene. The toluene was dried on molecular sieves (3Å) and stored under a dry argon atmosphere. Polycaprolactone (PCL) (M_W 80000 g mol^{-1}) and light mineral oil were purchased from Aldrich. Isostearyl isostearate (DUB ISIS®) was provided by Stéarinerie Dubois (Boulogne-Billancourt,

France), Miglyol 812 was from Condea Chemie GmbH (Hamburg, Germany), and the silicone oil was 200 Fluid® of 20 cSt viscosity from Dow Corning.

10.2.2
Synthesis of Block Copolymers (PCL-b-PEO)

The synthesis of PCL-b-PEO block copolymers was carried out with a coordinated anionic polymerization of ε-caprolactone from PEG methyl ether (MPEG) and TEA. The synthesis of such copolymers is well documented, and proceeds usually with stannous octanoate as initiator and MPEG in stoichiometric amounts. In the present studies $AlEt_3$ was used as initiator, in a [MPEG]/[TEA] ratio of ≈ 10, in order to synthesize the alcoholate, taking advantage of the rapid transfer reaction between alcoholate and alcohol. This reaction is a versatile one-step process for the synthesis of functional low molar mass polymers from oxygenated heterocycles [43–45], and was readily adapted to the synthesis of block copolymers using the MPEG block as transfer agent [46]. This reaction also allowed the synthesis of polymer materials using less metal atoms than for the process using stannous octanoate. As it is difficult to eliminate the residual metal compounds, even after precipitation of the polymer, the initiator (metal catalyst) is generally not removed on reaction completion. The final material was found to contain fewer metal species in the present case, while the number-average DP (m) was the ratio of caprolactone to MPEG transfer agent:

$$m = \frac{[\text{caprolactone}]}{\text{MPEG}}$$

Four commercially available MPEGs of number-average molar mass $M_n = 350$, 750, 2000 and 5000 g mol^{-1} (corresponding to number-average DPs of $n = 7.6$, 16.7, 45.5 and 113.3, respectively) were used as transfer agents. The reaction was performed in toluene under dry argon in a 250 ml round-bottomed flask. MPEG was dissolved in dry toluene by heating. Part of the toluene was distilled out in order to eliminate any residual water as an azeotrope; residual water was mainly contained in MPEG. After cooling to room temperature, the aluminum alkoxide was formed by adding TEA according to a MPEG: TEA ratio of ≈ 10. ε-Caprolactone was added after a 15 min interval and the polymerization was then performed at 60 °C for 2 h. After complete conversion of the monomer, the block copolymer was precipitated in cold heptane at -18 °C. The polymer was dried at room temperature under vacuum, weighed, and characterized using ^{1}H NMR and MALDI-ToF mass spectrometry.

10.2.3
Methods

Particles sizes were measured using either dynamic or classical light scattering methods. These two light-scattering techniques are complementary as they are sensitive to different size-ranges but differ widely in their basic principle of operation. Thus, small-angle light scattering (SALS) is based on measurement of the time-averaged scattered intensity as a function of the scattering angle; according to the explored angular domain, particle diameters are measured between 100 nm and 2 mm. Dynamic light scattering (DLS) is a measurement of the Brownian motion of the particles that is related to their hydrodynamic diameter; the upper limit of diameters that can be measured is 2 µm. DLS was performed using a Malvern® Zetasiser 3000HS instrument operating at a wavelength of 633 nm and a 90° scattering angle. The suspensions were diluted to approximately 10^{-4}, so that the count rate was of the order of 200 kHz. The autocorrelation signal was analyzed by means of the method of cumulants, giving the z-average diameter of the particles. Characterization by static light scattering was performed using a Coulter® LS 230 (Beckman Coulter). The dispersion was diluted inside the instrument in such a way that the obscuration was 12% according to the supplier's instructions. The size distribution was calculated from the Mie theory, according to a suitable optical model. The real parts of the refractive indices of the optical models used were 1.33 for water, 1.45 for PCL, 1.455 for isostearyl isostearate, 1.45 for Miglyol 812, 1.467 for mineral oil and 1.40 for silicone oil; the imaginary part of the refractive index was zero for all compounds.

10.2.4
Emulsification of Oils or PCL

Oil or PCL emulsions were prepared using the nanoprecipitation process. The process consisted of mixing an organic phase in acetone into water. A typical example of the organic phase contained 0.2 g of oil or PCL and 0.1 g of block copolymer in 25 ml of acetone. The organic solution was rapidly injected into 50 ml of water; the mixture was seen immediately to turn 'milky' as the O/W emulsion formed, after which the acetone was evaporated under reduced pressure. By following this process an emulsion containing 0.6% dispersed phase was obtained; the sizes of the emulsion particles were measured using static and dynamic light scattering methods, as described above.

10.3
Results and Discussion

The chemical structures of copolymers used in the following procedures are listed in Table 10.1; these are abbreviated as PCL(m)-b-PEO(n), where m and n are the number-average DPs of the PCL and PEO blocks, respectively (see Section 10.1.1.1). The hydrophile–lipophile balance (HLB) numbers of the copolymer emulsifiers were estimated from the definition of Griffin, using the number-average masses of the PEO block and the full copolymer:

$$\text{HLB} = 20 \times \frac{M_n \text{ PEO}}{M_n \text{ copolymer}}$$

All copolymers listed in Table 10.1 were soluble in acetone. The only water-soluble copolymer was PCL(10)-b-PEO(113) which, indeed, was the compound with the highest HLB number among the series.

Table 10.1 Chemical structure and HLB numbers of polymeric emulsifiers.

n PEO	m PCL	HLB
8	10	4.6
17	5	11.2
17	10	7.8
45	20	9.3
113	10	16.2
113	20	13.6
113	25	12.6
113	30	11.7
113	40	10.3

10.3.1
Emulsions of PCL by Spontaneous Emulsification

10.3.1.1 Fabrication of the Emulsions

The PCL emulsions were prepared using the 'spontaneous emulsification' process, better known in the case of the emulsification of a polymer instead of a liquid oil as the 'nanoprecipitation' process. A PCL solution in acetone was emulsified without energy supply by dispersing it into a larger volume of water. Depending on its solubility, the copolymer emulsifier was introduced either in the organic phase together with PCL, or dissolved in the aqueous phase.

The block copolymer must be compatible with the emulsification process (in this case nanoprecipitation). As polymers are known to adsorb quite slowly to the surface of particles [3, 47], the emulsification is quite difficult and might even be unsuccessful. However, some well-known polymer emulsifiers such as poly(vinyl alcohol) or poly(ethylene oxide)/(propylene oxide) block copolymers (Pluronic®) allow the preparation of suspensions of polyester nanoparticles. Thus, the aim of this part of the study was to identify which PCL-b-PEO block copolymers were useful, and how they could be used properly for the nanoprecipitation of PCL.

Although block copolymers can be incorporated into either the aqueous or organic phase, it is common practice to use the emulsifier in the solvent in which it is soluble – in this case the continuous phase of the emulsion. According to the Bancroft rule, water-soluble emulsifiers (with high HLB-values) allow the stabilization of O/W emulsions, while the reverse holds for oil-soluble emulsifiers (low HLB-values). The Bancroft rule is empirical and holds for conventional processes and formulations – that is, for emulsions made using high-shear mechanical processes with classical nonionic emulsifiers. The present case proved to be an exception to this empirical

Table 10.2 Properties of aqueous suspensions of PCL nanoparticles stabilized with various PCL-b-PEO block copolymers incorporated in the organic phase. The recipe was: [PCL] = 0.4%, [PCL-b-PEO]/[PCL] = 1/2.

Copolymer	HLB	Appearance of the suspension	Particle diameter (nm)
PCL(10)-b-PEO(8)	4.6	Immediate precipitation	
PCL(5)-b-PEO(17)	11.2	Precipitate + milky suspension	
PCL(10)-b-PEO(17)	7.8	Precipitate + milky suspension	
PCL(20)-b-PEO(45)	9.3	Milky suspension	280
PCL(10)-b-PEO(113)	16.2	Milky suspension	256
PCL(20)-b-PEO(113)	13.6	Milky suspension	204
PCL(25)-b-PEO(113)	12.6	Milky suspension	229
PCL(30)-b-PEO(113)	11.7	Milky suspension	220
PCL(40)-b-PEO(113)	10.3	Milky suspension	210

rule, however, because the emulsifiers were not classical and the emulsification process by means of spontaneous emulsification directed the emulsion towards O/W, whichever type of emulsifier was used.

As all of the block copolymers were soluble in acetone, they were all incorporated into the organic phase. Stable emulsions could be prepared in the favorable cases listed in Table 10.2, and in such cases the mixture turned milky when the acetone phase was poured into the aqueous phase. No solid precipitate was observed at either the bottom or the top of the suspension. Moreover, the final properties of the emulsion were unaffected by the rate of addition of the organic phase.

The PCL(10)-b-PEO(113) copolymer, which was soluble both in water and acetone, allowed a comparison to be made with regards to the phase-solubilization of the emulsifier. Thus, it was possible to prepare stable emulsions with the PCL(10)-b-PEO(113) block copolymer, whichever phase it was introduced in. There was, however, a slight difference in the particle size: given the same recipe ([PCL] = 0.4%, [PCL-b-PEO]/[PCL] = 1/2), the particle size was 256 nm when the block copolymer emulsifier was introduced into the organic phase, but only 210 nm when it was introduced into the aqueous phase. Therefore, as a general rule, the block copolymer emulsifier should be better incorporated in the organic phase when the nanoprecipitation process is applied and, indeed, this has proved to be a specific property of the nanoprecipitation process.

With regards to the optimum choice of block copolymer emulsifier, observations relating to the stability of dispersions on completion of their preparation are listed in Table 10.2. Stable milky suspensions could be prepared only with those copolymers having a PEO chain of molar mass between 750 and 2000 g mol^{-1} ($17 < n < 45$), while decreasing the length of the PEO block caused a partial flocculation, which most often occurred immediately. Hence, the block copolymer PCL(10)-b-PEO(17) was close to the limit, as it allowed the complete emulsification of PCL, although a slow coagulation did occur following the preparation of emulsions. The limit of the PEO block length was estimated at 1000 g mol^{-1} ($n = 22$); any partial flocculation that took place upon using too-short PEO block lengths generally occurred immediately after emulsification.

The length of the PCL block appeared to be less important. Although it was clear that a too-small hydrophobic portion would be disastrous with respect to stabilization of the suspensions (because such an emulsifier would not be adsorbed at the surface of the PCL particles), this situation was not encountered in the series investigated. Even PCL(10)-b-PEO(113), which is soluble in water, allowed an efficient stabilization. Finally, the HLB-value was found not to provide a satisfactory means of predicting stability. Despite the general trend being that a minimum HLB was required, this limit was not well defined; for example, an efficient stabilization was not ensured by PCL(5)-b-PEO(17) of HLB 11.2, but *was* ensured by PCL(40)-b-PEO(113) of HLB 10.3. It is likely that block copolymers with a longer PCL chain would work equally well.

10.3.1.2 Particle Sizes

In the case of stable suspensions, the sizes of the particles have been measured using DLS. For a given recipe ([PCL] = 0.4%, [PCL-b-PEO]/[PCL] = 1/2), the particle size was found to depend less on the type of copolymer emulsifier (Table 10.2). Although the particle sizes were between 200 and 300 nm in all instances, the size was better controlled by the emulsification process than by the emulsifier content. The one formulation parameter that does allow control of the particle size in a large domain is the PCL concentration in the organic phase (acetone) [11].

The choice and concentration of emulsifier allowed control of the particle size across the narrow range available. It should be noted here that the behavior of the most hydrophilic block copolymer PCL(10)-b-PEO(113), which is water-soluble, differed from the four other polymers of the series (PCL(m)-b-PEO(113)) (with $m = 20$ to 40), all of which are insoluble in water. Thus, water solubility appears to be a relevant parameter with regards to the control of particle size. In the case of the water-soluble copolymer PCL(10)-b-PEO(113), increasing the polymeric surfactant content in the formulation caused the particle size to decrease (Figure 10.1). This is an expected trend that is readily observed with classical surfactants, but not with copolymers which are water-insoluble. No comparison can be made at this point with classical surfactants, because the latter with low HLB-values will stabilize W/O, but not O/W emulsions.

As a final important observation, emulsifiers that were water-insoluble were able to stabilize O/W emulsions quite well, despite this contradicting the Bancroft rule (water-soluble surfactant gives O/W emulsion; oil-soluble surfactant gives W/O emulsion). Hence, the nanoprecipitation process forced the emulsion type to be O/W, such that there was no chance for a W/O emulsion to form. However, the emulsions were stable and did not phase-invert. The same experiment with classical surfactants of low HLB would have produced an unstable O/W emulsion that would rapidly phase-invert into a W/O emulsion, possibly coexisting with a nonemulsified part of pure water.

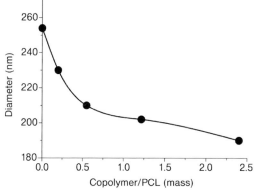

Figure 10.1 Particle size as a function of the amount of PCL(10)-b-PEO(113) copolymer emulsifier introduced in the aqueous phase. Although the emulsion prepared without emulsifier was not stable, its size could be measured immediately after preparation.

10.3.1.3 Stability of the Emulsions

Classically, the colloidal stability of dispersions has been evaluated by measuring the particle size against storage time over four months (Figure 10.2). Typically, the particle size did not vary significantly over this period at 4, 20 and 40 °C. Emulsions that were stable immediately after their preparation showed excellent long-term stability. However, in the case of unstable emulsions prepared with unsuitable block copolymer emulsifiers, coagulation occurred very rapidly after emulsification. No significant differences in emulsion stability were identified between the block copolymers PCL(x)-b-PEO(113) with $10 < x < 40$.

All of the emulsions prepared showed excellent long-term stability. However, in order to identify any differences in stabilization efficiency between the block

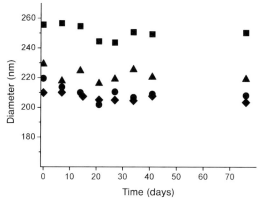

Figure 10.2 Assessment of colloidal stability: Particle size plotted against storage time at 20 °C. $n(PEO) = 113$; $m(PCL) = 10$ (■); 25 (▲); 30 (●); and 40 (♦).

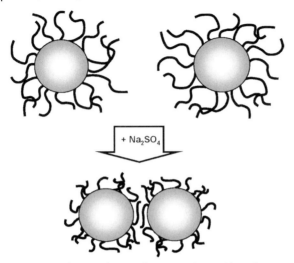

Figure 10.3 The PEO chains collapse onto the particle surface as sodium sulfate is added; the steric stabilization decreases and the colloidal suspension undergoes coagulation. Rapid coagulation regime is reached when the thickness of the PEO layer is less than the range of the attractive forces between the particles.

copolymer emulsifiers, a more severe stability test was employed. Here, stability was assessed by monitoring the rate of destabilization following the addition of salt (Na_2SO_4); the latter caused flocculation (coagulation) by collapsing the adsorbed polymer layer onto the particle surface (Figure 10.3). The presence of sodium sulfate in the aqueous solution causes 'deswelling' of the PEO coils that leads to coil collapse. PEO is present under good solvent conditions in pure water, whereas Θ-solvent conditions are achieved in 0.5 M Na_2SO_4.

The coagulation rate was measured by recording the absorbance of dispersions as a function of time at different concentrations of Na_2SO_4. This classical method of colloid chemistry [48] is analogous to an accelerated aging experiment. Initially, each of the tested emulsions was stable in pure water, but as Na_2SO_4 was added the initially slow coagulation rate was gradually increased as a function of the amount added. This was termed the 'slow coagulation regime'. Above a salt concentration referred to as the 'critical coagulation concentration' (CCC), the coagulation rate was maximal and no longer dependent on the amount of electrolyte present; this was the 'rapid coagulation regime'. For the slow coagulation regime, the coagulation rate was expressed as the ratio of rapid coagulation rate to actual coagulation rate, termed the 'stability ratio', W [48]. The CCC was read on the log-log plot of W against electrolyte concentration (see Figure 10.4).

The CCC is a measurement of colloidal stability: a very stable emulsion can only be coagulated with the help of a very large concentration of salt, and the CCC is therefore large. The observed CCC values were similar (~190 g l^{-1}) for all polymeric emulsifiers having a PEO chain of 5000 g mol^{-1}, whatever the length of the PCL chain. However, the CCC was slightly lower for a copolymer having 10 PCL units. Lastly, the CCC was

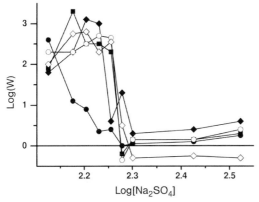

Figure 10.4 Dependence of the stability ratio W on electrolyte concentration for emulsions prepared with the block copolymer introduced in the organic phase. ● PCL(10)-b-PEO(113); ■ PCL(20)-b-PEO(113); ◆ PCL(25)-b-PEO(113); ○ PCL(30)-b-PEO(113); ◇ PCL(40)-b-PEO(113).

significantly lower for formulations where the copolymer emulsifier has been introduced in the aqueous phase. Such difference was also noted for the amounts of block copolymer adsorbed at the surface of particles, with adsorption being stronger when the copolymer was introduced in the organic phase. These findings show that the emulsification process is important with regards to the final stability of the emulsion, and leads to two conclusions: (i) that the polymeric emulsifier should preferably be introduced in the dispersed phase; and (ii) the block copolymer PCL(10)-b-PEO(113) which is soluble in water does not stabilize the emulsion with the same efficiency as block copolymers having larger PCL contents.

10.3.2
Emulsions of Various Oils by Spontaneous Emulsification

As the process of spontaneous emulsification works for both solid and liquid oils [11], it is not limited to PCL and can be used for the emulsification of liquid oils on the same grounds. It is, nevertheless, important that the hydrophobic block makes favorable interactions with the oil, as weak interactions would result in low adsorption and poor stabilization of the interface. Therefore, the emulsification properties are expected to depend on the type of oil through swelling of the PCL block by the oil. It is also expected that oils which are good solvents of PCL might be stabilized as emulsions by PCL-b-PEO emulsifiers. With regards to the emulsification of PCL, the latter is under Θ-solvent conditions in the PCL core of the emulsion droplets. Several different oils have been investigated to verify this point, including those used in pharmaceutical and cosmetic emulsions, such as mineral oil, silicone oil, isostearyl isostearate and the triglyceride Miglyol 812.

Stable emulsions of mineral oil and silicone oil could not be prepared with the PCL-b-PEO copolymers, as either the emulsification failed or the emulsions coagulated rapidly. In all trials with mineral oil the main part of the oil was found to float at the surface of the aqueous phase, which contained a dilute emulsion of the remaining part of the oil. Emulsification also failed with silicone oil, although the results were slightly different; typically, the emulsion creamed very quickly, leaving an almost clear aqueous phase at the bottom of the sample. It was presumed that emulsions of silicone oil could be formed and would coagulate rapidly, accelerating the creaming; however, the coalescence was much slower because the silicone oil was quite viscous. As a result, there was a long interval before the pure silicone oil phase became separated on top of the creamed emulsion. The same phenomenon was also expected with the mineral oil, but coalescence occurred rapidly because the light mineral oil was very fluid. It appears that cases of failure result from a successful emulsification followed by rapid coagulation and coalescence.

The emulsification was successful with isostearyl isostearate and Miglyol 812. Thus, stable emulsions could be prepared with the PCL(10)-*b*-PEO(17), PCL(10)-*b*-PEO(113) copolymers. The average diameters were between 200 and 400 nm, in the same range as for emulsions of PCL. The copolymer PCL(10)-*b*-PEO(8) with a short hydrophilic block did not stabilize the emulsions. As with the PCL emulsions, a minimum length of the PEO block is necessary, and the limit is slightly shorter with the liquid oils. The results of the investigations are summarized in Table 10.3.

Table 10.3 Properties of aqueous emulsions of various oils stabilized with the PCL-*b*-PEO block copolymers incorporated in the organic phase. The recipe was: [oil] = 0.4%, [PCL-*b*-PEO]/[oil] = 1/2.

Oil	Copolymer	Emulsification	Particle diameter (nm)	Stability
PCL	PCL(10)-*b*-PEO(8)	Failed	–	Unstable
	PCL(10)-*b*-PEO(17)	Partial	290	Slow coagulation
	PCL(10)-*b*-PEO(113)	Successful	256	Stable
Mineral oil	PCL(10)-*b*-PEO(8)	Failed	–	Unstable
	PCL(10)-*b*-PEO(17)	Failed	–	Unstable
	PCL(10)-*b*-PEO(113)	Failed	–	Unstable
Silicone oil	PCL(10)-*b*-PEO(8)	Failed	–	Unstable
	PCL(10)-*b*-PEO(17)	Successful	500	Slow coagulation and creaming
	PCL(10)-*b*-PEO(113)	Successful	370	Slow coagulation and creaming
Isostearyl isostearate	PCL(10)-*b*-PEO(8)	Failed	–	Unstable
	PCL(10)-*b*-PEO(17)	Successful	410	Slow coagulation and creaming
	PCL(10)-*b*-PEO(113)	Successful	290	Stable
Miglyol 812	PCL(10)-*b*-PEO(8)	Failed	–	Unstable
	PCL(10)-*b*-PEO(17)	Successful	310	Stable
	PCL(10)-*b*-PEO(113)	Successful	280	Stable

10.4 Conclusions

Block copolymer emulsifiers of the PCL-*b*-PEO type were found to be efficient for the preparation and stabilization of aqueous emulsions of PCL and various oils. Such block copolymers are readily available as their synthesis is quite simple and could easily be scaled up. Compared to the well-known ethylene oxide/propylene oxide triblock copolymers, the main benefit of using PCL-*b*-PEO block copolymers is the biodegradability of the PCL block.

Systematic studies as a function of block length have shown that a sufficiently long PEO block is required, typically $1000\,\mathrm{g\,mol^{-1}}$ ($n=22$). The copolymers that show negative results do not allow the emulsification of PCL by the spontaneous emulsification process, or they give rise to unstable suspensions that coagulate and settle very quickly. Even when the copolymer is soluble in water it is better introduced in the organic phase, against the Bancroft rule. Given a PEO block length, a block copolymer with a sufficiently short PCL block to be water-soluble provides a lesser steric stabilization of the colloidal suspensions. The typical particle sizes are 200 to 300 nm for all copolymer emulsifiers, with $n=45$ or 113. These suspensions are stable during several months of storage and offer good resistance to coagulation by electrolytes. Various liquid oils can be emulsified in the same way; only those oils that cause the PCL block to swell produced stable emulsions, however, and the emulsions of mineral oil and silicone oil were unstable. The behavior of block copolymers appears similar with regards to PCL and liquid oils. The shortest PEO block length to allow stabilization of the emulsions was $n\approx 20$ in every case. It should be noted that the same limit with regards to PEO block length was found for the stabilization of aqueous suspensions of mineral particles (calcium carbonate) by PEO diphosphonate dispersants [49, 50].

The successful use of those polymer surfactants that are insoluble in water is also of great interest, as all emulsifier macromolecules occur at the surface of the particles. Consequently, the efficiency is maximal and there is no residual surfactant left in the aqueous phase. The surfactant molecules in solution in the aqueous phase are bioavailable, and are responsible for possible biological effects such as hemolysis and irritancy.

The present copolymers are good candidates for use in the formulation of emulsions made not only from PCL but also from other polymers and various oils. Although, the present studies were performed only with the spontaneous emulsification process, it is likely that such polymeric emulsifiers are well-suited for their implementation in different emulsification processes. The strong adsorption behavior is a clear advantage, and the absence of any bioavailable surfactant in the aqueous phase is a direct consequence of this. Nonadsorbed polymer which remain in the aqueous phase is also problematic as its presence increases the viscosity of the continuous phase of the emulsion. For example, emulsions prepared with high concentrations of poly(vinyl alcohol) as emulsifier are quite viscous, the consequence being a larger droplet size for a given emulsification process [4, 5, 51]. Even water-soluble block copolymers would not greatly increase the viscosity of the aqueous phase because they self-associate as micelles in solution.

References

1 Piirma, I. (1992) *Polymeric Surfactants*, Marcel Dekker, New York.
2 Tadros, T.F. (2001) Polymeric surfactants, in *Handbook of Applied Surface and Colloid Chemistry* (ed. K. Holmberg), Wiley, New York. Chapter 16, pp. 373–384.
3 Riess, G. (2003) *Prog. Polym. Sci.*, **28**, 1107–1170.
4 Murakami, H., Kawashima, Y., Niwa, T., Hino, T., Takeuchi, H. and Kobayashi, M. (1997) *Int. J. Pharm.*, **149**, 43–49.
5 Guinebretière, S., Briançon, S., Lieto, J., Mayer, C. and Fessi, H. (2002) *Drug Develop. Res.*, **57**, 18–33.
6 Davies, J.T. and Haydon, D.A. (1957) Proceedings, 2nd International Congress on Surface Activity, Volume 1, 417–425.
7 Davies, J.T. and Rideal, E.K. (1963) *Interfacial Phenomena*, 2nd edition, Academic Press, pp. 360–367.
8 Ruschak, K.J. and Miller, C.A. (1972) *Ind. Eng. Chem. Fundam.*, **11**, 534–540.
9 Miller, C.A. (1988) *Colloids Surfaces*, **29**, 89–102.
10 Vitale, S.A. and Katz, J.L. (2003) *Langmuir*, **19**, 4105–4110.
11 Ganachaud, F. and Katz, J.L. (2005) *Chem. Phys. Chem.*, **6**, 1–8.
12 (a) Montasser, I., Fessi, H., Briançon, S. and Lieto, J. Patent FR 2806005 (2000); (b) Montasser, I. Fessi, H., Briançon, S. and Lieto, J. US Patent 2003059473 (2001); (c) Montasser, I. Fessi, H. Briançon, S. and Lieto, J. WO 0168235 (2001).
13 Montasser I. Briançon S. Lieto J. and Fessi H. 2000 *J. Pharm. Belg.*, **55**, 155–167.
14 Moinard-Checot, D., Chevalier, Y., Briançon, S., Fessi, H. and Guinebretière, S. (2006) *J. Nanosci. Nanotechnol.*, **6**, 2664–2681.
15 Matijević E. 1993 *Chem. Mater.*, **5**, 412–426.
16 Edlund U. and Albertsson A.-C. (2002) *Adv. Polym. Sci.*, **157**, 67–112.
17 Albertsson A.-C. and Varma I.K. (2003) *Biomacromolecules*, **4**, 1466–1486.
18 Wang S. Cai Q. and Bei J. (2003) *Macromol. Symp.*, **195**, 263–268.
19 Couvreur, P., Grislain, L., Lenaerts, V., Brasseur, F., Guiot, P., and Bernachi A. 1986 in *Polymeric Nanoparticles and Microspheres* (eds. P. Guiot and P. Couvreur), CRC Press.
20 Arshady R. (1999) (ed.), *Microspheres, Microcapsules & Liposomes*, Citus Books, London, Volume 1.
21 Benita S. (1996) (ed.), *Microencapsulation. Methods and Industrial Applications*, Marcel Dekker, New York.
22 European Pharmacopoeia, 5th edition, version 5.5.2005.
23 Gref, R., Domb, A., Quellec, P., Blunk, T., Müller, R.H., Verbavatz, J.M. and Langer, R. (1995) *Adv. Drug Delivery Rev.*, **16**. 215–233.
24 Holmberg, K., Tiberg, F., Malmsten, M. and Brink, C. (1997) *Colloids Surfaces A*, **123–124**, 297–306.
25 Shi, B., Fang, C., You, M.X., Zhang, Y., Fu, S. and Pei, Y.Y. (2005) *Colloid Polym. Sci.*, **283**, 954–967.
26 Heald, C.R., Stolnik, S., Kujawinski, K.S., De Matteis, C., Garnett, M.C., Illum, L., Davis, S.S., Purkiss, S.C., Barlow, R.J. and Gellert, P.R. (2002) *Langmuir*, **18**, 3669–3675.
27 Riley, T., Heald, C.R., Stolnik, S., Garnett, M.C., Illum, L., Davis, S.S., King, S.M., Heenan, R.K., Purkiss, S.C., Barlow, R.J., Gellert, P.R., and Washington, C. (2003) *Langmuir*, **19**, 8428–8435.
28 Vangeyte, P., Gautier, S. and Jérome, R. (2004) *Colloids Surfaces A*, **242**, 203–211.
29 Letchford, K., Zastre, J., Liggins, R. and Burt, H. (2004) *Colloids Surfaces B*, **35**, 81–91.
30 Allen, C., Yu, Y., Maysinger, D. and Eisenberg, A. (1998) *Bioconjugate Chem.*, **9**, 564–572.
31 Allen, C., Han, J., Yu, Y., Maysinger, D. and Eisenberg A. (2000) *J. Control. Release*, **63**, 275–286.
32 Rösler, A., Vandermeulen, G.W.M. and Klok, H.-A. (2001) *Adv. Drug Delivery Rev.*, **53**, 95–108.

33 Ravi Kumar, M.N.V., Kumar, N., Domb, A.J. and Arora, M. (2002) *Adv. Polym. Sci.*, **160**, 45–117.

34 Soo, P.L., Luo, L., Maysinger, D. and Einsenberg, A. (2002) *Langmuir*, **18**, 9996–10004.

35 Hu, Y., Jiang, X., Ding, Y., Zhang, L., Yang, C., Zhang, J., Chen, J. and Yang, Y. (2003) *Biomaterials*, **24**, 2395–2404.

36 Ravenelle, F. and Marchessault, R.H. (2003) *Biomacromolecules*, **4**, 856–858.

37 Ahmed, F. and Discher, D.E. (2004) *J. Control. Release*, **96**, 37–53.

38 Shuai, X., Ai, H., Nasongkla, N., Kim, S. and Gao, J. (2004) *J. Control. Release*, **98**, 415–426.

39 Jeong, Y.-I., Kang, M.-K., Sun, H.-S., Kang, S.-S., Kim, H.-W., Moon, K.-S., Lee, K.-J., Kim, S.-H. and Jung, S. (2004) *Int. J. Pharm.*, **273**, 95–107.

40 Mahmud, A. and Lavasanifar A. (2005) *Colloids Surfaces B*, **45**, 82–89.

41 Meier, M.A.R., Aerts, S.N.H., Staal, B.B.P., Rasa, M. and Schubert, U.S. 2005 *Macromol. Rapid Commun.*, **26**, 1918–1924.

42 Wang, F., Bronich, T.K., Kabanov, A.V., Rauh, R.D. and Roovers, J. (2005) *Bioconjugate Chem.*, **16**, 397–405.

43 Jacquier, V., Miola, C., Llauro, M.-F., Monnet, C. and Hamaide, Th. (1996) *Macromol. Chem. Phys.*, **197**, 1311–1324.

44 Delaite-Miola, C., Hamaide, Th. and Spitz, R. (1999) *Macromol. Chem. Phys.*, **200**, 1771–1778.

45 Pantiru, M., Iojoiu, C., Hamaide, Th. and Delolme, F. (2004) *Polym. Int.*, **53**, 506–514.

46 Iojoiu, C., Hamaide, Th., Harabagiu, V. and Simionescu, B.C. (2004) *J. Polym. Sci. Part A. Polym. Chem.*, **42**, 689–700.

47 Gohy, J.-F., (2005) *Adv. Polym. Sci.*, **190**, 65–136.

48 Overbeek, J.Th.G., (1952) in *Colloid Science* (ed. H. R. Kruyt), Volume I, Chapter VII, Elsevier, Amsterdam, pp. 278–301.

49 Mosquet, M., Chevalier, Y., Le Perchec, P. and Guicquero, J.-P. (1997) *New J. Chem.*, **21**, 143–145.

50 Mosquet, M., Chevalier, Y., Brunel, S., Guicquero, J.-P. and Le Perchec, P. (1997) *J. Appl. Polym. Sci.*, **65**, 2545–2555.

51 Loxley, A. and Vincent, B. (1998) *J. Colloid Interface Sci.*, **208**, 49–62.

11
Routes Towards the Synthesis of Waterborne Acrylic/Clay Nanocomposites

Gabriela Diaconu, Maria Paulis, and Jose R. Leiza

11.1
Introduction

Polymer/clay nanocomposites exhibit often outstanding mechanical, thermal, barrier and flammability properties [1–3] that have made them very attractive in the development of a new class of materials, the so-called 'nanocomposites'. The enhanced properties of nanocomposite materials, as compared with conventional materials, are derived from the small size of the clays and the high aspect ratio of the clay platelets that, upon exfoliation, may render polymeric materials with enormous surface interaction areas between the polymer and the clay. In order to synthesize these materials it is necessary to disperse the clay mineral homogeneously into the host polymer matrix (hence, compatibility between polymer and clay is a key parameter), and furthermore to delaminate (exfoliate) the clay platelets which, due to electrostatic interactions, are naturally forming stacks. A variety of different routes and alternatives have been proposed to achieve this goal, including polymer/pre-polymer intercalation from solution, melt intercalation and *in situ* polymerization [4]. In this chapter, only waterborne polymer/clay nanocomposites are considered, and therefore only routes towards the synthesis of such materials by means of *in situ* polymerization methods are discussed.

Here, montmorillonite (MMT) clays were used to produce waterborne polymer/clay nanocomposites. Natural sodium montmorillonite (Na-MMT) is hydrophilic and it is well known [5] that, when dispersed in water at concentrations below 0.5 wt%, clay platelets are completely exfoliated (the individual platelets are randomly dispersed in the continuous phase). It has been shown recently [6], by using small angle X-ray scattering (SAXS) that at higher concentrations (1–1.5 wt%) Na-MMT also presented a completely exfoliated structure in water with clay platelets at average distances above 150 nm. However at concentrations above 3 wt% a certain amount of platelets tend to interact, most likely in pairs separated at distances of 14–16 nm. Therefore, *ab initio* emulsion polymerization (either nucleating the particles by micellar nucleation or, in the absence of micelles, by homogeneous nucleation) in the

Emulsion Science and Technology. Edited by Tharwat F. Tadros
Copyright © 2009 WILEY-VCH Verlag GmbH & Co. KGaA, Weinheim
ISBN: 978-3-527-32525-2

presence of pristine Na-MMT should allow the production of an exfoliated waterborne polymer/clay nanocomposite, provided that the clay does not jeopardize the polymerization in the particles nor the stability of the polymer particles. In principle, this route is both precise (Na-MMT does not require any modification) and simple (typical batch, semibatch and continuous reactors and large-sized industrial-scale reactors can be used). It also explains the large number of studies using this route to produce polymer/MMT nanocomposites [7–12]. However, aspects such as the role of the clay on the kinetics and microstructural properties of the polymers produced under these conditions have not been carefully addressed. In addition, the production of stable and coagulum-free waterborne nanocomposites with high solids contents (>50 wt%) and clay loadings of 3–4% based on the polymer has not been demonstrated. The rheology of these latexes, and their comparison with counterparts produced by blending processes, also remains to be investigated.

The main advantage of the *emulsion polymerization route* is that clay platelets are spontaneously exfoliated in the water phase along the polymerization, and it might be assumed that they remain as that in the final latex (there is no direct proof of that as far as we know). However, the compatibility of the clay and the polymer upon film formation, and the morphology achieved, will depend on the nature of the polymer (and surfactants) used in the polymerization and the film formation process itself, and differences have been reported for similar systems [13–16].

Another route to produce waterborne polymer/clay nanocomposites, at the same time increasing interaction/anchorage between the polymer and the clay platelets, is that of *miniemulsion polymerization*. This has been reported to be suitable for incorporating water-insoluble compounds to the reaction loci [17], and therefore it should be feasible also for incorporating organically modified clays into the polymer particles. For this purpose, the clay minerals must be modified to render them sufficiently hydrophobic so as to increase their compatibility with the monomer/polymer mixture and improve their dispersion in the polymer particles. In order to hydrophobize clay minerals, several different alternatives have been presented. Typically, organic compounds such as cationic surfactants (including primary, secondary, tertiary and quaternary ammonium ions or phosphonates) have been used to render the clay surface hydrophobic by a simple ion-exchange reaction [18–20]. Commercial organically modified clays produced using this method are currently available. Alternatively, the organic cations can bear a reactive group in one end (double bond, initiator fragment or a capped living radical that might be activated by appropriate conditions) [21–26] that might favor polymerization inside the clay platelets. This would facilitate exfoliation of the clays, compatibility of the clay with the polymer matrix, and also enhance the final dispersion of the clay in the film.

Clearly, for the above-mentioned process to be reliable, droplet nucleation should be the main nucleation mechanism, ultimately rendering a waterborne polymer/clay nanocomposite containing particles with exfoliated/intercalated clay platelets. Therefore, the size of the monomer miniemulsion droplets and that of the modified clays should be adequate to carry out the polymerization under such conditions; in other words, the size of the monomer droplets needs to be sufficiently large so as to accommodate the clay inside the monomer droplet, or the size of the clay small enough to fit into the droplets. This is not always achieved, however, and intermediate situations might arise. In any case, other issues such as interaction of the surfactant with the clay, the stability of the monomer miniemulsions under the presence of clay, the amount of clay used or the occurrence of secondary nucleation mechanisms, might jeopardize the success of synthesizing polymer/clay nanocomposites by miniemulsion polymerization.

The better compatibility of the polymer matrix with the clay can be considered as the main advantage of the miniemulsion polymerization route. As a consequence, the clay platelets will be within the polymer particles and hence, during film formation, they will be more constrained against migration and reorganization; this would ensure a more homogeneous dispersion of the clay platelets in the film. If this was achieved, the route might lead to nanocomposites with better mechanical and barrier properties than nanocomposites produced by emulsion polymerization; however, this remains to be demonstrated.

In the present studies we have explored both routes for the production of waterborne polymer/clay nanocomposite coatings made from an acrylic polymer (methyl methacrylate (MMA)/butyl acrylate (BA) = 50/50, w/w). Pristine Na-MMT was used for the emulsion polymerization, while for the miniemulsion route two organically modified MMTs were employed. On the one hand, commercially available organically modified MMT (Cloisite 30B; Southern Clays, USA) was used, while on the other hand a reactive organic compound was synthesized and ionically exchanged with Na-MMT to produce an organically modified clay. The reactive group was a living cationic oligomer (methyl methacrylate/styrene = 90/10, w/w) produced with a nitroxide agent (SG1) and a cationic initiator (AIBA) in aqueous solution.

11.2
Experimental

11.2.1
Materials

The monomers – MMA, BA and styrene (S) were all obtained from Quimidroga (Spain) and used as received, without further purification. Sodium lauryl sulfate (SLS; Aldrich) was used as the emulsifier both in emulsion and miniemulsion polymerization processes. Stearyl acrylate (SA; Aldrich) was used as the costabilizer (hydrophobe) in the miniemulsion polymerization process. Potassium persulfate (KPS; Aldrich) was used as a thermal initiator for the emulsion polymerization. *Tert*-butyl hydroxyperoxide (TBHP; Panreac) and ascorbic acid (AsAc; Aldrich) were used as redox initiators for the miniemulsion polymerization.

Na-MMT and Cloisite 30B were provided by Southern Clay Products Inc. (Texas, USA). Na-MMT has a cationic exchange capacity (CEC) of 92.6 mEq. per 100 g clay. Cloisite 30B (C30B) is a natural MMT modified with a quaternary ammonium salt: methyl tallow bis-2-hydroxyethyl quaternary ammonium with a cationic exchange capacity of 90 mEq. per 100 g clay. X-ray diffraction (XRD) analysis showed interlayer spaces of 1.15 and 1.85 nm, respectively.

An organically modified Na-MMT clay was synthesized in our laboratory by exchanging the naturally occurring Na^+ cations of Na-MMT clay with a (living oligomer) that contains one quaternary ammonium group in one end and one capped living radical (nitroxide radical) in the other end. The cationic macroinitiator was synthesized in aqueous phase by polymerizing MMA and S (90:10 molar ratio, $[M]_T = 4.3 \times 10^{-2}$ M) in the presence of AIBA (2,2′-azo-bis(2-amidinopropane) dihydrochloride; Aldrich) and SG1 (N-tert-butyl-N-(1-diethylphosphono-2,2-dimethylpropyl) nitroxide; kindly supplied by ARKEMA) at a molar ratio SG1:AIBA = 1.1 at 90 °C. It should be noted that the amount of monomer used was below the solubility of both monomers [27]. The number average molecular weight of the oligomer measured by gel-permeation chromatography (GPC) was 2660 g mol^{-1}, which corresponds to a chemical structure shown in Scheme 11.1, with an average amount of 23 monomer units (MMA/S = 90/10).

The cationic exchange was carried out under the following conditions: Na-MMT (10.0 g) was dispersed in water (200 g) and the cationic macroinitiator (12 mEq.) was added to the clay solution and kept for 3 h under mechanical stirring (700 rpm) at room temperature. The dispersion was precipitated and washed several times with water, after which a powder was recovered and used in miniemulsion polymerizations (as described below). The wide-angle X-ray diffraction (WAXD) patterns of both organically modified clays used in these studies are presented in Figure 11.1, and showed that the basal space increased from the d(001) = 1.15 nm of the pristine clay to d(001) = 1.85 nm for the Cloisite 30B and d(001) = 1.53 nm for the macroinitiator MMT. This indicated that the reactive species were successfully intercalated in the interlaminar space of the clay platelets. A thermogravimetric analysis of the organically modified clays showed that the onset decomposition temperatures were 174 °C [28] for Cloisite 30B and 238 °C for the macroinitiator modified MMT, while the residual masses were 73% [28] and 34.5%, respectively. This indicated that the macroinitiator-modified Na-MMT clay was exchanged at 77% of CEC, whereas Cloisite 30B was fully exchanged.

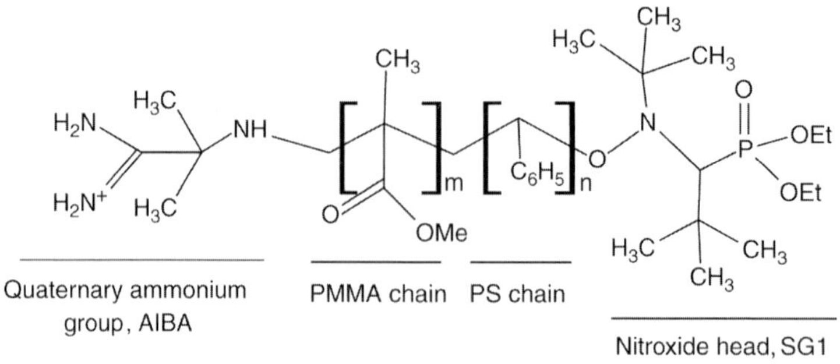

Scheme 11.1 The chemical structure of the cationic macroinitiator used to modify the Na-montmorillonite clay (MMA/S: 90/10).

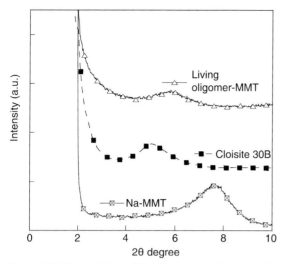

Figure 11.1 Wide-angle X-ray diffraction patterns of the Na-MMT and the organically modified clays.

11.2.2
Synthesis of Waterborne (MMA-BA)/MMT Nanocomposites by Emulsion Polymerization

All reactions were carried out in a 1 liter stirred-tank reactor equipped with a jacket, reflux condenser, sampling device, nitrogen inlet, two feeding inlets and a stainless steel anchor stirrer equipped with two blade impellers rotating at 250 rpm. The reactor temperature, monomer and initiator feed flow rates were controlled by an automatic control system (Camile TG, Biotage). Waterborne poly (MMA-co-BA)/MMT nanocomposite latexes with 30 wt% solids content and 3 wt% Na-MMT based on monomers were synthesized by seeded semibatch emulsion polymerization.

The formulation used for the preparation of waterborne nanocomposites by seeded semibatch emulsion polymerization (PCN1) is shown in Table 11.1. The typical procedure for the synthesis of these nanocomposite latexes was as follows: first, an aqueous dispersion containing all Na-MMT of the recipe was stirred at 250 rpm and heated at 75 °C for 14 h. The monomers, emulsifier and initiator were then added to this aqueous phase, and the seed with 20 wt% solids content (MMA/BA = 50/50 wt%) was synthesized batchwise at 75 °C for 1 h. The remaining amount of comonomer and an aqueous solution of KPS were fed continuously into the reactor in two different streams for 3 h. The reactor contents were then allowed to polymerize for another hour at 75 °C. The final solids content of the latex was 30 wt%, and the content of Na-MMT was 3 wt% based on monomer. For the sake of comparison, a blank emulsion latex (BE) was synthesized following the above-described synthesis procedure, but without Na-MMT.

It is worth noting that the same nanocomposite latexes can be produced without pretreatment of the Na-MMT aqueous solution at 75 °C for 14 h [6, 26].

Table 11.1 Formulation used for the seeded semibatch emulsion copolymerization of MMA/BA in the presence of Na-MMT. Solids content: 30 wt%.

Reagent	Seed initial charge[a]	Feeding streams[b]	
		F1	F2
MMA (g)	67.5	67.5	–
BA (g)	67.5	67.5	–
SLS (g)	5.4	–	–
Na-MMT (g)	8.1	–	–
KPS (g)	0.675	–	0.675
H$_2$O (g)	550	–	30

[a] Seed prepared in batch, at 75 °C for 1 h.
[b] Feeding time 3 h.

11.2.3
Synthesis of Waterborne (MMA-BA)/MMT Nanocomposites by Miniemulsion Polymerization

The miniemulsions were prepared as follows. The oil phase was prepared by dissolving the costabilizer (SA) and the organically modified clay (O-MMT clay) in the monomers (MMA and BA; 50/50 wt%). This mixture was stirred for 15 min at 1000 rpm using a magnetic stirrer. The aqueous phase was prepared by dissolving the emulsifier in water. Both phases (aqueous and oil phase) were brought together and mixed for 15 min at 1000 rpm. The dispersion was sonicated using a Branson Sonifier 450 (operating at 8-output control and 80% duty cycle) for 15 min in an ice bath and under magnetic stirring. The pH of the miniemulsion was adjusted to 10 using an aqueous solution of ammonia and boric acid. (Note: if the pH of the dispersion was acidic, then unstable latexes and coagulum were formed upon polymerization.) The nanocomposite latexes with a solids content of 30 wt% were synthesized batchwise. The formulation used to prepare waterborne nanocomposite latexes (PCN2 and PCN3) by batch miniemulsion polymerization is presented in Table 11.2. The typical procedure for the synthesis of nanocomposite latex PCN2 was as follows (for PCN3 the procedure was slightly different, and is described below): the miniemulsion prepared as above was charged in the reactor (the set-up described for the emulsion polymerizations was also used here) and the temperature raised to 70 °C under nitrogen flow (15 ml min^{-1}). On reaching the reaction temperature, aqueous solutions of the redox components (TBHP/AsAc = 2 : 1 molar ratio) were fed to the reactor in two separate streams for 3 h. The reactor contents were then allowed to polymerize for another hour at 70 °C. The miniemulsion polymerization reaction carried out in presence of the macroinitiator modified Na-MMT clay (PCN3) was performed at 90 °C. The miniemulsion, when stabilized with 2 wt% of emulsifier (SLS) based on monomer, was kept at 90 °C for 4 h under nitrogen flow (15 ml min^{-1}), without any further initiator addition. The purpose of this stage was to

Table 11.2 Formulation used for the batch miniemulsion copolymerization of MMA/BA in presence of organically modified clay.[a] Solids content: 30 wt%.

Reagent	Initial charge	Feeding streams[b]	
		F1	F2
MMA (g)	112.5	–	–
BA (g)	112.5	–	–
SA (g)	6.75	–	–
SLS (g)	4.5; 9[c]	–	–
O-MMT clay (g)[d]	9	–	–
TBHP (g)	–	1.125	–
AsAc (g)	–	–	0.5625
H$_2$O (g)	425	50	50

[a]Polymerization with the macroinitiator MMT clay was carried out at a lower scale (1/7).
[b]Feeding time 3 h.
[c]6.75 for macroinitiator MMT and 9 for C30B.
[d]O-MMT clay: Cloisite 30B or macroinitiator MMT.

initiate polymerization in the interlayer of the clays by extending the capped chains. The redox initiators system was then fed continuously into the reactor for 1 h. For the sake of comparison, a blank miniemulsion latex (BM) was synthesized following the procedure described for run PCN2, but without adding the organically modified clay.

A summary of all experiments carried out, together with an indication of the type of process employed and the percentages and types of clay used in each experiment is provided in Table 11.3.

11.2.4
Characterization and Measurements

Polymer particle size and monomer droplet size were measured using dynamic light scattering (DLS) with a Coulter N4 Plus instrument, in unimodal analysis. For this

Table 11.3 Summary of the nanocomposite latexes with 30 wt% solids content synthesized by means of emulsion and miniemulsion polymerization.

Run	Process	Type of clay	Clay content (wt%)[a]
BE	Emulsion/semibatch	–	–
PCN1	Emulsion/semibatch	Na–MMT	3
BM	Miniemulsion/batch	–	–
PCN2	Miniemulsion/batch	Cloisite 30B	4
PCN3	Miniemulsion/batch	Macroinitiator MMT	3

[a]Based on the total amount of the monomer in the recipe.

analysis, a fraction of the latex (or miniemulsion) was diluted with deionized water (saturated with monomers in the case of miniemulsion droplet size measurement). The reported particle size (droplet size) values represented an average of three repeated measurements. Particle size distribution of the final latexes was measured by means of a disc centrifuge photosedimentometer (Brookhaven BI-DCP). The extent of conversion was measured by gravimetric analysis.

The gel contents of the samples were measured via conventional Soxhlet extraction, using tetrahydrofuran (THF) as solvent. A glass-fiber disk was impregnated with a few drops of latex and the extraction carried out for 24 h under reflux conditions. The gel remained in the glass fiber, whereas the polymer soluble fraction was recovered from the THF solution. The amount of gel was calculated using the following equation:

$$\text{gel}(\%) = \frac{w_{gel} - (w_{total} x_{clay})}{w_{total} - (w_{total} x_{clay})} \times 100$$

where w_{gel} is the amount of insoluble polymer that remained in the glass fiber, w_{total} is the whole polymer sample, and x_{clay} is the fraction of clay content based on monomer in the formulation.

The molecular weight distribution (MWD) and average molecular weights ($\bar{M}_w$) of the soluble polymer fraction were determined using GPC with a size-exclusion chromatography (SEC) instrument consisting of a pump (Waters 510), three columns (Styragel HR2, HR4 and HR6) and a differential refractometer as detector. The soluble polymer fraction obtained by Soxhlet extraction in THF for 24 h under reflux conditions was concentrated, filtered (filter pore size 0.45 µm; Albert) and then analyzed by SEC. Polystyrene standards were used to calibrate the equipment, and the absolute molecular weights were calculated using the Mark–Houwink–Sakurada constants ($k_{MMA} = 14.3 \times 10^{-5}$ dl g^{-1}, $\alpha_{MMA} = 0.71$ and $k_{BA} = 12.3 \times 10^{-5}$ dl g^{-1}, $\alpha_{BA} = 0.7$).

WAXD analyses were performed using a Philips PW 1729 generator connected to a PW 1820 (Cu K$_\alpha$ radiation with $\lambda = 0.154056$ nm) at room temperature. The range of diffraction angles was $2\theta = 2$–$12°$ with a scanning rate of $0.02°$ per 3 s. The (001) basal spacing of the clay, d, was calculated using the Bragg equation: $n\lambda = 2d\sin\theta$. In order to perform the WAXD analysis, the films cast from the latex were rinsed several times with water to remove the SLS and to avoid its peaks at $7°$, $5°$ and $2.5°$ in the WAXD patterns.

Small-angle X-ray scattering experiments were carried out on the SAXS instrument constructed at the European Synchrotron Radiation Facility (ESRF) in Grenoble (France), using the Synchrotron Radiation of the beamline BM16. The beamline is equipped with a 2048 × 2048 (2 × 2 binned) pixels position-sensitive, two-dimensional marCCD detector with an active surface area of 165 mm × 165 mm. The measurements were made using a monochromatic X-ray at wavelength $\lambda = 0.726$ nm (energy $E = 17.068$ KeV) and at a sample-to-detector distance of 3.93 m. The samples were calibrated to the diffraction peaks of silver behenate.

The morphology of the nanocomposite films was studied using transmission electron microscopy (TEM), with two microscopes: a Hitachi 7000FA at 75 kV (UPV-EHU, San Sebastian) and a Jeol 1010 (Universitat de Barcelona, Serveis Cientificotècnics). The dried samples were cryosectioned with a Leica UltraCut FCS or Leica EM UC6 cryoultramicrotome, using a Diatome 35° diamond knife. The ultra-thin sections were placed on a 300-mesh formvar-coated copper grid and observed with TEM. For observation of the latex particles a negative staining was used: one drop of latex was placed on the formvar and carbon-coated copper grid. After cleaning off the excess latex, the grid was immersed in an aqueous uranyl acetate solution (2%) for 90 s. Any excess of solution was rinsed free and the grids were dried before TEM observations.

11.3
Results and Discussion

11.3.1
Waterborne Nanocomposites by Emulsion Polymerization

The polymerization rates for runs PCN1 and BE were similar (data not shown). For PCN1, the instantaneous conversion was not affected by the presence of clay, the processes evolved under rather starved conditions (>92% in both cases), and full conversions were achieved. The particle size of the seed was 60 nm in both cases, and the final particle sizes were 76 and 72 nm, respectively. In the feeding stage, the particle size increased linearly with conversion (the time evolutions of the particle size plots are not presented here), which indicated that no secondary nucleation had occurred during the monomer addition period. Chen et al. [29], in an *ab initio* emulsion polymerization of styrene in the presence of 1 wt% Na-MMT, showed the particle size for the run with Na-MMT to be smaller compared to the counterpart without Na-MMT at the same SLS concentration. In contrast to the findings of these authors, in the present studies the presence of clay had very little influence on the nucleation of the seed polymer particles, most likely because of the high concentration of surfactant employed during the seed polymerization ([SLS] = 36 mM) and consequently on the subsequent polymerization in semibatch.

The average molecular weight (M_w), polydispersity index (PDI) and gel content for BE and PCN1 latexes are listed in Table 11.4, while GPC traces of the latexes BE and PCN1 are shown in Figure 11.2.

Table 11.4 The M_w, PDI and the gel content of runs BE and PCN1.

Run	Mw (g mol^{-1})	PDI (M_w/M_n)	Gel content (%)
BE	1.08×10^6	10.9	<1
PCN1	8.93×10^5	12.2	<1

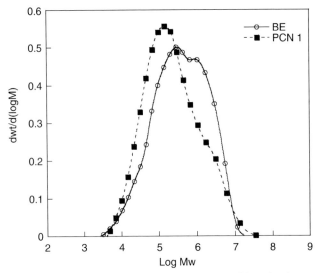

Figure 11.2 Gel-permeation chromatography traces of the sol polymer fraction of runs BE and PCN1.

Although the molecular weights were similar for both latexes, they were slightly higher for run BE without clay. However, as can be seen in the GPC traces (Figure 11.2), both showed a shoulder at higher molecular weights, which should be attributed to chain transfer to the polymer and to the high polymer concentrations in these starved semibatch polymerizations. However, the level of gel polymer was negligible (see Table 11.4), in agreement with polymerizations carried out under similar conditions by Elizalde et al. [30] and Gonzalez et al. [31], without clay. This was due to the higher reactivity of the MMA monomer and the lower tendency of MMA-terminated radicals towards coupling which would prevent network and gel formation.

The WAXD diffraction patterns of the film cast from the latex PCN1, and that of the Na-MMT clay, are shown in Figure 11.3. The characteristic peak of the Na-MMT powder clay appeared at $2\theta = 7.63°$, which corresponds to an interlayer space of $d = 1.15$ nm. The nanocomposite film PCN1 showed no such peak, which indicated that the basal spacing of the Na-MMT had been extended beyond the separation that could be detected with WAXD (≈ 4 nm); namely, that the ordered structure of the clay had been lost during the polymerization.

It is well known that TEM can provide useful information in a localized area on the morphology, structure and spatial distribution of the dispersed phase of nanocomposites. Typical TEM images for the nanocomposite film cast from the latex PCN1 are shown in Figure 11.4, at different magnifications. The individual layers, as well as zones with more than one clay layer (these represent the intercalated structures) that are well distributed into the polymer matrix can be distinguished. These TEM images confirm that a mixture of intercalated and exfoliated structure was obtained for

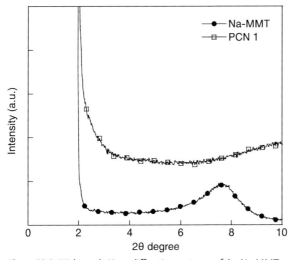

Figure 11.3 Wide-angle X-ray diffraction patterns of the Na-MMT clay and nanocomposite latex film PCN1.

nanocomposite PCN1, which is in agreement with the WAXD pattern. It should be noted that a similar degree of clay exfoliation was achieved when the reaction was carried out without any previous thermal pretreatment of the aqueous clay solution [6, 26].

11.3.2
Waterborne Nanocomposites by Miniemulsion Polymerization

As discussed in Section 11.1 waterborne polymer/clay nanocomposites can be also prepared by miniemulsion polymerization using hydrophobic MMT clay. In these studies, two different hydrophobic MMTs were used, namely a commercial organically

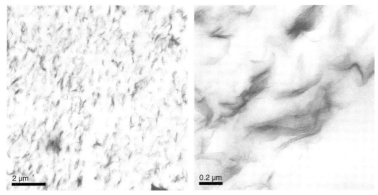

Figure 11.4 Transmission electron microscopy images of the nanocomposite films PCN1, at different magnifications.

modified MMT (Cloisite 30B) and a MMT modified with a cationic macroinitiator. A summary of the nanocomposites prepared by miniemulsion polymerization using different organically modified clays is provided in Table 11.3. The study aim was to incorporate the hydrophobic clay platelets (150–200 nm wide) into the monomer droplets, and to delaminate the clay by polymerization within the interlayer space. However, the surfactant concentration required to produce miniemulsion droplet sizes >300 nm (so as to fully engulf clay platelets in the droplets) yielded unstable miniemulsions. The minimum amount of surfactant (SLS) required to produce stable latexes with clay contents between 2 and 4 wt%, and with negligible amount of coagulum, was 4 wt%, based on the monomer. The stability of the latexes was also greatly improved when a high pH was used in the reaction medium, most likely because the edges-to-face agglomeration was avoided. Under these conditions, the miniemulsion droplet sizes were smaller than the size of the clay platelets, such that the clay could not be incorporated into the droplet (i.e. not engulfed). The location of the hydrophobic clay platelets after miniemulsification was not clear, although SAXS measurements of the clay dispersed in the monomer mixture, and of a miniemulsion containing clay (30% monomer dispersed phase and 3 wt% of clay based on monomer) may shed some light on this aspect (see Figure 11.5). From this figure it can be seen that the Cloisite 30B clay swelled in the presence of the MMA/BA mixture, from 1.85 nm (as calculated using the Bragg equation applied to the WAXD peak) for the Cloisite 30B to approximately 4 nm (calculated using the equation $d = 2\pi/q$, where q is the position of the SAXS peak).

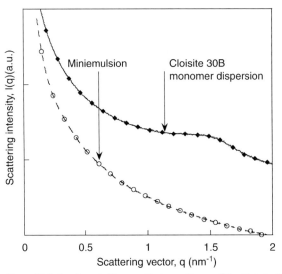

Figure 11.5 Small-angle X-ray scattering patterns of the Cloisite 30B monomer dispersion and miniemulsion containing Cloisite 30B. The amount of clay in both dispersions is approximately 1 wt%.

Table 11.5 Miniemulsion droplet size, final particle size and the final conversion of the miniemulsion polymerization reactions.

Run	Modified clay	Miniemulsion droplet size (nm)	Final particle size (nm)	Final conversion (%)
BM	–	108	93	>99
PCN2	Cloisite 30B	210	200	>99
PCN3[a]	Macroinitiator MMT	144	180	89

[a] A small amount of coagulum (1 wt%) was obtained at the end of the reaction.

After sonication and formation of the miniemulsion, however, the SAXS pattern showed no peak; this not only indicated that the ordered structure had been lost, but also that the miniemulsion droplets were no longer spherical. Otherwise, the SAXS scattering profile would have shown fringes rather than the continuously decreasing pattern which is typically attributed to disk-like or rod-like structures [32]. Therefore, due to the relatively hydrophobic character of Cloisite 30B, a preferential droplet surface location of the platelets could be suggested.

As can be seen in Table 11.5, light-scattering measurements showed that the droplet size of the miniemulsion containing clay was larger than the counterparts without modified clays, due most likely to the presence of clay on the droplet surface. The final particle sizes for runs BM and PCN2 were smaller than those of the initial miniemulsion droplet, which indicated that some particles had been produced through mechanisms other than droplet nucleation. By using a parking area of 131 Å^2 per molecule for SLS on MMA/BA copolymers, the total surface area that could be covered by the emulsifier (SLS) was calculated for runs BM, PCN2 and PCN3. The total droplet surface areas (A_D) and the surface area that the emulsifier could cover (A_E) for runs BM, PCN2 and PCN3 are listed in Table 11.6. From these data it was observed that the total surface area that could be stabilized by the emulsifier was greater than that of the miniemulsion droplets for all runs. Taking into account that the CMC of SLS was 7 mM [33], micelles were presented in the medium for these two cases, and it is likely therefore that micellar nucleation occurred during these polymerizations. Both, latexes BM and PCN2 were coagulum-free. For run PCN3, the larger particle size of the final latex could be attributed to

Table 11.6 The droplet surface area (A_D) and the area coverable by the emulsifier (A_E).

Run	d_d (nm)[a]	A_D (Å^2)	A_E (Å^2)
BM	108	5.26×10^{24}	6.2×10^{27}
PCN2	210	2.9×10^{24}	5.8×10^{27}
PCN3	144	4.17×10^{24}	2.2×10^{27}

[a] d_d is the miniemulsion droplet size.

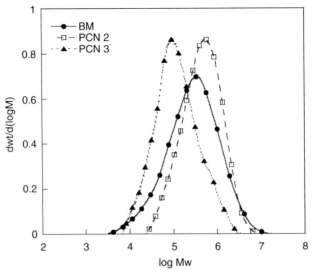

Figure 11.6 The molecular weight (M_w) distribution for runs BM, PCN2 and PCN3.

the particle flocculation that occurred during polymerization (1.2% coagulum was measured in the latex).

The miniemulsion polymerizations occurred rapidly, with full conversion being reached in 60 min for runs BM and PCN2. It should be noted that the PCN2 latex was stable and presented no coagulum, indicating that all of the Cloisite 30B clay added to the formulation had been incorporated into the final latex. The polymerization process for run PCN3 was different from that of runs BM and PCN2, due to the living character of the modifier employed in the clay. Here, the polymerization temperature was 90 °C and the redox initiator couple was only added after 4 h of polymerization. During this stage the conversion reached 15%, indicating that some polymer had been formed within the interlayer space of the clay. After the 4 h polymerization period, the redox couple was fed for 1 h, after which the final latex showed a small amount of coagulum (most likely due to the lower amount of emulsifier of 2 wt% used to stabilize the miniemulsion droplets), and the conversion achieved was 89%.

The average molecular weights, PDIs and gel contents for latexes BM, PCN2 and PCN3 are listed in Table 11.7. Interestingly, the molecular weights were less than those obtained in emulsion polymerization (the distributions were also narrower). Furthermore, the three latexes presented a gel polymer. Such gel formation should be less likely than in a semibatch emulsion polymerization process carried out under starved conditions, such as those used in these studies for emulsions [34, 35]. The reason for gel polymer formation in latexes BM and PCN2 most likely lies in the semibatch addition and the redox initiator couple used in the polymerization. As mentioned in Section 11.2, although the redox couple was fed to the reactor for 3 h, a 100% monomer conversion had been achieved by the first hour. During the

Table 11.7 The molecular weights, PDIs and gel contents of runs BM, PCN2 and PCN3.

Run	M_w (g mol^{-1})	PDI (M_w/M_n)	Gel content (%)
BM	6.3×10^5	6.1	5.0
PCN2	7.2×10^5	2.8	11.2
PCN3	2.2×10^5	3.4	17.6

remaining 2 h, therefore, highly active radicals were generated which abstracted hydrogen from the polymer backbone, and this led (at termination) to the formation of gel. This result was largely comparable to that reported by Ilundain and colleagues [36] in the post-polymerization of acrylic latexes. These authors found that, when this redox couple was used in the post-polymerization, the gel content of the latexes increased to a greater extent than when KPS was used. The even higher gel content obtained for run PCN3 should be attributed to the higher reaction temperature, which led to an enhanced chain transfer to the polymer [37].

The polymer obtained after extraction in THF for 24 h was analyzed with GPC (see Figure 11.6). For all the cases, a relatively broad unimodal molecular weight distribution was obtained.

In order to shed more light on the location of the clay platelets in the final nanocomposite latexes, SAXS measurements of the liquid latexes were performed. Both, the blank miniemulsion latex (BM) and the nanocomposite latexes (PCN2 and PCN3) were analyzed (Figure 11.7). The SAXS trace of latex BM showed the presence of fringes, in agreement with the scattering intensity reported for spherical polymer

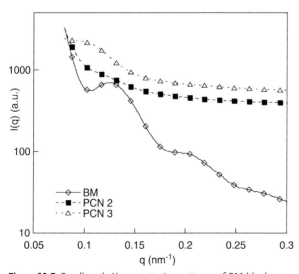

Figure 11.7 Small-angle X-ray scattering patterns of BM blank miniemulsion latex and nanocomposite latexes PCN2 and PCN3.

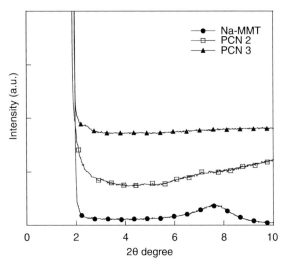

Figure 11.8 Wide-angle X-ray diffraction patterns of the nanocomposite latex films PCN2 and PCN3.

particles [38]. In contrast, the SAXS scattering for PCN2 and PCN3 did not show fringes but a rather a continuously decreasing pattern. This type of scattering has been attributed to disk-like or rod-like structures [32] which, in the present latexes, might have indicated that the clay platelets were located on the surface of the particles, thus avoiding the scattering of the spherical particles.

The WAXD diffraction patterns of Na-MMT clay and of the nanocomposite latex films, PCN2 and PCN3, prepared by batch miniemulsion polymerization, are shown in Figure 11.8. The WAXD patterns showed no diffraction peaks for the nanocomposite films PCN2 and PCN3, which indicated that the platelets were separated by at least 4 nm.

The TEM images of the nanocomposite latex films of PCN2 and PCN3 are shown in (Figures 11.9a and 11.10a, respectively. Upon nanocomposite film formation,

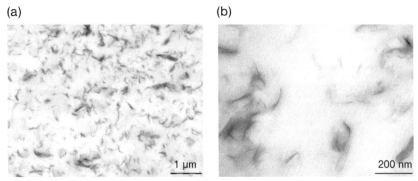

Figure 11.9 Transmission electron microscopy images of nanocomposite latex PCN2 film at different magnifications.

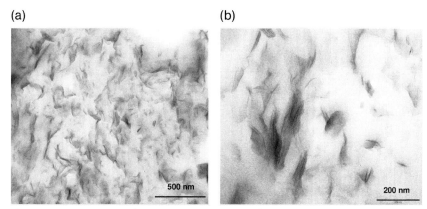

Figure 11.10 Transmission electron microscopy images of nanocomposite latex PCN3 film at different magnifications.

both organically modified clays were well dispersed in the polymer matrix (Figures 11.9a and 11.10a), and individual layers as well as zones consisting of more than one clay layer could also be clearly distinguished (Figures 11.9b and 11.10b). The structures shown in these TEM images were in agreement with the WAXD patterns of Figure 11.8.

The TEM images of negatively stained PCN2 and PCN3 latexes are shown in Figure 11.11. Unfortunately, the clay platelets could not be distinguished in these images, although irregular, nonspherical particle shapes were seen that may be a consequence of the clay presence. Clearly, the location of MMT platelets within the latex requires further TEM analysis and investigation.

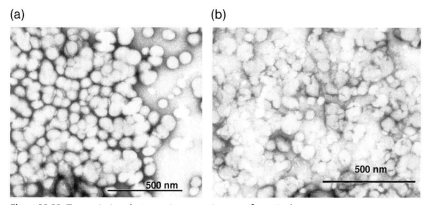

Figure 11.11 Transmission electron microscopy images of negatively stained nanocomposite latexes PCN2 (a) and PCN3 (b).

11.4
Conclusions

In these studies, waterborne acrylic polymer/clay nanocomposites were prepared by both emulsion and miniemulsion polymerization processes. In the former process, pristine Na-MMT was used as it is readily exfoliated in water phase. However, for the latter process organically modified clays were used in order to incorporate the clay to the monomer droplets and to produce the nanocomposite latex. The use of a reactive modifier in the clay did not reveal any major advantage over the commercial counterpart, and both routes allowed the synthesis of waterborne nanocomposites with well dispersed and intercalated/exfoliated structures of the clay in the polymer matrix on film formation. The nanocomposites prepared by either route showed similar improvements in mechanical and barrier properties (not presented in this work), despite having used pristine Na-MMT for the emulsion polymerization and organically modified montmorillonite for the miniemulsion. This fact, together with the ease of emulsion polymerization, might designate this route as the preferred means to produce waterborne polymer clay nanocomposites. Nonetheless, the miniemulsion route might be preferred for the production of polymers that cannot be synthesized by conventional emulsion polymerization, yet still have enhanced mechanical and barrier properties (e.g., hybrid alkyd/acrylic, urethane/acrylic, etc.).

Acknowledgments

G.D. acknowledges the Marie Curie fellowship (HPMT-CT-2001–00227) for financial support. Financial support was also provided by the Ministerio de Educacion y Ciencia of Spain (MAT 2003-01963) and the Basque Government (ETORTEK 2005). The authors wish to thank to Drs Francois Fauth and Ana Labrador for making the SAXS measurements at the Spanish CRG beamline at the European Synchrotron Radiation Facility in Grenoble, and Carmen Lopez Iglesias from Servicios Cientifico-Tecnicos, Universidad de Barcelona, Spain and Jose Ramos from the University of the Basque Country, Spain.

References

1 Yano, K., Usuki, A. and Okada, A. (1997) *J. Polym. Sci., Part A: Polym. Chem.*, **35**, 2289.
2 Koo, C.M., Kim, S.K. and Chung, I.J. (2003) *Macromolecules*, **36**, 2748.
3 Ray, S.S., Okamoto, K. and Okamoto, M. (2003) *Macromolecules*, **36**, 2355.
4 Ray, S.S. and Okamoto, M. (2003) *Prog. Polym. Sci.*, **28**, 1539.
5 Van Olphen, H. (1991) *An Introduction to Clay Colloid Chemistry*, 2nd edition, Krieger Publishing Company, Malabar.
6 Diaconu, G., Paulis, M. and Leiza, J.R. (2008) *Polymer*, **49**, 2444.
7 Lee, D.C. and Jang, L.W. (1996) *Journal of Applied Polymer Science*, **61**, 1117.
8 Noh, M.H. and Lee, D.C. (1999) *Journal of Applied Polymer Science*, **74**, 2811.
9 Bandyopadhyay, S. and Giannelis, E.P. (2000) *Polym. Mater. Sci. & Eng.*, **82**, 208.
10 Choi, Y.S., Choi, M.H., Wang, K.H., Kim, S.O., Kim, Y.K. and Chung, I.J. (2001) *Macromolecules*, **34**, 8978.

11 Negrete-Herrera, N., Letoffe, J.M., Putaux, J.L., David, L. and Bourgeat-Lami, E. (2004) *Langmuir*, **20**, 1564.
12 Zhang, Z., Zhao, N., Wei, W., Wu, D. and Sun, Y. (2005) *Studies in Surface Science and Catalysis*, **156**, 529.
13 Kim, Y.K., Choi, Y.S., Wang, K.H. and Chung, I.J. (2002) *Chem. Mater.*, **14**, 4990.
14 Wang, C., Wang, Q. and Chen, X. (2005) *Macrom. Mater. Eng.*, **290**, 920.
15 Li, H., Yu, Y. and Yang, Y. (2005) *Eur. Polym. J.*, **41**, 2016.
16 Park, B.J., Kim, T.H., Choi, H.J. and Lee, J.H. (2007) *J. Macrom. Sci. B: Phys.*, **46**, 341.
17 Asua, J.M. (2002) *Prog. Polym. Sci.*, **27**, 1283.
18 Kwolek, T., Hodorowics, M., Stadnicka, K. and Czapkiewics, J. (2003) *J. Coll. Int. Sci.*, **254**, 14.
19 Fornes, T.D., Hunter, D.L. and Paul, D.R. (2004) *Macromolecules*, **37**, 1793.
20 Kozak, M. and Domka, L. (2004) *J. Phys. Chem. Solids*, **65**, 441.
21 Weimer, M.W., Chen, H., Diannelis, E.P. and Sogah, D.Y. (1999) *J. Am. Chem. Soc.*, **121**, 1615.
22 Fan, X., Xia, C. and Advincula, R.C. (2005) *Langmuir*, **21**, 2537.
23 Zhang, B.Q., Pan, C.Y., Hong, C.Y., Luan, B. and Shi, P.J. (2006) *Macrom. Rapid Commun.*, **27**, 97.
24 Di, J. and Sogah, D.Y. (2006) *Macromolecules*, **39**, 1020.
25 Konn, C., Morel, F., Beyou, E., Chaumont, P. and Bourgeat-Lami, E. (2007) *Macromolecules*, **40**, 7464.
26 Diaconu, G., Asua, J.M., Paulis, M. and Leiza, J.R. (2007) *Macrom. Symp.*, **259**, 305.
27 Lesko, P.M. and Sperry, P.R. (1995) in *Emulsion Polymerization and Emulsion Polymers* (eds. P.A. Lovell and M.S. El-Aasser), John Wiley & Sons, New York, p. 620.
28 Cervantes, J.M., Cauich-Rodriguez, J.V., Vazquez-Torres, H., Garfias, L.F. and Paul, D.R. (2007) *Thermochimica Acta*, **457**, 92.
29 Chern, C.S., Lin, J.J., Lin, Y.L. and Lai, S.Z. (2006) *European Polymer Journal*, **42**, 1033.
30 Elizalde, O., Arzamendi, G., Leiza, J.R. and Asua, J.M. (2004) *Ind. Eng. Chem. Res.*, **43**, 7401–7409.
31 Gonzalez, I., Asua, J.A. and Leiza, J.R. (2007) *Polymer*, **48**, 2542–2547.
32 Svergun, D.I. and Koch, M.H.J. (2003) *Rep. Prog. Phys.*, **66**, 1735.
33 Van Os, N.M., Haak, J.R. and Rupert, L.A.M. (1993) *Physico-Chemical Properties of Selected Anionic, Cationic and Nonionic Surfactants*, Elsevier, Amsterdam.
34 Li, D., Sudol, E.D. and El-Aasser, M.S. (2006) *J. Appl. Polym. Sci.*, **102**, 4616.
35 Gonzalez, I., Paulis, M., de la Cal, J.C. and Asua, J.M. (2007) *Macromol. React. Eng.*, **1**, 635.
36 Ilundain, P., Alvarez, D., Da Cunha, L., Salazar, R., Barandiaran, M.J. and Asua, J.M. (2003) *J. Polym. Sci. A: Polym. Chem.*, **41**, 3744.
37 Gonzalez, I., Leiza, J.R. and Asua, J.M. (2006) *Macromolecules*, **39**, 5015.
38 Megens, M., van Kats, C.M., Bösecke, P. and Vos, W.L. (1997) *Langmuir*, **13**, 6120.

12
Preparation Characteristics of Giant Vesicles with Controlled Size and High Entrapment Efficiency Using Monodisperse Water-in-Oil Emulsions

Takashi Kuroiwa, Mitsutoshi Nakajima, Kunihiko Uemura, Seigo Sato, Sukekuni Mukataka, and Sosaku Ichikawa

12.1
Introduction

A wide variety of amphiphilic lipid molecules, such as phospholipids, can form vesicles that are quasispherical shells composed of lipid bilayer membranes. Vesicles of micrometer-scale diameters are called *giant vesicles* (GVs). As GV membranes are structurally more similar to cellular membranes than are the membranes of smaller vesicles, a number of biological, medical and physico-chemical studies using GVs have been conducted [1–5]. Notably, GVs can entrap various materials in their internal water phases, which are separated from the external environment by lipid bilayer membranes. More specifically, GVs can entrap not only water-soluble small molecules but also large biomolecules and particulate materials, because the internal volumes of GVs are larger than those of smaller vesicles [6–8]. Therefore, the potential applications of GVs, including delivery systems for drugs, genes, nutrients and cosmetic materials [6, 9, 10] and microreactor systems [11–15], have developed remarkably over the past two decades.

The utilization of GVs is often limited by their preparation methods. Methods reported for preparing GVs involve electroformation [16, 17], gentle hydration [18–20], evaporation of two-phase phospholipid solutions [21], and the use of microfluidic channel devices [8, 22]. Although some of these methods have been used widely in laboratories, certain problems are still associated with GV preparation, including: (1) a difficulty of size control of GVs; (2) a low efficiency of entrapment of hydrophilic materials in the internal water phases of GVs; and (3) a low productivity of size-controlled GVs capable of entrapping hydrophilic materials. The size of GVs is important in order to quantitatively control their chemical and biological behaviors *in vitro* or *in vivo*. Additionally, when GVs are used as carriers and microreactors, various functional compounds such as drugs, genes and enzymes must be entrapped in their active forms, and with a high efficiency, into GVs. Finally, we have been unable to identify a method which is applicable for the mass production of size-controlled GVs with a high entrapment efficiency.

Emulsion Science and Technology. Edited by Tharwat F. Tadros
Copyright © 2009 WILEY-VCH Verlag GmbH & Co. KGaA, Weinheim
ISBN: 978-3-527-32525-2

Taking these points into consideration, we have previously developed a novel method for preparing GVs [23, 24] (also T. Kuroiwa et al., unpublished results). This technique, which is referred to as the 'lipid-coated ice droplet hydration method', uses monodisperse water-in-oil (W/O) emulsions obtained by the microchannel (MC) emulsification technique [25, 26]. The method enables us to control the size of GVs and to improve the entrapment yield of hydrophilic materials, because size-controlled water droplets in the starting W/O emulsions become the internal water phases of the GVs. In this study, we investigated the preparation characteristics of GVs created by using the new method. Monodisperse W/O emulsions containing water droplets of different sizes were prepared by MC emulsification and used as starting materials for GV preparation. The relationship between the diameters of the starting water droplets and the obtained GVs was studied. The entrapment efficiency of the GVs was also evaluated through the entrapment of a water-soluble fluorescent dye, *calcein*. The mechanistic characteristics of GV formation by means of this method are also discussed from a technical viewpoint.

12.2
Materials and Methods

12.2.1
Materials

Sorbitan mono-oleate (Span 80), cholesterol (Chol) and stearylamine (SA) were purchased from Wako Pure Chemical Industries (Osaka, Japan). Phosphatidylcholine (PC) from egg yolk was obtained from Funakoshi Co., Ltd (Tokyo, Japan). Calcein was purchased from Sigma (St Louis, MO, USA). All other chemicals were from commercial sources and were either analytical or extra-pure grades.

12.2.2
Preparation of W/O Emulsions Using MC Emulsification

Hexane containing Span 80 (3 wt.%) and SA (0.1 wt.%) was used as a continuous phase for preparing W/O emulsions. To prevent the solubilization of prepared water droplets into this continuous hexane phase, the phase was saturated with water before use. Tris–HCl buffer (50 mM, pH 9) containing 0.4 mM calcein was used as the dispersed phase.

Table 12.1 Dimensions of MC plates used in the studies.

MC plate	MC width (μm)	MC depth (μm)	Terrace length (μm)
MC-A	5	2	15
MC-B	16	2	30
MC-C	16	4	30

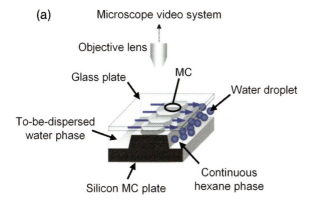

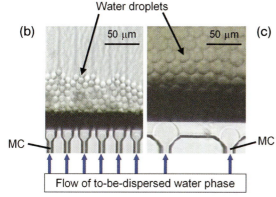

Figure 12.1 (a) Schematic illustration of microchannel (MC) emulsification and emulsification behavior of water-in-oil (W/O) emulsions prepared with (b) MC-A and (c) MC-C plates.

The experimental equipment consisted of a MC module, a silicon MC plate, a glass plate, a syringe pump supplying the continuous phase, a reservoir to feed the dispersed phase and a microscope video system to observe the emulsification behavior. In this study, we used three types of MC plate with MC dimensions [26], as listed in Table 12.1. The to-be-dispersed water phase was pressed into the continuous hexane phase through the MCs, after which the monodisperse water droplets were formed (Figure 12.1a) and recovered at the outlet of the MC module. For preparing W/O emulsions, the silicon MC plates and a glass plate were hydrophobized with octadecyltriethoxysilane to prevent the MCs from being wetted by the dispersed phase [27, 28].

12.2.3
Formation of GVs

The GVs were formed by means of the monodisperse W/O emulsions, as described previously [23, 24]. The vesicle-formation procedure consisted of two processes, namely surfactant replacement and lipid hydration. First, Span 80 in the starting W/O emulsion was replaced by vesicle-forming lipids (this process is referred to as 'surfactant replacement'). In order to avoid water droplet coalescence during surfactant replacement, the water droplets in the W/O emulsion were frozen by liquid nitrogen and kept frozen throughout the process. Sedimentation of the frozen water droplets was followed by suction removal of the supernatant containing Span 80, after which hexane containing a mixture of vesicle-forming lipids (PC, Chol, and SA at a molar ratio of 5:5:1) was added. This surfactant replacement procedure was repeated several times to reduce the surfactant concentration to approximately 0.003 wt.% [24]. On completion of the surfactant replacement process the concentrations of water droplets and lipid mixture were adjusted to a prescribed value. To adjust the water concentration in the sample, the water content in the W/O emulsion was measured using a Karl Fischer coulometer (model E-684; Metrohm, Herisau, Switzerland). After adjusting the concentrations of the water droplets and lipid mixture, hexane was evaporated under reduced pressure while the water droplets were kept frozen at $-7\,^\circ$C. Portions (0.2 ml) of Tris–HCl buffers (pH 9, 50 mM) containing 10 wt% glycerol and/or large unilamellar vesicles (FAT-VET$_{50\mathrm{nm}}$) (T. Kuroiwa et al., unpublished results) were added to hydrate the lipid layers and form GVs; this process is referred to as 'lipid hydration'. An additional, external water phase (Tris–HCl buffer, pH 9, 50 mM) was added for dispersing the GVs.

12.2.4
Measurement of Droplet and Vesicle Diameters

The diameters of the water droplets of W/O emulsions and GVs were measured from images taken with the microscope video system using WinROOF image analysis software (Mitani Corporation, Fukui, Japan). A number-average diameter (μm) and coefficient of variation (%; the percentage of standard deviation of diameters to the average diameter) were determined from diameter measurements of at least 100 droplets.

12.2.5
Determination of Entrapment Yield

The entrapment yield of the fluorescent marker, calcein, was determined fluorometrically by a modification of the method used to determine the trapped volumes of liposomes [29], as reported previously [23]. First, a 3 ml portion of a GV suspension was placed into a quartz cuvette and the fluorescence intensity (F_{total}) measured. Cobalt chloride (CoCl$_2$) solution (10 mM; 30 μl) was added to the suspension to quench the fluorescence of nonentrapped calcein, after which the fluorescence intensity of the calcein entrapped within the GVs (F_{in}) was measured. Finally, Triton X-100 solution (10 wt%, 0.3 ml) was added to the mixture to disrupt the GVs and to quench all fluorescence of calcein in the mixture; the resultant fluorescence intensity (F_q) was then measured. The entrapment yield of calcein was calculated from the following equation:

$$\text{(Entrapment yield, \%)} = (F_{in}r_1 - F_q r_2)/(F_{total} - F_q r_2) \times 100 \quad (12.1)$$

Here, r_1 and r_2 are dilution volume factors due to the addition of $CoCl_2$ and Triton X-100 solutions. The excitation and emission wavelengths for fluorescence measurement of calcein were 490 nm and 520 nm, respectively.

12.3 Results and Discussion

12.3.1 Preparation of GVs Using Monodisperse W/O Emulsions

In this study, the size of GVs was adjusted by controlling the water droplet diameter of monodisperse W/O emulsions, using the MC emulsification technique. In MC emulsification, the diameter of water droplets can be varied by using different MC plates with different channel dimensions. The emulsification behaviors of W/O emulsions prepared with different MC plates (MC-A and MC-C, see Table 12.1) are shown in Figure 12.1. As shown in Figure 12.1b and c, monodisperse water droplets with different diameters were successfully produced by varying the dimensions of the MC plates. The droplet sizes and coefficient of variation (CV) values of the emulsions recovered from the MC module after a one-day emulsification differed slightly from those observed in the MC module just after formation: notably, the average diameters became smaller, and the CV values became larger. These observed changes in diameter and monodispersity might have been caused by dissolution of the dispersed water phase into the continuous phase. Furthermore, the extent of this dissolution might have been affected by operation conditions such as emulsification temperature and storage time prior to recovery. Here, we used recovered emulsion droplets with average diameters of 6.6, 10.4 and 13.4 µm (obtained with MC-A, MC-B and MC-C, respectively) as starting materials for GV preparation.

Two microphotographs of the GVs prepared from the W/O emulsion obtained with MC-A are shown in Figure 12.2. Most of the GVs were spherical, with the diameter of each GV typically being several micrometers and comparable with that of the water droplets in the original W/O emulsions. These results indicated that the diameter of GVs could be controlled by means of W/O emulsions with arbitrary sizes. Fluorescent microscopic observations confirmed that the GVs contained calcein, a water-soluble fluorescent marker, in their internal water phases (Figure 12.2b). In the case of the sample shown in Figure 12.2, the entrapment yield of calcein into the internal water phase was approximately 36%, which is a relatively high entrapment yield for hydrophilic molecules in GVs. Therefore, our strategy for GV preparation – that is, the utilization of monodisperse water droplets as a starting material for the GVs' internal water phase – appeared to be effective for obtaining GVs with a controlled size and a high entrapment yield.

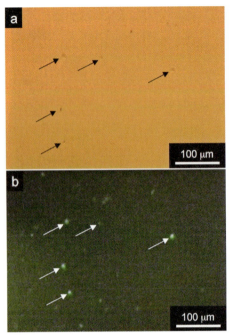

Figure 12.2 Microphotographs of prepared giant vesicles (GVs) taken in (a) brightfield and (b) fluorescence modes. The arrows point toward the focused GVs. The positions of each GV were slightly different between photographs (a) and (b) because of movement of the vesicles during observation.

12.3.2
Size Control of GVs and Entrapment of a Hydrophilic Molecule into GVs

Giant vesicles of varying diameters were prepared from the W/O emulsions with different mean droplet diameters. Figure 12.3 shows a group of fluorescent microscopic images of the W/O emulsions used as the starting materials for GV preparation (Figure 12.3a and b, obtained with MC-A and C, respectively) and their resulting GVs (Figure 12.3c and d). These images show that the sizes of the obtained GVs reflected those of the water droplets of the original W/O emulsions. The diameter distributions of the original water droplets and the prepared GVs shown in Figure 12.3 are depicted in Figure 12.4. The average diameters of GVs prepared from water droplets with mean diameters of 6.6 and 13.4 µm were 7.2 µm and 11.1 µm, respectively. In both cases, more than 50% of the prepared GVs were in the range of the diameter of the original water droplets, although the size distributions of both GVs became broader (CV values: 5–10% for the water droplets, 30–45% for the GVs). As GVs can be fractionated by simple centrifugation and membrane filtration, without any specialized equipments [30], the size distribution of these GVs might be narrowed further through simple fractionation.

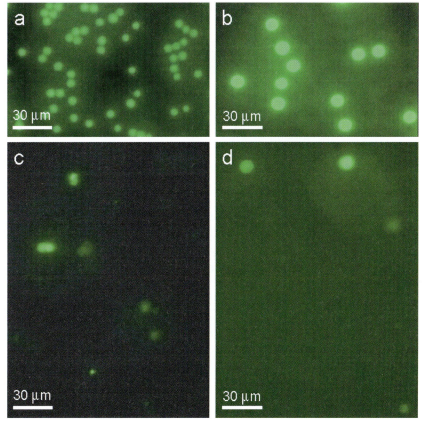

Figure 12.3 Fluorescent microphotographs of calcein-containing W/O emulsions with water droplet diameters of (a) 6.6 μm and (b) 13.4 μm, and the resultant giant vesicles (c and d, respectively).

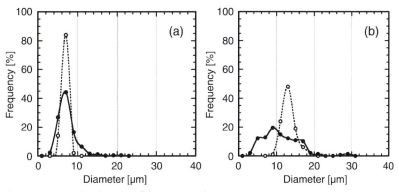

Figure 12.4 Size distributions of the W/O emulsion water droplets and resultant giant vesicles shown in Figure 12.3. The W/O emulsions were prepared with (a) MC-A and (b) MC-C plates.

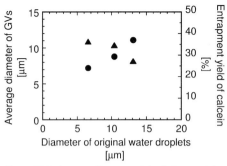

Figure 12.5 Average diameters (●) of giant vesicles and their entrapment yields for calcein (▲) as a function of diameter of the original water droplets in W/O emulsions.

The average diameters of the GVs and their entrapment yields for calcein were plotted against the average diameters of the original water droplets in the W/O emulsions (Figure 12.5). The diameter of the GVs increased with increasing diameter of the original water droplets. However, the increase was not linear: the average diameters of the GVs became smaller than those of their original water droplets as the water droplet diameter increased. By contrast, the entrapment yield of calcein decreased with increasing diameter of the original water droplets. These results suggest that GVs of larger diameters are more disrupted in their closed-membrane structure than are smaller-diameter GVs, thus decreasing the ability of larger GVs to retain calcein. Furthermore, such a disrupted GV structure indicates that the lipid layers covering larger water droplets are less stable during GV preparation than those covering smaller droplets, perhaps because of the lipid membranes' sensitivity to shear stress.

The entrapment yield of the GVs as a function of increasing ratio of lipids to internal water phase (L/W ratio) [24] (also T. Kuroiwa *et al.*, unpublished results) in a W/O emulsion after the surfactant replacement with 6.6 μm droplets is shown in Figure 12.6. As reported previously [24], the entrapment yield of calcein was increased with an increasing L/W ratio. This result can be explained by the increase

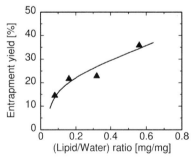

Figure 12.6 The entrapment yield of calcein as a function of L/W ratio in a W/O emulsion containing 6.6 μm water droplets.

in thickness of the lipid membranes on the surface of the frozen water droplets induced by an increasing L/W ratio: a larger amount of lipids covering the same surface area of water droplets led to thicker, more stable lipid bilayers surrounding internal water phase, resulting in a higher entrapment yield.

12.3.3
Formation Characteristics of GVs

The results shown in Figures 12.5 and 12.6 indicate that both the increase in GV diameter and the decrease in L/W ratio (i.e. the decrease in GV membrane thickness) lowered the relative stability of the lipid layers against shear force caused by the external fluid during GV preparation. For a thin spherical membrane shell such as a vesicle, the pressure difference required to disrupt the closed membrane structure is expressed by Laplace's law [31]:

$$\Delta P_L = 4\tau/d \tag{12.2}$$

where ΔP_L is the pressure difference between the inside and outside of the vesicle required to induce disruption (in Pa), τ is the membrane lysis tension (in $N\,m^{-1}$), and d is the diameter of the sphere (in m).

Additionally, the vesicles with a diameter of the order of the energy-dissipating eddy size in the flow should experience a velocity difference, u, between opposite sides of an individual vesicle, and this leads to a local pressure difference, ΔP_F. This pressure difference is represented by the following equation, according to Bernoulli's law:

$$\Delta P_F \propto \rho_e u^2 \tag{12.3}$$

where ρ_e is the density of the external water phase (in $kg\,m^{-3}$) and u is the local velocity difference of the flow in which the vesicles are suspended (in $m\,s^{-1}$).

Disruption of a vesicle will occur when ΔP_F exceeds ΔP_L. Therefore, the mechanical stability of vesicles in a flow can be related to the ratio of both pressure differences:

$$\Delta P_F/\Delta P_L \propto (\rho_e u^2)/(\tau/d) = \rho_e u^2 d/\tau \tag{12.4}$$

According to the Kolmogorov theory for isotropic turbulence, u is proportional to $d^{1/3}$ under fixed experimental conditions (i.e. constant energy density). Therefore,

$$\Delta P_F/\Delta P_L \propto d^{5/3} \tag{12.5}$$

Furthermore, if the membrane lysis tension τ of a vesicle is assumed to be proportional to the membrane thickness [32–34], then τ can also be assumed to be proportional to the amount of lipids per unit surface area of vesicles in the present experiments. Then, we have

$$\Delta P_F/\Delta P_L \propto 1/m_{lipid,s} \tag{12.6}$$

where $m_{lipid,s}$ is the amount of lipids per unit surface area of vesicles (in kg m^{-2}). By combining Equations 12.5 and 12.6, the following relationship is obtained:

$$\Delta P_F/\Delta P_L \propto d^{5/3}/m_{lipid,s} \qquad (12.7)$$

Thus, the quantity $d^{5/3}/m_{lipid,s}$ can be related to the mechanical stability of GVs in the suspension, and the dependency of the entrapment yield on the experimental parameters (i.e. the diameter of GVs and the L/W ratio) can be discussed based on Equation 12.7. Figure 12.7 shows, graphically, the relationship between the value of $d^{5/3}/m_{lipid,s}$ and the entrapment yield. Although the number of experimental results is limited, a clear correlation is observed between the wide range of diameters (6.6–36.1 μm) and L/W ratios (0.08–1.6 mg mg^{-1}), including our previous result [23]. This relationship indicates that the relative magnitude of shear force and membrane strength is one of the key parameters for improving the GV preparation process, and the value of $d^{5/3}/m_{lipid,s}$ is a useful index to predict the effect of operational factors on the preparation of GVs. The conditions that provide a low $d^{5/3}/m_{lipid,s}$ value should be preferable for efficient entrapment of water-soluble molecules and also for the control of vesicle size.

On the basis of the above considerations, the possible patterns of GV formation are summarized as a flowchart in Figure 12.8. In our method, water droplets in the original W/O emulsions (Figure 12.8a) were frozen and then converted to lipid-coated ice droplets through the surfactant replacement and solvent evaporation (Figure 12.8b). At this stage, it was confirmed by microscopic observation that most of the ice droplets after evaporational removal of hexane were still spherical and maintained their original diameters [23]. Therefore, changes in the diameter of GVs and decreases in entrapment yields most likely occurred mainly during the following processes: lipid hydration, melting of ice droplets, and the recovery of GVs in a suspension (Figure 12.8c). If the initial shape and size of a water droplet were to be maintained throughout the GV preparation process, then a GV with a diameter similar to that of its original water droplets should be obtained and, in principle, the

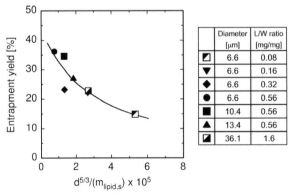

Figure 12.7 Correlation of the entrapment yield of calcein with the value of $d^{5/3}/m_{lipid,s}$. The data represented by the symbol ◪ ($d = 36.1$ μm, L/W ratio = 1.6 mg mg^{-1}) are from Ref. [23].

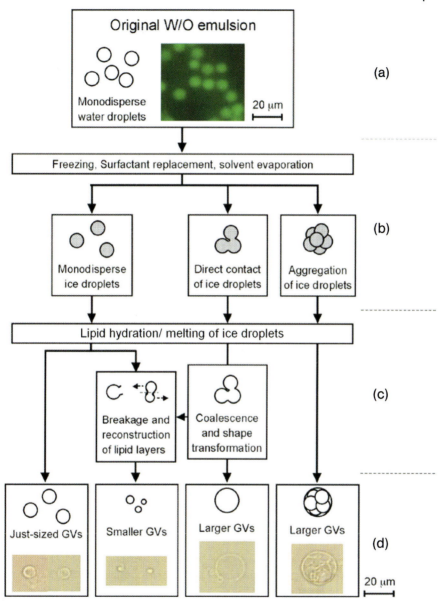

Figure 12.8 Schematic flowchart of possible mechanisms of giant vesicle formation in the present experiments. (a) Fluorescence microphotograph and schematic illustration of a starting W/O emulsion prepared with an MC-B plate (average droplet diameter 10.4 μm). (b) Schematic illustrations of frozen emulsion droplets after evaporative removal of the continuous hexane phase. (c) Schematic illustrations of lipid hydration and giant vesicle formation. (d) Brightfield microphotographs and schematic illustrations of the resulting giant vesicles.

resultant entrapment yield should be near-optimal. In our experimental observation, however, four types of GV were observed following the preparation procedures (Figure 12.8d): (i) GVs with diameters similar to those of the original water droplets; (ii) GVs smaller than the original water droplets; (iii) GVs larger than the original water droplets; and (iv) larger, multiple GVs (vesicle-in-vesicle structure).

Smaller GVs could be formed by the reconstruction of lipid membranes after partial breakage due to insufficient coverage of the surface of water (ice) droplets with lipid layers, shear force by flow of the external water phase, or both. We believe that these processes were the main causes of the decreased entrapment yield. Additionally, we have previously reported that the L/W ratio affects the entrapment yield of calcein in GVs prepared by this method (T. Kuroiwa *et al.*, unpublished results). As a higher L/W ratio should give a thicker lipid membrane around the ice droplets following solvent evaporation, a higher L/W ratio could therefore reduce the possibility of insufficient lipid coverage of droplets and prevent the breakage of GVs due to shear stress (Figure 12.8c). This hypothesis might explain why a higher entrapment yield was obtained at a higher L/W ratio (see Figure 12.6). Furthermore, the addition of small unilamellar vesicles to the external water phase and hydration under low temperature (below $0\,°C$) are also effective methods for improving the entrapment yield (T. Kuroiwa *et al.*, unpublished results), since small vesicles may protect against vesicle breakage, and low temperatures minimize the deformation of the frozen water droplets. Therefore, prevention of GV breakage is one possible strategy both for controlling the size of the GVs and for improving entrapment yields.

As in other cases of GV formation, large GVs with a single internal water phase were most likely formed through the coalescence of water droplets (Figure 12.8d). In order for two or more droplets to aggregate, they must contact each other directly after the lipid hydration process. By contrast, the aggregation of ice droplets followed by an overall hydration of the droplet aggregates resulted in large GVs that had multiple internal water phases (vesicle-in-vesicle structure). As shown in Figure 12.8d, such multiple GVs seemed to comprise a cluster of GVs with diameters similar to those of the water droplets in the original W/O emulsion. The formation of these two types of large GV would not accompany lipid membrane breakage and subsequent leakage of the internal compartments of the water droplets, and thus would not lead to a direct decrease in the entrapment yield. However, large GVs would be more easily broken by shear-flow stress. Therefore, the coalescence and aggregation of droplets might have caused the breakage of formed GVs and consequently led to a decreased entrapment yield. Improving the hexane continuous-phase removal process could effectively prevent direct contact and aggregation of the water droplets and thus provide a more uniform GV diameter, as well as a higher entrapment yield.

12.4
Conclusions

In these studies, we have investigated the preparation characteristics of GVs with a controlled size and a high entrapment yield of a hydrophilic molecule, calcein, by the lipid-coated ice droplet hydration method. W/O emulsions with different water droplet diameters were prepared using the MC emulsification technique. The GV preparation process consisted of MC emulsification, surfactant replacement, solvent removal and lipid hydration with an external water phase. This method has simultaneously permitted control of both the GV diameter and the entrapment of calcein in a high yield. The average diameter of the obtained GVs increased with the increasing average diameter of the water droplets in the original W/O emulsions. Therefore, the size of the GVs could be controlled by using W/O emulsions with different droplet diameters. However, the entrapment yield of calcein in the GVs increased with *a decreasing* original water droplet diameter and with an *increasing* ratio of lipids to internal water phase. The entrapment yield obtained experimentally could be correlated with the relative magnitude of disruptive force by flow of external water phase and mechanical strength of lipid membrane. These experiments have provided an index for the improvement of the GV preparation process. Further investigations will undoubtedly focus on the hydration process to improve both the uniformity of GV size and the entrapment yield.

Acknowledgments

These studies were supported financially by the Industrial Technology Research Grant Program from the New Energy and Industrial Technology Development Organization (NEDO) of Japan and by the Grant-in-Aid for Young Scientists (B) (No. 19760559) from the Ministry of Education, Culture, Sports, Science and Technology of Japan.

References

1 Menger, F.M. and Keiper, J.S. (1998) *Curr. Opin. Chem. Biol.*, **2**, 726.
2 Hotani, H., Nomura, F. and Suzuki, Y. (1999) *Curr. Opin. Coll. Interf. Sci.*, **4**, 358.
3 Döbereiner, H.G. (2000) *Curr. Opin. Coll. Interf. Sci.*, **5**, 256.
4 Fischer, A., Oberholzer, T. and Luisi, P.L. (2000) *Biochim. Biophys. Acta*, **1467**, 177.
5 Mally, M., Majhenc, J., Svetina, S. and Zeks, B. (2002) *Biophys. J.*, **83**, 944.
6 Antimisiaris, S.G., Jayasekera, P. and Gregoriadis, G. (1993) *J. Immunol. Methods*, **166**, 271.
7 Tresset, G. and Takeuchi, S. (2005) *Anal. Chem.*, **77**, 2795.
8 Tan, Y.C., Hettiarachchi, K., Siu, M. and Pan, Y.P. (2006) *J. Am. Chem. Soc.*, **128**, 5656.
9 Lasic, D.D. (2000) in *Giant Vesicles* (eds Luisi, P.L. and Walde, P.), John Wiley & Sons, Chichester, Chapter 2.
10 Nomura, S. and Yoshikawa, K. (2000) In: *Giant Vesicles* (eds P.L. Luisi and P. Walde), John Wiley & Sons, Chichester, Chapter 23.
11 Wick, R., Angelova, M.I., Walde, P. and Luisi, P.L. (1996) *Chem. Biol.*, **3**, 105.
12 Bucher, P., Fischer, A., Luisi, P., Oberholzer, T. and Walde, P. (1998) *Langmuir*, **14**, 2712.
13 Tsumoto, K., Nomura, S.M., Nakatani, Y. and Yoshikawa, K. (2001) *Langmuir*, **17**, 7225.
14 Fischer, A., Franco, A. and Oberholzer, T. (2002) *ChemBioChem*, **3**, 409.
15 Noireaux, V. and Libchaber, A. (2004) *Proc. Natl. Acad. Sci. USA*, **101**, 17669.

16 Angelova, M.I. and Dimitrov, D.S. (1986) *Faraday Discuss. Chem. Soc.*, **81**, 303.

17 Angelova, M.I. (2000) in *Giant Vesicles* (eds Luisi, P.L. and Walde, P.), John Wiley & Sons, Chichester, Chapter 3.

18 Hub, H.H., Zimmermann, U. and Ringsdorf, H. (1982) *FEBS Lett.*, **140**, 254.

19 Mueller, P., Chien, T.F. and Rudy, B. (1983) *Biophys. J.*, **44**, 375.

20 Magome, N., Takemura, T. and Yoshikawa, K. (1997) *Chem. Lett.*, **26**, 205.

21 Moscho, A., Orwar, O., Chiu, D.T., Modi, B.P. and Zare, R.N. (1996) *Proc. Natl. Acad. Sci. USA*, **93**, 11443.

22 Lorenceau, E., Utada, A.S., Link, D.R., Cristobal, G., Joanicot, M. and Weitz, D.A. (2005) *Langmuir*, **21**, 9183.

23 Kuroiwa, T., Nakajima, M., Sato, S., Mukataka, S. and Ichikawa, S. (2007) *Membrane*, **32**, 229.

24 Sugiura, S., Kuroiwa, T., Kagota, T., Nakajima, M., Sato, S., Mukataka, S., Walde, P. and Ichikawa, S. (2008) *Langmuir*, **24**, 4581.

25 Kawakatsu, T., Kikuchi, Y. and Nakajima, M. (1997) *J. Am. Oil Chem. Soc.*, **74**, 317.

26 Sugiura, S., Nakajima, M., Iwamoto, S. and Seki, M. (2001) *Langmuir*, **17**, 5562.

27 Kawakatsu, T., Trägårdh, G., Trägårdh, C., Nakajima, M., Oda, N. and Yonemoto, T. (2001) *Colloids Surf. A*, **179**, 29.

28 Sugiura, S., Nakajima, M., Ushijima, H., Yamamoto, K. and Seki, M. (2001) *J. Chem. Eng. Jpn.*, **34**, 757.

29 Oku, N., Kendall, D.A. and Macdonald, R.C. (1982) *Biochim. Biophys. Acta*, **691**, 332.

30 Kato, K., Walde, P., Mitsui, H. and Higashi, N. (2003) *Biotechnol. Bioeng.*, **84**, 415.

31 Mui, B.L.S., Cullis, P.R., Evans, E.A. and Madden, T.D. (1993) *Biophys. J.*, **64**, 443.

32 Kwok, R. and Evans, E. (1981) *Biophys. J.*, **35**, 637.

33 Hallett, F.R., Marsh, J., Nickel, B.G. and Wood, J.M. (1993) *Biophys. J.*, **64**, 435.

34 Akashi, K., Miyata, H., Ito, H. and Kinoshita, K. Jr. (1996) *Biophys. J.*, **71**, 3242.

13
On the Preparation of Polymer Latexes (Co)Stabilized by Clays*
Ignác Capek

13.1
Introduction

During recent years, montmorillonite (MMT) has attracted great industrial and academic interest because of its high aspect ratio of silicate nanolayers, its high surface area, and its wide application in the field of polymer materials [1]. Polymer/clay nanocomposites have been extensively studied in recent years, because they often exhibit physico-chemical properties that are dramatically different from their micro- and macrocomposite counterparts [2]. These polymer nanocomposites have attracted great interest because of their dramatically improved properties compared to conventional composites [3]. Polymer/clay nanocomposites consist of clay nanolayers dispersed in a polymeric matrix [4].

As is well-known, polymer/clay nanocomposites are materials of increasing interest because of their structural or functional behaviors [5]. When compared to pristine polymers, polymer/clay nanocomposites possess many desirable properties, such as enhanced thermal and mechanical properties [6, 7], improved barrier properties [8], enhanced barrier characteristics [9], increased moduli and strengths [10], high heat distortion temperatures [11], decreased thermal expansion coefficients [12], reduced gas permeabilities [8], decreased absorption in organic liquids [13], enhanced ionic conductivities [14, 15] and reduced flammability [16]. Interesting electrical and electrochemical properties of poly(ethylene oxide)-smectites have been extensively reported [17]. These improved properties depend on the nanostructural configuration and interfacial bonding between the clay and the matrix (polymer...) [18]. Because of these many advantages, such as excellent mechanical properties, good gas barrier and flame retardation, polymer/clay nanocomposites have in recent years been the subject of intense investigation [19, 20].

The composites have attracted the attention of many research groups, with the consequent development of polymer/silicate nanocomposites with excellent properties. With regards to microstructure, the clay particles are dispersed in the polymer matrix in either intercalated or exfoliated states [21]:

*A list of abbreviations is given at the end of this chapter.

Emulsion Science and Technology. Edited by Tharwat F. Tadros
Copyright © 2009 WILEY-VCH Verlag GmbH & Co. KGaA, Weinheim
ISBN: 978-3-527-32525-2

- *Intercalated nanocomposites* are generally obtained when the polymer is located between the silicate layers; even though the layer spacing increases, there remain attractive forces between the silicate layers such that the layers are stacked with uniform spacing. In the intercalated composites, although the polymer chains are inserted into the interlayer space of the stacking silicate platelets, the silicate layers remain well ordered, despite the basal space being greatly expanded.

- In *exfoliated composites*, the discrete clay layers are randomly dispersed in the continuous polymer matrix. Exfoliated nanocomposites are favored when there is a need to enhance the material's properties; they are formed when the layer spacing is increased to the point where there are no longer sufficient attractions between the silicate layers to maintain a uniform layer spacing. In particular, exfoliated polymer/clay hybrids offer improved mechanical and thermal properties due not only to the homogeneous dispersion of the clay in polymer matrix but also to the large interfacial area of the clay layers. In the past, it has generally been considered that exfoliated polymer/silicate nanocomposites had better mechanical properties than intercalated polymer/silicate nanocomposites, due mainly to the uniform dispersion of silicate layers in the polymer matrix. In order to prepare exfoliated polymer/silicate nanocomposites, a variety of methods has been developed, including melt intercalation [22], *in situ* polymerization [23], curing systems [24] and sol–gel techniques [25].

As is well-known, the dispersion scale of silicate in the polymer matrix is largely responsible for the efficiency of clay as a reinforcing filler [26]. Accordingly, the process and mechanism of highly exfoliated polymer/clay nanocomposites have been focal points in areas of both fundamental and applications research [27, 28]. Exfoliated nanocomposites can be prepared in three main ways: (1) solution intercalation; (2) melt intercalation; and (3) *in situ* intercalative polymerization [29]. Whether conducting fundamental research studies or meeting the ever-increasing commercial demands for these materials, highly exfoliated polymer/MMT nanocomposites obtained in a direct and low-cost manner are becoming increasingly attractive [30].

Since Friedlander and Grink [31] first reported a slight expansion of the d_{001} spacing of clay galleries upon intercalation, both intercalated and exfoliated polystyrene (PSt)/clay nanocomposites have been the subjects of extensive investigations. The melt intercalation of PSt into an organically modified sodium bentonite (a layered, mica-type silicate) was also observed, and this led to an approximate 25% increase in the spacing between the silicate layers [32]. In addition, Hoffmann *et al.* [33] reported a correlation between the morphology and rheology of exfoliated PSt nanocomposites, based on organophilic layered silicates such as fluoromicas. In these studies, the intercalated PSt/clay nanocomposites – which were prepared by the polymerization of styrene in organophilic and hydrophilic clays [34] – and exhibited better thermal stability than PSt.

Intercalated polystyrene/clay nanocomposites have been prepared by the *in situ* polymerization of styrene in the presence of an organophilic clay [34]. Likewise,

Akelah *et al.* [35] synthesized intercalated PSt/clay nanocomposites using a short-chain reactive surfactant and solvents, to facilitate intercalation. Qutubuddin and coworkers [36] synthesized exfoliated PSt/clay nanocomposites via an *in situ* polymerization of styrene and a reactive organoclay, while Weimer *et al.* [37] prepared exfoliated PSt/clay nanocomposites by anchoring a living free radical initiator inside the clay galleries. The synthesis of intercalated PSt/clay nanocomposites was also achieved using a polymer melt intercalation [38]. For example, Hasegawa *et al.* [39] synthesized PSt nanocomposites with a melt intercalation by using an organoclay prepared by ion exchange with a protonated, amine-terminated PSt. Exfoliated PSt/clay nanocomposites were also prepared by the melt blending of a styrene–vinyloxazoline copolymer with an organophilic clay [33]. Several research groups have reported the preparation of PSt/clay nanocomposites [33, 40–42] and poly(methyl methacrylate) (PMMA)/clay nanocomposites, based on their high market volume [43–46], although most of these nanocomposites demonstrate the intercalated form of clay.

Some pioneering studies have been conducted with nylon/clay hybrid composites by the research team at Toyota, to create lightweight materials [47]. For example, nylon-6/silicate [23] nanocomposites produced by the Toyota group opened a novel field of materials to both research workers and engineers, mainly because these new composites exhibited balanced mechanical properties such as tensile strength and modulus.

Both, polymer/clay nanocomposites with polyaniline [48] and a styrene–acrylonitrile copolymer [49] have been introduced as candidate materials for drybase electrorheological (ER) fluids. In one study, when dispersed polarizable colloidal suspensions were placed in an external electric field, their rheological properties were abruptly altered due to structural changes in the particle aggregation of the chain- or column-like structures [50].

Nanoparticles such as layered silicates (e.g. naturally occurring bentonites) generally have a thickness of 1 nm, but a width in the submicrometer to micrometer range. These platelet-like particles have been extensively studied during the past few years as filler materials for the formation of polymer/clay nanocomposites, thereby combining the beneficial properties of both materials, including low density, flexibility, excellent moldability, high tensile strength, heat stability and chemical resistance [51].

Polymer-layer silicate nanocomposites (PLSN) can be prepared using four different methods: (1) solution intercalation; (2) *in situ* intercalative (solution and dispersion) polymerization; (3) polymer melt intercalation; and (4) exfoliation adsorption. Various methods based on simple mechanical mixing, bulk polymerization, solution polymerization, suspension and emulsion polymerizations have also been employed. Among the above-mentioned approaches, emulsion and suspension polymerizations are the most attractive for their ease of manipulation, low cost and environmental friendliness. With regards to the layered silicates, the surfactant and/or initiator interactions and the overall view of the dispersion polymerization process must take into account a variety of aspects, including the modification of polarity brought by about by the presence of clay in the dispersion media [52]; a possibility of the persulfate counterion being adsorbed onto the clay surface [53]; or an interaction of the sulfate ion on acid bridges in the center crystal of the aluminosilicate structure [52].

The synthesis of nanoclay composites from dispersed systems has attracted much interest in recent years, the main reasons for this being the encapsulation of layered silicates, the nanolayer stabilization, the *in situ* polymerization of the monomers in intergalleries of the layered silicates, and the possibility of obtaining hybrid particles and nanoparticles [54, 55].

The formation of a stable emulsion from two immiscible liquids in the micrometer or submicrometer range can normally only be achieved by the addition of surface-active compounds, such as surfactants. Organic latex particles and polymers are known to stabilize emulsions, suspensions, emulsion–suspension mixtures (suspoemulsions) and multiemulsions by assembly at the liquid–liquid interface. Moreover, inorganic particles have also been used to stabilize suspensions and emulsions [56]. Several attempts have been made to use these layered silicates for the *in situ* formation of nanocomposites by surfactant-stabilized emulsion polymerization [19, 43, 57–62].

When particles, rather than surfactant molecules, are used to stabilize the emulsion the term 'Pickering emulsion' is commonly used [63]. By reducing the dimensions of the stabilizing particles to the nanoscale, it is possible to generate emulsion droplet sizes down to 50 nm in size [64]. Indeed, the preparation even of a PSt-encapsulated Laponite composite system has been claimed by Sun *et al.* [60], via miniemulsion polymerization.

Dispersions of inorganic particles have been used to prevent agglomeration [56], while inorganic clay platelets have been used as stabilizers for emulsions and inverse emulsions [64]. In the case of emulsion polymerizations in the presence of clay platelets, so-called 'armored latex' were commonly synthesized. In a recent study conducted by Bon *et al.* [64], stable styrene oil-in-water (o/w) emulsions were prepared by using synthetic hectorites as a stabilizer in the absence of a surfactant. Following the polymerization of styrene, a Pickering dispersion of latex particles that was stabilized by synthetic clay platelets was obtained. Layered silicates have recently been reported by several research groups as stabilizers for water-in-oil (w/o) emulsions for cosmetics, drilling liquids and asphalt coatings [65]. Alternatively, Binks *et al.* [66] prepared an inverse emulsion stabilized by hydrophobically modified bentonite particles.

The emulsion polymerization of vinyl acetate and acrylic monomers with different stabilizing systems (anionic, nonionic and protective colloids) in the presence of MMT caused an enhancement of both the conversion and polymerization rates of the monomer [67, 68]. However, in some cases, where there was a high concentration of layered silicate in relation to the monomer (>12.5 wt%), a small decrease was observed in the rate of polymerization. The same authors also studied the conditions by which to obtain polymer/clay stable latexes, by controlling the dimensions of the particles and the polymer/clay nanocomposites through microemulsion polymerization [69]. Under specific conditions, it was found that hybrid nanolatexes were obtained.

13.2
Cloisite Clays and Organoclays

The exfoliation of pristine MMT is impeded by an electrostatic attraction between the negatively charged clay layer and cations, such as Na^+ and Ca^{2+}, in the gallery. Thus, a pretreatment is normally required when preparing exfoliated MMT nanocomposites. The opportunity to combine clays and a natural or synthetic polymer, at the nanometric level, seems to be an attractive way to develop new organic–inorganic hybrid materials that possess properties inherent to both types of component. The very large majority of polymer/natural silicate nanocomposites have shown an intercalated morphology. In the preparation of polymer/silicate nanocomposites, it is important to consider several points such as the surface character, the regular structure of pristine silicates [70], and the compatibility between silicates and polymers. For example, pristine clay is naturally hydrophilic (Figure 13.1), while its polymers are often hydrophobic [71]. This hydrophilic nature of pristine clay impedes its homogeneous dispersion in a polymer matrix. However, a brilliant suggestion was made that the silicate layers and polymers could be connected covalently by using modifiers containing reactive sites or, alternatively, to anchor the surface-active additives on the silicate surfaces [72]. However, this technique also included a step for the modification of pristine silicates. Consequently, a few organic modifiers [73] were adopted to modify silicate layers. Thus, the clay modified with alkylammonium facilitates its interaction with a polymer, because the alkylammonium renders the hydrophilic clay surface organophilic. Many research studies on the synthesis of nanocomposites have focused on the organic modification of layered silicates (OLS) [74–76].

A typical synthesis of polymer/clay nanocomposites involves the organic modification of a clay with an alkylammonium cation, and intercalation of a suitable monomer followed by *in situ* polymerization. The role of the alkylammonium cation is to improve the penetration of the organophilic monomers into the interlayer space, while the role of the monomer is to promote dispersion of the clay particles.

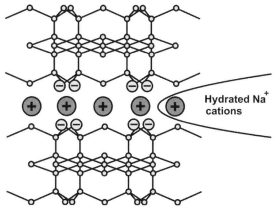

Figure 13.1 Natural Cloisite Na (NaMMT) with basal spacing d_{001} ca. 1.19 nm.

An increased basal spacing arises from the expansion of the interlayer space to accommodate the polymer and, as a result, the intercalation process of polymers can be distinguished by the difference in basal spacing [77]. The van der Waals attraction between the negatively charged silicate layers and positively charged alkylammonium cation causes major alterations to the interior of the clay. It has been speculated that the result of this attraction is the formulation of an electrical 'triple' layer, or two double layers, on both sides of the silicate layer. The densely packed layers of the natural clay can then be defined in terms of the 'triple' layer in the interlayer. A similar behavior is favored by the low concentration of modifier, or a modifier with small alkylammonium cations. In this case, the two double layers might be formed by a modification of clays by a high concentration of modifier, or a modifier with large alkylammonium cations (alkyl = tallow) or charged polymers, as a result of which the gap between layers is increased.

The basic clay is a Cloisite Na (Cl Na); this has no surfactant modification and can have the chemical formula of:

$$M_x[A_{4-x}Mg_x]Si_8O_{20}(OH)_4$$

where M is a monovalent charge-compensating cation. Natural Cl Na (NaMMT) has the formula $Na_{0.02}K_{0.02}Ca_{0.39}[Fe_{0.45}Mg_{1.10}Al_{2.51}][Si_{7.91}Al_{0.09}]O_{20}(OH)_4 \times nH_2O$.

The surfactants which are broadly used to prepare the organoclays are ammonium salts of tallow derivatives, and the organoclays can be divided into two families by the structures of these surfactants. The first family includes Cloisite 30B, 10A and 25A, the surfactants of which possess only a single long tallow chain; in the second family, which includes Cloisite 93A, 20A, 20B and 15A, the surfactants possess two tallow chains, or a swallowtail:

- Group (family) 1 (e.g. Cloisite 30B) has one tallow (T) containing about 65% hydrocarbon chain C_{18}, 30% C_{16}, and 5% C_{14}, plus two ethanol molecules ($M_w = 404.8$):

$$CH_3 - \overset{+}{N}(CH_2CH_2OH)(CH_2CH_2OH) - T$$

- Group (family) 2 (e.g. Cloisite 20B) has two hydrogenated tallows (HTs) containing about ~65% C_{18}, ~30% C_{16}, and ~5% C_{14}, plus two methyl groups ($M_w = 551.6$):

$$CH_3 - \overset{+}{N}(CH_3)(HT) - HT$$

The number of tallow chains – or 'tails' – has a profound effect on the properties of the nanocomposites. Cloisite 30B enjoys a unique feature in that its surfactant molecules possess pendant hydroxyl groups, which provides Cloisite with excellent properties of either intercalation or viscoelasticity. The basal spacing in these organoclays is increased as follows [78]:

Group 1: d_{001}(nm)/Cloisite: 1.17/Na, 1.85/30B, 1.86/15A and 1.92/10A
Group 2: d_{001}(nm)/Cloisite: 2.36/93A, 2.42/20A, 3.15/15A

The gallery heights of the organoclays in the swallowtail family are about one-half of a fully stretched chain length of an alkyl carbon chain. It is suggested that the tallow tails in the galleries adopt a near-fully extended conformation, which is tilted to the silicate surface [79]. A higher ion-exchange capacity leads to a larger d-spacing for both families, which is to be expected as more exchanged surfactant molecules bring about more crowdedness and hence a higher d-spacing.

Clays can be modified with classical cationic surfactants (e.g. alkylammonium salts). For example, Cloisite Na was organically modified with octadecyltrimethylammonium bromide (ODTMAB) and hexadecyltrimethylammonium bromide (HDTMAB) before intercalation of the styrene monomer [80]. The results of X-ray diffraction (XRD) studies have shown that the basal spacing d_{001} (i.e. the interlayer distance of a (001) reflection plane) is increased by increasing the alkyl chain length of the cationic surfactant:

- d_{001} (nm)/additive: 1.25/none, 1.93/HDTMAB, 2.23/ODTMAB

Densely packed alkyl chains of either HDTMAB or ODTMAB form the organic layer between silicate surfaces (or double layers). In addition, the cations of the surfactants project to the negatively charged silicate surfaces, while their alkyl (HD or OD) chains project to the middle of the layer where they form the hydrophobic organic matrix.

Cloisite Na was functionalized by vinylbenzyl-dimethyldodecylammonium chloride (VDAC) using a cationic exchange process, and the VDAC-Cloisite organoclays formed were characterized using XRD [61]. In order to avoid coagulation, the amount of added VDAC was one-half that of the Cloisite. The amount of surfactant required for coagulation is less than (but close to) the amount of clay × the cation-exchange capacity (CEC)/100 of Cloisite [81]. The XRD pattern of untreated Cloisite Na shows a broad peak centered at $2\theta = 8.9°$ (0.99 nm). In contrast, VDAC-Cloisite, prepared by the cationic exchange of Cloisite Na with VDAC in aqueous solution in the presence of styrene droplets [36], exhibits a sharp peak at $2\theta = 4.6°$. The d_{001} spacing of VDAC-Cl Na was increased from 0.99 nm for pristine Cloisite Na [43] to 1.92 nm in the dried state [36] (Fig. 13.1A):

- d_{001} (nm)/additive: 0.99 (8.9°)/none, 1.92 (4.6°)/VDAC

The d_{001} spacing with water is 1.55 nm, but in the dried state this was reduced to 0.99; that is, the ratio of d_{001}s with and without water (1.55/0.99) was 1.56. The ratio of the d_{001}-values for ODTMAB in water and VDAC without water (2.23/1.92) was only 1.16, which indicated that the interlayer spacing was increasing with increasing size of the surfactant. The presence of water also strongly favors the expansion of interlayer spacing, due to colloidal forces.

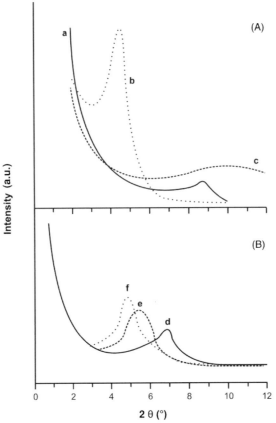

Figure 13.2 X-ray diffraction patterns. (A) (a) Cloisite Na (NaMMT) ($d_{001} = 0.99$ nm); (b) VDAC-Cl (1.92 nm); and (c) PSt/Cl nanocomposite particles (10.2°) [61]. (B) (d) pristine Cloisite Na ($d_{001} = 1.14$ nm); (e) Cl$_{AUA}$; and (f) Cl$_{CTAB}$ [59].

The XRD patterns shown in Figure 13.2A and B depict the limiting cases of untreated and VDAC-treated MMT. Curve c in Figure 13.2A shows the XRD pattern of a PSt/clay nanocomposite containing 5 wt% MMT [61]. For a Cloisite loading of 5 wt% or less, there is no peak corresponding to Cloisite Na (NaMMT) or VDAC-MMT (VDAC-Cl Na). However, there is a 'hump' which is centered at $2\theta = 10.2°$. The data in Figure 13.2A suggest that partially exfoliated PSt nanocomposites could be obtained via a one-step emulsion polymerization.

The zwitterion, amino-undecanoic acid (AUA), which was reported to be effective as a modifier for clay [59], possesses both an amino group ($-NH_2$) and a carboxylic group ($-COOH$) within its molecule, and can be protonated to form cationic $-NH_3^+$ in an acidic medium. Under acidic conditions the amino acid molecules can easily enter the Cloisite Na galleries to accomplish ion exchange with the interlayer cations, after which powder XRD patterns of the Cloisite Na and its AUA derivatives (Cl$_{AUA}$) and CTAB derivatives (Cl$_{CTAB}$) were measured (Figure 13.2B). Following cation exchange with AUA, SDS and CTAB, the d_{001} spacing was expanded to 1.72 nm for Cl_{AUA}, to 1.8 nm for Cl_{SDS}, and to 1.84 nm for Cl_{CTAB} [59]:

- d_{001} (nm)/clay: 1.14/Cl Na, 1.72/Cl_{AUA}, 1.8/Cl_{SDS}, 1.84/Cl_{CTAB}

These results indicate that AUA, SDS and CTAB are each intercalated into the interlayers of clay.

Donescu et al. have reported the dependence of clay mineral swelling capacities on the different monomer polarities [82, 83]. For Cloisite Na, the swelling capacity was found to be more than 100 wt% for acrylic monomers. Another important argument related to monomer–clay interaction relates to the expansion of d_{001} specific to clay galleries when the MMT is dispersed in the monomer phase [84]. In perspective clays, d_{001} is very much affected by the presence of different organic solvents [85], with the solubility parameters depending not only on dispersion and polar interactions but also on the hydrogen bonding related to the solvent. The most important parameter, however, is considered to be the dispersion, which is the determinant of clay exfoliation.

The d_{001} spacing of Cloisite Na (NaMMT) was followed by using XRD, as a function of the solvent type [86] (Figure 13.3):

- d_{001} (nm)/solvent: 1.62 ($2\theta = 5.46°$)/MeOH, 1.57 (5.64°)/EtOH, 1.40 (6.33°)/IPA, 1.21 (7.30°)/n-hexane, 1.19 (7.41°)/toluene, 1.18 (7.44°)/benzene, 1.17 (7.6°)/water

Liquids with high δ_h (hydrogen bonding component) values expand the d_{001} spacing of Cloisite more than liquids with low δ_h values. Moreover, the spacing of

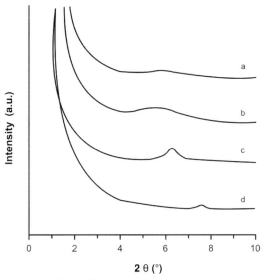

Figure 13.3 X-ray diffraction patterns of Cloisite Na (NaMMT) at different polar solvents: (a) 1.57 nm (5.64°)/EtOH; (b) 1.62 nm (5.5°) MeOH; (c) 1.4 nm (6.33°) IPA; (d) 1.17 (7.6°) water [86].

natural MMT is strongly related to δ_h rather than to δ_d (the dispersive component), because the liquids have almost the same δ_d values. Experiments using acrylonitrile (AN), n-butyl acrylate (BA), methyl methacrylate (MMA) and styrene (St), all of which have almost the same δ_h values, can be used to explain the effect of δ_p on the basal spacing expansion:

- d_{001} (nm)/unsaturated monomer: 1.81 nm ($2\theta = 4.87°$) BA, 1.66 nm (5.30°) AN, 1.60 nm (5.54°)/MMA, 1.20 nm (7.39°)/styrene

The basal spacing expansion depends on the polar components (δ_p) and hydrogen-bonding components (δ_h) of organic liquids. Liquids with high δ_h values showed an expansion from methyl alcohol to benzene, while monomers with high δ_p values exhibited an expansion from n-butyl acrylate to styrene. Monomers with high dipole moments showed large basal spacings, whereas those with low dipole moments showed smaller basal spacings.

The hydrogen (δ_h) and polar (δ_p) components of liquids are the primary parameters for the dispersion states, and for the basal spacing expansion. The degree of expansion by monomers reflects the expected structure of polymer/Cloisite nanocomposites: a large basal spacing produces exfoliated structures, while a small basal spacing produces interacted structures. The correlation between the basal spacing expansion and the structure of polymer/MMT nanocomposites is reasonable because the polymerization rates depend on the monomer concentrations. Because δ_h and δ_p both involve dipole–dipole interactions, monomer dipole moments are associated with the basal spacing expansion that is related to the structure of polymer/MMT nanocomposites.

The XRD peaks of the Cloisite dispersed in BA/styrene comonomers appear between the peaks of BA/MMT (4.87°) and styrene/MMT (7.39°) [86]. As the styrene content increases, the XRD peak shifts toward a high-angle position, approaching that of styrene/Cloisite (MMT). The BA/MMA/MMT is expected to have an XRD peak between the 2θ range of BA/MMT and MMA/MMT (5.54°), because the BA/MMA comonomers have dipole moment values between those of BA (2.31 D) and MMA (1.67 D). As the BA content in the comonomer increases, BA/MMA/MMT shows a decreasing basal spacing until the weight ratio of BA to MMA is 27 : 3. When this weight ratio is below 27 : 3, the basal spacing increases to that of BA/MMT. This increase could be due to two competitive interactions between: (1) MMT and the mixed monomer; and (2) the two monomers.

The peak of Cloisite Na dispersed in water occurs at 5.7°, which corresponds to the basal space of 1.55 nm in 2θ value, and means that the interlayer space of silicate is widened by water [87]. This d_{001} value differs somewhat from that of 1.17 reported by the same group in a previous model study [86]. Rather, a value of 1.55 is the result of a strong penetration of water into the hydrophilic layers of the clay. The Cloisite Na, when dispersed with a comonomer in water, shows a peak around 5.58°, with a d_{001}-space of 1.61 nm. The water-saturated clay does not favor absorption of the hydrophobic comonomers (styrene and MMA). The Cloisite Na, when dispersed with 2-acrylamido-2-methyl-1-propane sulfonic acid (AMPS) in water, shows a peak at about 3.8°, with a space of 2.32 nm. The surface-active AMPS molecules saturate and widen the interlayers of clay and, as a result, the Cloisite Na dispersion with AMPS and comonomer in water shows no peak. Rather, it indicates that a reactive surfactant AMPS, when intercalated in clay, attracts comonomers to widen the gap between the clay layers:

- d_{001} (nm)/additive: 1.55/water, 1.61/(water + comonomer), 2.32/(AMPS + water), no peak/AMPS + comonomer ($A_{0.3}M_{10}S_{10}T_5 = 0.3$ g AMPS, 10 g MMA (M), 10 g styrene (S), 5 wt% Cloisite Na (T) relative to 20 g comonomer (MMA and styrene))

The formulation of double layers with repulsive van der Waals forces might help to exfoliate the Cloisite matrix; the depletion forces generated by the absorption of comonomer may also contribute to such exfoliation.

The X-ray patterns for different contents of clay at fixed levels of AMPS were also investigated. The peaks for 1 and 3 wt% of clay appeared at 3.74° with a basal space of 2.36 nm, whereas the peak for 5 wt% clay was at 5.13° with a basal space of 1.72 nm. The d_{001}-spacing for 1 and 3 wt% was wider than that for 5 wt%; hence, the AMPS contents of 1 and 3 wt% clays were greater than for 5 wt% clay:

- d_{001} (nm)/additive: 2.32 (3.74°)/clay (1–3 wt%), 1.72 (5.13°)/clay (5 wt%)

$A_{0.3}M_{10}S_{10}T_5$ shows a peak at about 5.8° with the basal space of 1.52 nm, which indicates that the clays in the nanocomposite are intercalated and still have van der Waals force interactions between the layers (electric triple layer). In contrast, $A_{0.3}M_{10}S_{10}T_1$ and $A_{0.3}M_{10}S_{10}T_3$ show no peaks, indicating the exfoliation of clay up with to 3 wt% of silicates [87]:

- d_{001} (nm)/nanocomposite: 1.52 (5.8°)/$A_{0.3}M_{10}S_{10}T_5 \gg$ no peaks/$A_{0.3}M_{10}S_{10}T_1$ and $A_{0.3}M_{10}S_{10}T_3$

The interlayer space of pristine silicate is enlarged with water, showing (001) plane diffraction at 5.78° (d spacing of silicate layers: 1.55 nm) [88]. The diffraction peak of an aqueous dispersion of silicate containing styrene shows a (001) plane diffraction at 5.68° (1.58 nm), while the diffraction peak of aqueous dispersion of silicate with St and AMPS show at 4.28° (2.1 nm). Meanwhile, KPS, AN and AMPS also make the interlayer space of silicate even wider in the aqueous dispersion, and show no discernible diffraction patterns:

- d_{001} (nm)/additive(s): 1.55 (5.78°)/water, 1.58/(water + styrene), 2.1 (4.28°)/(water + styrene + AMPS), $\gg$ no peaks/(AN + water), (KPS + water), (AMPS + AN + water)

These results indicate that AN has a higher interaction with silicate layers than styrene, and the amount of AN in the interlayer space of silicate is higher than that of styrene. It also means that, as certain portions of the low-molecular-weight electrolyte (KPS; potassium peroxodisulfate, initiator), acrylonitrile and AMPS are present in the interlayer space of the silicate, polymerization can easily be induced in the interlayer space. That leads the exfoliated state of the polyacrylonitrile (PAN), (St-AN) copolymer/silicate nanocomposites. Monomers and initiators easily penetrate the interlayer spaces via the aqueous phase, while the preadsorbed monomers attract reactants to the interlayer spaces during the polymerization. The polymerization rate for the emulsion system correlates with approximately the first order of monomer concentration and the number of growing particles [89]; that is, the more monomers in the basal spacing the faster the polymerization rate.

The peak of pristine Cloisite Na (NaMMT) dispersed in water is detected at an angle of $5.7°$ (1.55 nm), and the NaMMT dispersed in MMA and water shows a peak around $3.0°$ (3.27 nm) [62]:

- d_{001} (nm)/additive: 1.55 (5.7°)/water, 3.28 (3°)/(water + MMA)

The latter has two peaks at angles of 3 and 6°. It is believed that the intercalation of MMA in pristine Cloisite Na layers causes the appearance of a peak at $3.0°$. In a water/MMA dispersion state, the gap of Cloisite Na appears to be widened to provide a volume sufficient enough for the surfactants to be intercalated into the layers, but the interaction between layers protects the layer gap against exfoliation. Both surfactants (AMPS and NaDBS) are intercalated into silicate layers before polymerization [62].

The peaks for the Cloisite Na with AMPS appear at $7.5°$ (1.2 nm) and $3.8°$ (2.3 nm): the peak at $7.5°$ corresponds to the peak for pristine Cloisite Na, while the peak at $3.8°$ is caused by the intercalation of AMPS into the silicate layers. In the case of dodecylbenzenesulfonic acid sodium salt (NaDBS, D), two peaks appear at $2.9°$ (3.0 nm) and $6.6°$ (1.4 nm), and the basal spaces are thought to become wider than those for AMPS. Even though water is removed, both surfactants remain in the silicates. Thus, it is concluded that the peak at $3.0°$ of $D_{2.5}M_{20}T_1$ is caused by the intercalation of NaDBS which is not fully extracted from Cloisite Na [62].

In the presence of $NaHCO_3$ [a buffer used to maintain a constant (pH-related) decomposition rate of the peroxodisulfate initiator], the d_{001} spacing was slightly expanded to 1.2 nm for Cloisite Na (Figure 13.4A). In the presence of both 2.66 mM $NaHCO_3$ and 9 mM SDS, the two peaks appeared at $6.75°$ and $4.3°$, which correspond to d_{001} spacings of 1.18 and 1.8 nm, respectively [90]:

- d_{001} (nm)/additive: 1.15/none, 1.2/$NaHCO_3$, 1.18 and 1.8/($NaHCO_3$ and SDS)

These data indicate that the penetration of $NaHCO_3$ slightly increases the d_{001} spacing due to a partial neutralization of the free ions within the Cloisite. The depression of van der Waals forces leads to a partial expansion of the Cloisite matrix. The presence of SDS micelles is accompanied by an expansion of the d_{001} spacing (1.8°). This is most likely due to penetration of the SDS molecules into the clay interlayers, and the formation of inverse SDS micelles and mixed Cloisite/SDS intercalated aggregates. The degree of expansion by the present SDS and $NaHCO_3$ solution is much less than with the aqueous solution of NaDBS (D).

The organization and exfoliation of clay (15A and 20A) platelets were studied in cyclohexane and xylene (both solvents form inverse emulsions) (Figure 13.4B) [91]. For the Cloisite 20A (after drying), a strong peak appeared at $2\theta = 3.48°$, corresponding to the basal spacing (d_{001}) of 2.54 nm, while a secondary weak peak was observed at $2\theta = 7.08°$. In the 2 wt% dispersion of clay in cyclohexane (CH) only a very weak peak at $2\theta = 3.48°$ was observed, indicating that the clay platelets may be in a partially exfoliated state in the organic medium:

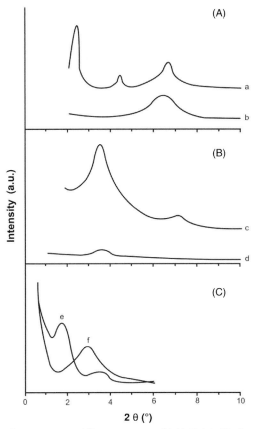

Figure 13.4 X-ray diffraction spectra. (A) (a) Cloisite Na dispersed in aqueous 2.66 mM NaHCO$_3$ and 9 mM SDS; (b) Cloisite Na dispersed in aqueous 2.66 mM NaHCO$_3$ [90]. (B) (c) Cloisite 20A after drying; (d) 2 wt% Cloisite 20A dispersed in cyclohexane [91]. (C) (f) nanoparticles of Cloisite 15A; (e) mixture of xylene + 5 vol% Cloisite 15A [92].

- d_{001} (nm)/additive: 2.54 (3.48°) and 1.25 (7.08°)/none, 2.55 (3.48°)/cyclohexane, 2.8 (3.1°)/2 wt% dispersion of clay in cyclohexane (crosslinked composite nanoparticles), ≫ no peaks/2 wt% dispersion of clay in cyclohexane (uncrosslinked composite nanoparticles)

The XRD analysis of a 5 vol% Cloisite 15A (Cl 15)/xylene dispersion did not show any evidence for complete exfoliation of the clay, although a significant increase in the interlayer distance of the clay, from 3.15 to 5.28 nm (2θ = 1.67°) was observed (Figure 13.4C) [92]:

- d_{001} (nm)/additive: 3.15/none, 5.28 (1.67°) and 2.5 (3.5°)/xylene

The penetration of xylene into the organically modifier Cloisite leads to an expansion of the interlayer spacing. This increase in the interlayer distance is discussed in terms of the formation of a percolating network structure. Because of the large specific surface area of Cloisite nanoparticles upon exfoliation, a percolating network structure may form at very low volume fraction of Cloisite in a liquid (xylene), leading to a significant increase in the melt viscosity of the xylene/Cloisite nanodispersion.

The basal spacing increases from 1.25 to 1.33 for Cl Na and Cl Cu respectively [80]. The addition of polyaniline (PAn) suppresses the effect of cation type on the basal spacing:

- d_{001} (nm)/Cloisite (MMT) type: 1.25/Na, 1.49/(Na + PAn), 1.33/Cu, 1.48/(Cu + PAn)

The location of larger Cu cations in the interlayers may increase the charge of the electrical layers, while the presence of a polymer can lead to the depletion effect.

Compared with the weight loss of pristine Cl Na, the weight loss 16.1% of Cl_{AUA} in the temperature range 200 to 800 °C is attributed to the decomposition of AUA [59]. As the calculated value on the basis of the CEC of clay is 20.1%, it can be see that the ion exchange reaction was satisfactory.

Exfoliation of the clay after ion exchange with AUA was greatly facilitated in an alkaline medium. When the pH of the Cl_{AUA} aqueous dispersion was adjusted to 8.5 and 10, a major change was observed in the viscosity after stirring overnight. It was reported that, at pH 7, the suspension existed as a Newtonian fluid and the viscosity was very low. However, at pH 10 the viscosity of the Cl_{AUA} suspension at a low shear rate was increased by about four orders of magnitude compared to the suspension at pH 7, and a pronounced shear-thinning behavior was observed. Such as vast increase in viscosity is usually associated with a very large increase in the number of dispersed particles, while the shear dependence is associated with an increase in the aspect ratio of the dispersed particles [93]. In other words, the change in rheological behavior with the rise in pH indicates the occurrence of exfoliation of the layered silicate in an alkaline medium.

It is reasonable to suggest that, in the alkaline medium the $-COOH$ groups of the interlayer amino acid molecules are ionized to form anions with $-COO^-$ terminal groups. The thus-derived anionic $-COO^-$ groups then interact repulsively with each other and also with negatively charged clay layers, forcing the layers apart and leading to an exfoliation of the clay into individual layers. The existence of exfoliated clay layers in the Cl_{AUA} suspension in the alkaline medium was confirmed by using atomic force microscopy (AFM); this showed the lateral dimension of the Cloisite particles to be in the range of 100–200 nm, while the height of 1–2 nm was consistent with one to two elementary silicate layers.

The sodium ions in naturally occurring clays can be exchanged with organic cations to make the platelets dispersible in an organic medium (Figure 13.5). The long alkyl chain of the quaternary ammonium compounds, $RH_3N^+ X^-$ (where R denotes alkyl chains and X mostly halogen anions), that are adsorbed onto the surface of the hydrophobized clay surface enables a swelling and exfoliation of the platelets in an apolar medium. Mixing the clay platelets that are dispersed in the oil phase with the monomer aqueous solution results in a slight increase in the turbidity

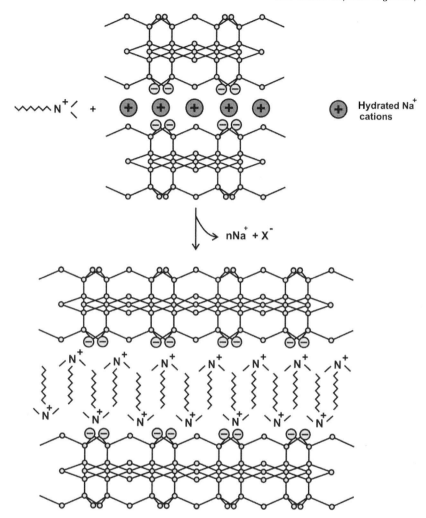

Figure 13.5 Exchange of sodium ions in naturally occurring clays by organic cations (R_4N^+ X^-) and swelling and exfoliation of clays by apolar solvents and polymers (figure continued on p. 258).

of the apolar phase. Although, the hydrophobically modified clay platelets are dispersed in the organic phase, they tend to adsorb a small portion of water and transport to the oil phase. This phenomenon was also observed by Binks et al. [66] and later by Voorn et al. [91]. This increase in turbidity indicates that some of the particles formed in the oil phase are able to scatter light. The exfoliation and dispersibility of the organically modified clays in cyclohexane was investigated with dynamic light scattering (DLS) [91]. First, the organically modified clay dispersed in cyclohexane showed a broad distribution, most likely due to an association of the clay platelets which form large aggregates (from ca. 100 nm to 3 μm). Following several

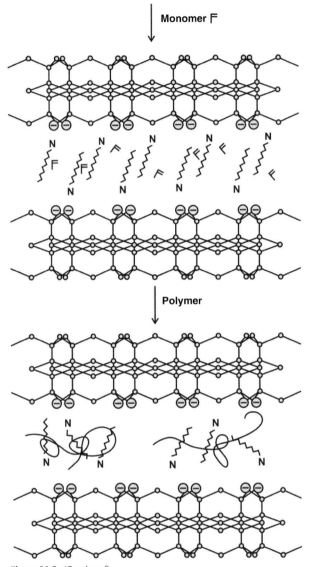

Figure 13.5 (Continued)

ultrasonication treatment steps, the stacked clay platelets further lowered the average particle size to ~500 nm. However, the increased concentration of the modified clay (Cloisite 20A) led to a slight decrease in the average particle size:

- d (nm)/Cloisite 20A (wt%): 700/0.27; 600/0.49; 500/0.69 wt%

In natural Cloisite (Cl_N), a content of 1–4% iron, located mainly in the octahedral lattice positions and existing in two oxidation states – Fe(II) and Fe(III) – leads to a paramagnetic response with the mass magnetic susceptibility $\chi \approx (1–5) \times 10^{-6}$ emu/(g Oe) [94] (Cloisite density $\chi \approx 2.5$ g cm^{-3}). Synthetic Cloisite (Cl$_S$), which commonly lack iron, are diamagnetic ($\chi < 0$) [94]. Callaway and McAtee have reported that such susceptibility does not correlate directly with the amount of Fe in the layered silicates [95], but may also be related to the presence of additional paramagnetic atoms in the lattice or separate phases (Mn^{2+}, Ti^{3+}) as well as the existence of paramagnetic centers intrinsic to the lattice (stable hole traps due to ionizing radiation or grinding). A chemical analysis of the base Cl$_S$ used for organic modification indicates that the mean composition of a unit cell of Cl$_S$ is (Si$_{7.8}$Al$_{0.2}$O$_{16}$)(Al$_{2.96}$Fe$_{0.45}$Mg/Ca$_{0.46-0.13}$O$_4$(OH)$_4$)Na$_{0.66}$ (CEC = 91 mEq. per 100 g), while that of Cl$_N$ is (Si$_{7.72}$Al$_{0.28}$O$_{16}$)-(Al$_{2.7}$Fe$_{0.39}$Mg/Ca$_{0.91-0.13}$O$_4$(OH)$_4$))Na$_{1.19}$ (CEC = 145 mEq. per 100 g).

The paramagnetic response of Cl$_S$, however, does not account for the opposite alignment of the Cl$_N$. The superconducting vibrating sample magnetometer (VSM) data at 300 K for Cl$_N$ and Cl$_S$ are summarized in Figure 13.6. The magnetic mass susceptibility χ [$\chi_{NC} = 7.08$ emu/(g Oe) $\times 10^{-6}$, $\chi_{SC} = 6.12$ emu/(g Oe) $\times 10^{-6}$] is in good agreement with data published elsewhere [94]. However, the hysteresis, which is significantly more pronounced for Cl$_N$, indicates the presence of antiferro- and ferrimagnetic phases in both Cl$_S$ and associated remnant magnetization M$_r$ (M$_{r,N}$ = 0.0055 emu cm^{-3}, M$_{r,S}$ = 0.0015 emu cm^{-3}). The remnant magnetization, M$_r$ (intercept at H = 0 of the linear extrapolation of the high-field magnetization curve), of Cl$_N$ is significantly larger than that of Cl$_S$ (M$_{r,N}$ = 0.0055 emu cm^{-3}, M$_{r,S}$ = 0.0015 emu cm^{-3}), which is indicative of a much larger contribution of antiferro- and/or ferrimagnetic components. Investigations at different temperatures show an increase in magnetization from 300 to 5 K, as expected (Bloch's law), and a Curie temperature, T_C, for the majority antiferro/ferrimagnetic phase in both systems greater than 500 °C.

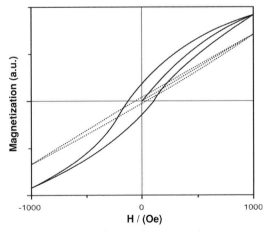

Figure 13.6 Superconducting vibrating sample magnetometer data of Cl$_N$ (solid) and Cl$_S$ (dashed) at 300 K showing hysteresis at M$_{mag}$ = H = 0 [96].

13.3
Radical Polymerization

13.3.1
Solution/Bulk Polymerization

Water- and oil-soluble polymers and copolymers can be synthesized by different routes. Among the most important requirements for the technological applications of water-soluble polymers are their high solubility in water, simple and cheap route of synthesis, adequate molecular weight and molecular weight distribution, chemical stability, high affinity for one or more metal ions, and selectivity for the metal ion of interest. The most frequently used synthetic procedures are addition polymerization, especially radical polymerization, and the functionalization of polymer backbones through polymer-analogous reactions.

During recent years, much attention has been paid to layered clay/polymer nanocomposites as advanced plastic materials prepared by *in situ* intercalative polymerization. The nanocomposite produced from PMMA/clay has been the most widely studied, while MMT is used widely as a layer silicate [97]. Here, the main studies have focused on the layer structure changes caused by the use of different intercalative agents and different polymerization conditions. Clearly, the polymerization of monomers will also be affected by the silicates added to the system, which in turn will affect the layer structure and, ultimately, the properties of the nanocomposites formed.

Uskov reported that MMA can graft onto air-dried sodium bentonite under vibrodisintegration [98] whilst, at the same time, no homopolymer was found in the polymerization. This finding indicates that the bentonite has an initiation effect on the polymerization of vinyl monomers. Solomon systematically studied the effect of different minerals on the polymerization of styrene [99], and showed that the polymerization has characteristics expected of both a radical and a cationic reaction. It was also found that the aluminum atom in the octahedral coordination of the mineral surface was responsible for the catalyst activity. In the radical polymerization of MMA, the aluminosilicates and magnesium silicates were seen to play an inhibitory role [100], which was in contrast to the styrene system.

The rate of bulk MMA polymerization was observed to increase slightly with increasing amounts of Cloisite 6A (Cl 6A, BPO = 0.04 mol dm^{-3}, 60 °C) [101]:

- $R_p \times 10^{-4}$ mol dm^{-3} s^{-1}/Cl 6A (wt%): 4.6/0, 4.6/0.1, 4.7/0.2, 4.8/0.4

When the dibenzoylperoxide (BPO) concentration is small, then with an increasing Cl 6A content the R_p will rise to a maximum and then decrease; however, when the BPO concentration is relatively large, the value will increase continually. This indicates that the effect of the organically modified Cl (OCl) content on R_p is related to the BPO concentration, and R_{p0} is affected by the combination of BPO with OCl.

The classic radical polymerization theory gives the following equation for the rate of polymerization [102]:

$$R_p = k_p[M][R^{\bullet}] = k_p(2fk_d/k_t)^{0.5}[M]^{1.0}[I]^{0.5} = (k_p/k_t^{0.5})[M]^{1.0}R_i^{0.5} \quad (13.1)$$

where [M], [R$^\bullet$] and [I] are the concentrations of monomer, radical and initiator, respectively, k_p is the propagation rate constant, f the initiator efficiency, k_d the initiator decomposition constant and k_t the termination rate constant. When the termination of the chain is a bimolecular process, the reaction order n on [I] is 0.5. In the case of monomolecular termination, $n = 1$. For present MMA/Cl systems, n varies with increasing clay amounts from 0, 0.1, 0.2 and 0.4 wt% to 0.48, 0.47, 0.52 and 0.56 wt%. The results indicate that there is an increasing fraction of more stable growing radicals formed by chain transfer to Cl matrix (encapsulation) or increasing contribution of monomolecular termination.

The increased rate of polymerization can be discussed in terms of the rate of initiation and decreased rate of bimolecular termination [103]:

1. Initiation

$$I - k_{d1} \rightarrow R^\bullet \tag{13.2}$$

$$I + Cl - k_{d2} \rightarrow Cl^*(+ R^\bullet) \tag{13.3}$$

$$I + Cl^* - k_{d3} \rightarrow R^\bullet(+ Cl) \tag{13.4}$$

This step increases the decomposition of initiator and encapsulation of reaction loci in Cl

$$R^\bullet + M \rightarrow M^\bullet \tag{13.5}$$

$$Cl^* + M \rightarrow ClM^\bullet \tag{13.6}$$

where Cl^* (MMT*) represents the reaction loci of Cloisite (montmorillonite). [Cl^*] was assumed to be in proportion to the concentration of OCl and initiator.

2. Propagation

$$\begin{aligned} M_i^\bullet + M - k_p &\rightarrow M_{i+1}^\bullet \\ ClM^\bullet + M - k_{pc} &\rightarrow ClM_{i+1}^\bullet \end{aligned} \tag{13.7}$$

The growth of oligomer and polymer chains can proceed in two zones: outside of the Cl, and within the monomer-saturated Cl (encapsulation of polymer chains) matrix.

3. Termination

a) Bimolecular termination:

$$\begin{aligned} M_i^\bullet + M_i^\bullet - k_{t1a} &\rightarrow P \\ ClM_i^\bullet + M_i^\bullet - k_{t1b} &\rightarrow P_{Cl} \end{aligned} \tag{13.8}$$

The classical bimolecular termination of two growing radicals proceeds in the reaction vessel. The entry of primary and oligomer radicals into the palisades of Cl^* terminates the polymer growth.

b) Monomolecular termination:

$$Cl^*(+M) - k_{t2} \to P_{Cl,de} \tag{13.9}$$

$$M_i^\bullet + Cl^* - k_{t2} \to P + Cl^* \tag{13.10}$$

The immobilization (by precipitation, crosslinking, ...) of growing radicals within the Cl palisade leads to the deactivation of chain growth.

c) Chain transfer:

$$M_i^\bullet + P_{Cl} - k_{t3a} \to P + Cl^* \tag{13.11}$$

$$X + Cl^* - k_{t3b} \to X^\bullet + Cl \tag{13.12}$$

The interaction of growing radicals with Cl and active Cl with additive (X) can lead to the chain transfer reactions.

The rate of polymerization can be written as follows

$$R_p = -d[M]/dt = k_p[M][M^\bullet] \tag{13.13}$$

The induction periods (IP) appeared in the bulk polymerization of MMA and St initiated by BPO in the presence of clay (Cloisite 6A) [103]. This was attributed to the impurities (metal salts, etc.) in the clay or the aluminum on the crystal edges, which reacted with the primary radical. More clay contained more deactive spots; that is, the IP was prolonged with increasing Cloisite 6A content (BPO = 0.04 mol dm^{-3}):

- IP(min)/Cloisite 6A (wt%): 9/0.1, 11/0.2, 14/0.4, 22/0.8

This behavior indicates that the interior of Cl forms part of the reaction system – that is, it contains the initiator, monomer and solvent. The primary radicals formed by decomposition of the BPO are deactivated by the radical scavenger Clay molecules. The impurities would not lead to a linear dependence of IP on the Cl concentration. IP can be discussed in terms of the reduction of metal cations with radicals [104]. The disappearance of metal cations is followed by the start of polymerization.

Clays are known to be free-radical scavengers and traps [100]. The clay minerals inhibit the free radical reactions by adsorption of the propagating or initiating radicals to the Lewis acid surface. The radicals then either undergo bimolecular termination or form carbocations by electron transfer to the Lewis acid site. Minerals containing higher contents of aluminosilicates are more effective retarders or inhibitors. Polymerization of MMA in the presence of clay led to an increase in the molecular weight of PMMA; this behavior may be discussed in terms of the decreased number of reaction loci.

The relative ratios of the polymerization rates in the presence ($R_{p,MMT}$) and absence (R_p) of Cloisite 6A at [BPO] = 0.04 mol dm^{-3} vary with the content (wt%) of Cloisite 6A as follows:

- {$R_{p,MMT}/R_p$}/Cloisite 6A (wt%): 1.06/0.1, 1.2/0.2, 1.3/0.4, 1.36/0.8

$R_{p,MMT}$ was seen to increase monotonically with increasing Cloisite 6A content. These results were discussed in terms of increased contribution of monomolecular mode of chain termination [103]. The presence of Fe(II) and Fe(III) ions favor the contribution of the redox system BPO/Fe(II), which is known to increase the rate of initiation as well as the rate of polymerization. The metal nanoparticles formed by the reduction of metal ions with primary radicals can take part in the catalysis of the polymerization. Thus, in the first step of this reaction the radicals are captured by metal ions, whereas in the second step the formed metal nanoparticles favor the growth events.

13.3.2
Radical Polymerization in Micellar Systems

In very dilute aqueous solutions, emulsifiers dissolve and exist as monomers; however, when their concentration exceeds a certain minimum – the so-called critical micelle concentration (CMC) – they associate spontaneously to form aggregates, or 'micelles' (Figure 13.7).

At low concentrations, the emulsifier dissolves as free monomers; however, as soon as the emulsifier concentration exceeds the CMC, the monomer concentration remains roughly constant and the emulsifier aggregates into micelles. In aqueous solutions, at concentrations not too large with respect to the CMC (e.g. in the range 1 CMC to 10 CMC), ionic emulsifiers form spherical or close-to-spherical micelles [105]. Depending on the proportion of suitable components and the hydrophilic–lipophilic balance (HLB) value of the surfactant used, the formation of micelles or microdroplets can be in the form of oil-swollen micelles dispersed

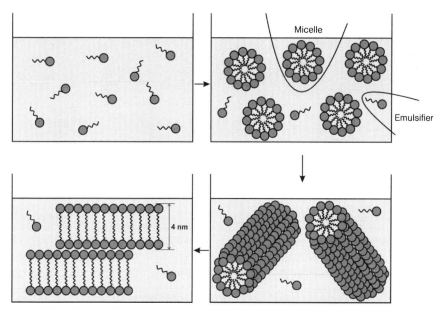

Figure 13.7 Schematic illustration of spherical, rod-like and bilayers direct micelles.

in the aqueous phase, as water-swollen micelles dispersed in oil, as the w/o micelles, or as reverse micelles [89]. Micelles are responsible for many processes, including: (1) an enhancement of the solubility of organic compounds in water or inorganic compounds in oil; (2) the catalysis of many reactions; (3) the alteration of reaction pathways, rates and equilibria; and (4) the reaction loci for the production of polymer products. The primary influence of micelles is to concentrate all reactants in or near the micelles. The extent of solubilization, the ionic charge of micelle and its shape are also important factors. The formation of o/w micelles is driven by strong hydrophobic interactions of the hydrophobic tail of the emulsifier molecules, whereas the formation of w/o micelles is driven by weak electrostatic interactions of polar groups of emulsifier molecule.

It was observed that, the titration of a coarse emulsion by a coemulsifier leads in some cases to the formation of a transparent microemulsion. Microemulsion formation involves:

- A large increase in the interface (e.g. a droplet of radius 120 nm will disperse ca. 1800 microdroplets of radius 10 nm, causing a 12-fold increase in the interfacial area.
- The formation of a mixed emulsifier/coemulsifier film (complex) at the oil/water interface, which is responsible for a very low surface tension (γ_i).

Microemulsions (monomer-swollen micellar solution, micellar emulsions, or spontaneous transparent emulsion) are dispersions of oil and water prepared with emulsifier and coemulsifier molecules. In many respects, they are small-scale versions of emulsions, being homogeneous on a macroscopic scale but heterogeneous on a molecular scale. They consist of oil and water domains which are separated by emulsifier monolayers [106].

Microemulsions act as attractive media for polymerization reactions. Indeed, polymerization in microemulsions represents a new technique that allows the preparation of ultrafine latex particles within the size range $10\,nm < d < 50\,nm$, and with a narrow size distribution [107, 108]. However, the formulation of polymerizable microemulsions is subject to severe constraints, due in large part to the high emulsifier level (ca. 10%) needed to achieve their thermodynamic stability. This fact, together with the requirement of high polymer contents in most applications, raises the problem of maintaining specific emulsifier–coemulsifier, monomer–emulsifier and monomer–coemulsifier interactions, which are disrupted in the presence of large amounts of polymer and tend to destabilize the polymer microemulsion, or to produce large-sized polymer particles.

A proposed mechanism for microemulsion polymerization consists of the following steps (Figure 13.8) [108, 109]:

- The particles are nucleated by the capture of radicals from the aqueous phase for both water-soluble and oil-soluble (partly soluble in water) initiators.
- The microemulsion droplets that did not capture radicals serve as reservoirs to supply the monomer and emulsifier to the polymer particles.

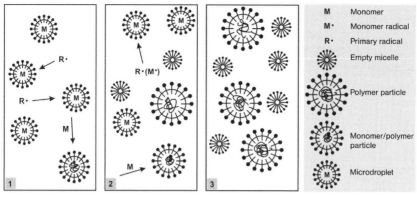

Figure 13.8 Microemulsion polymerization mechanism [109].

- The polymer particles compete with the microemulsion droplets in capturing radicals. However, owing to the much larger surface area provided by microemulsion droplets, the radical flux to the polymer particles is still smaller.
- The continuous nucleation of polymer particles during polymerization results from the very high number of monomer-swollen micelles or microdroplets.
- The ratio of monomer to emulsifier or the monomer concentration at the reaction loci decreases with increasing conversion. The result of these two opposing effects is the appearance of maximal rate at ca. 10–20% conversion.
- Light-scattering measurements prove the presence of both the microdroplets and mixed micelles (monomer-starved microdroplets). The ratio of microdroplets to mixed micelles decreases with increasing conversion [108].

Emulsions are thermodynamically unstable, and exhibit flocculation and coalescence unless significant energetic barriers to droplet interactions are present. Emulsions degrade towards phase separation via mass transfer, and other mechanisms. When an o/w emulsion is created by the application of a shear force to a heterogeneous phase containing surfactants and additives, a distribution of droplet sizes results. The interdroplet mass transfer (Ostwald ripening [110]) determines the fate of this distribution because of the higher Laplace pressure. If the small droplets are not stabilized against diffusional degradation they will be lost, thereby increasing the average droplet size. Emulsions are sensitive to coarsening phenomena such as coalescence and Ostwald ripening, as their thermodynamically most stable state is the completely demixed form. Coalescence can often be prevented by a careful choice of stabilizers. Ostwald ripening involves the movement of oil molecules from small to large droplets (Figure 13.9) [111] and is, therefore, the process whereby large droplets grow at the expense of small ones because the solubility of a material within a droplet increases as the interfacial curvature increases [112]. Unstable emulsions become more stable with respect to the Ostwald ripening process by the addition of small amounts of a hydrophobic additive, which distributes preferentially in the dispersed phase [113].

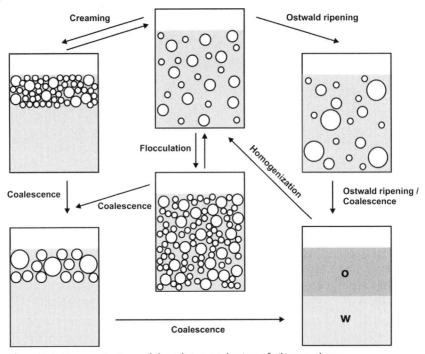

Figure 13.9 Homogenization and degradation mechanism of oil/water phases.

Their small droplet size and high kinetic stability make miniemulsions suitable for the efficient delivery of active ingredients. Unlike microemulsions, which require a high concentration of emulsifiers for their preparation (usually in the range of 10–30 wt%), miniemulsions can be prepared at moderate emulsifier concentration (in the range of 4–8 wt%). Thus, the droplet size distribution of nanoemulsions is a complex function of the breaking and coalescence of droplets, droplet degradation by monomer diffusion, and the presence or the absence of an emulsifier, coemulsifier (or surface-active agent) or hydrophobe. The stability of an o/w (nano)emulsion is directly connected with the transport of oil through the aqueous phase. The addition of an initiator to the stable monomer droplets leads to particle nucleation and the formation of polymer particles. The droplets are small enough to benefit from all the advantages of conventional emulsion polymerization process, such as high rates of polymerization, the depressed bimolecular termination of propagating radicals, and high molecular weights of the final polymers (Figure 13.10) [114]. Miniemulsion polymerization allows the incorporation of water-insoluble materials such as resins, organic pigments and polymers. The additive seed allows control of the particle number and particle size during the production process. Furthermore, miniemulsion polymerizations and copolymerizations carried out with acrylic and methacrylic monomers in the presence of unsaturated alkyd resins lead to the production of stable hybrid latex particles that contain grafted and crosslinked alkyd resin/acrylic products as the

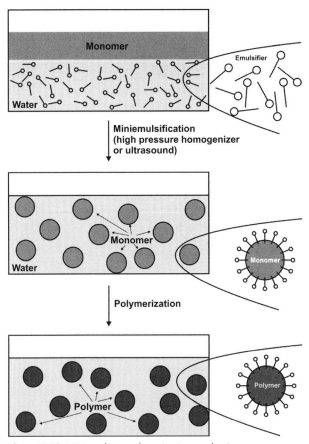

Figure 13.10 Miniemulsion polymerization mechanism.

coating polymer [115]. In this reaction, the multifunctional resin acts as both the hydrophobe and the costabilizer of the miniemulsion.

A polymerization using an emulsion technique has significant advantages for obtaining polymer/clay hybrids [57, 116]. One such important advantage is the use of water as a dispersion medium. Because the swelling property of NaMMT is caused by hydration of water to the interlayer inorganic cation (Na^+), this in turn leads to the insertion of micelles into the galleries of layered silicates. However, in the presence of limited amounts of water, NaMMT forms gels which contain isolated silicate layers and aggregate of several layers. Therefore, reducing the size of aggregated NaMMT before the polymerization reaction is considered to be an essential requirement to allow the fine and homogeneous dispersion of clay particles in the aqueous medium of an emulsion system.

For the preparation of water-borne polymers [e.g. (PSt, PMMA)/MMT nanocomposites], this represents a direct means of incorporating the clay particles into the formulation of an emulsion polymerization. Simple recipes and a batch process can

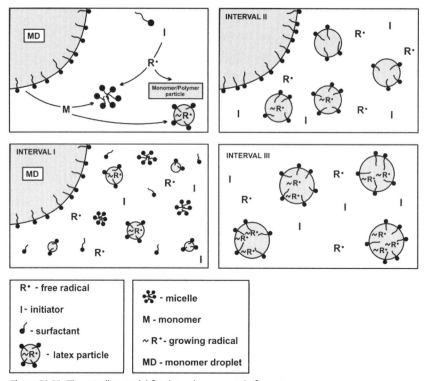

Figure 13.11 The micellar model for the polymer particle formation in emulsion polymerization [120].

be used to produce the polymer/MMT nanocomposites. Emulsion polymerization involves the propagation reaction of free radicals with monomer within the monomer-swollen polymer particles dispersed in the continuous aqueous phase. The discrete hydrophobic particles are stabilized by a surfactant such as the anionic sodium dodecyl sulfate (SDS), nonionic Tween, or cationic cetyl trimethyl ammonium bromide (CTAB). When the level of surfactant is greater than its CMC, the polymerization kinetics can be adequately predicted by the Smith–Ewart theory (Figure 13.11) [117–120].

Particle nuclei are generated via the capture of radicals by micelles, termed the 'micelle nucleation'. Emulsion polymerization involves the dispersion of a relatively water-insoluble monomer (e.g. styrenes, alkyl methacrylates) in water with the aid of emulsifiers, followed by the addition of a water-soluble (e.g. ammonium peroxodisulfate, APS) or oil-soluble (e.g. dibenzoyl peroxide, BPO) initiator. The APS-initiated emulsion polymerization is a two step process:

1. The formation of primary radicals and their transformation to the surface-active oligomeric radicals through the addition of monomer units to the growing radical.

2. The entry of oligomeric (surface-active) radicals into the monomer-swollen micelles (micellar nucleation) or the precipitation of growing radicals (homogeneous nucleation) from the aqueous phase [120–123]:

- decomposition of initiator in the aqueous phase:

$$I(APS) \rightarrow 2\,R^{\bullet}. \tag{13.14}$$

- water-phase propagation:

$$R^{\bullet} + M \rightarrow RM^{\bullet} \rightarrow RM_n^{\bullet} \rightarrow RM_z^{\bullet} \tag{13.15}$$

- entry of surface-active oligomeric radical ($RM_z^{\bullet}$) into the monomer-swollen micelle or polymer particle:

$$RM_z^{\bullet} + \text{particle (micelle)} \rightarrow \text{active particle} \tag{13.16}$$

Here, $R^{\bullet}$ is the charged primary radical derived from peroxodisulfate initiator (I), M is the monomer in the water phase, $RM^{\bullet}$ and $RM_n^{\bullet}$ are growing radicals, and $RM_z^{\bullet}$ is the surface-active radical with a high degree of hydrophobicity. The surface-active radical enters the polymer particle or monomer-swollen micelles, and begins the polymerization. The particle nuclei are generated via the capture of radicals by micelles, termed the 'micelle nucleation'.

Variations of the conversion and polymerization rates with time in the emulsion polymerization of unsaturated monomers are shown in Figure 13.12.

Both, homogeneous nucleation [121–123] and coagulative nucleation [124] are considered to be operative in polymerization systems where the surfactant concentration is below its CMC, and which contain the hydrophilic monomer(s) and interactive particles. In the former case, the oligomeric radicals do not enter the polymer particles but rather propagate until they reach a critical degree of polymerization, P_{crit} ($RM_z^{\bullet}$), whereupon they become insoluble and form primary particles

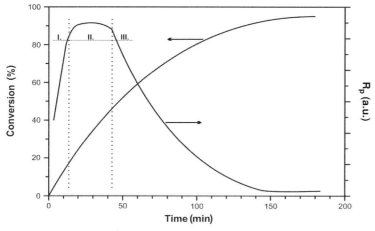

Figure 13.12 Variations of conversion and polymerization rate with time [124].

($RM_j^\bullet$) that grow by absorbing the monomer and the emulsifier-precipitation of growing radicals from the aqueous phase:

$$RM_n^\bullet + M \rightarrow RM_z^\bullet \rightarrow RM_j^\bullet \text{(primary particle)} \quad (13.17)$$

The unstable primary particles ($RM_j^\bullet$) aggregate between themselves and so form larger and more stable particles, or they may flocculate with larger, previously formed particles. In the absence of an emulsifier, the flocculation of surface-active oligomers governs the particle nucleation – that is, the oligomeric radicals will self-nucleate to form primary particles. The restricted termination of growing radicals within the monomer/polymer particle leads to a very rapid polymerization and the formation of large polymers.

Here, R_p is linearly proportional to the number of particles nucleated per liter of water (N_p) and the average number of radicals per particle (n) [119]:

$$R_p = k_p[M]_p(nN_p/N_A) \quad (13.18)$$

where k_p is the propagation rate constant, $[M]_p$ is the concentration of monomer in the particles, and N_A is Avogadro's number.

In aqueous solutions without a polymer, emulsifiers are known to self-aggregate into micelles at a well-defined CMC. Here, the micelles solubilize oil or water in the micelle volume, thereby introducing heterogeneity into the local concentration of the reactants. Likewise, polymer micelle-like aggregates begin to form along the polymer chain at a critical aggregation concentration (CAC). The CAC is thus an analogue of the CMC, but occurs in the solution with an added polymeric compound. Characteristically, the CAC is always lower than the CMC of the corresponding emulsifier (surfactant) [125, 126]. A lower CAC is especially pronounced in solutions of polyelectrolytes with an opposite charge to the emulsifier. The emulsifier often interacts cooperatively with polymers at the CAC, forming micelle-like aggregates within the polymer. Noncooperative association between the emulsifier and polymer is characterized by a simple partitioning of the emulsifier between the polymer and the aqueous phase. The addition of emulsifiers to aqueous solutions of amphiphilic polymers can either induce or break up interpolymer aggregation [127]. The emulsifier can interact cooperatively with polymers at the CAC, forming micelle-like aggregates within the polymer. Emulsifiers with a relatively long tail bind to the amphiphilic copolymers simply by partitioning between the aqueous phase and the polymer (noncooperative association) [128].

In the miniemulsion polymerization of styrene, in the presence of laponite, the silicate layers can be located not only on the surface of the particle but also inside the emulsion–latex particle, depending on the nature of the initiator [60]. In the suspension polymerization of styrene, in the presence of different organophilic MMTs, it is likely that silicates will also locate inside the particle [129]. In the suspension polymerization of MMA, in the presence of organophilic MMT, the clay layers are situated both inside and on the surface of the latex particles [84], and in this case, the kinetics of the monomer polymerization is much influenced by the presence of clay. Van Herk et al. have reported the presence of clays inside the latex particle by using a coupling system based on titanium or silica precursors [130].

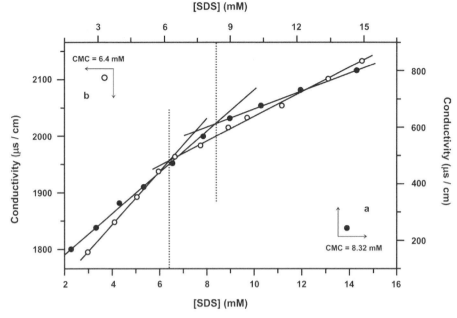

Figure 13.13 Specific conductivity of SDS as a function of the bulk SDS concentration. (a) SDS; (b) SDS + 1 wt% Cloisite Na [90].

The addition of 1 wt% Cl Na reduces the CMC of SDS from 8.3 mM (without Cl Na) to 6.4 mM (Figure 13.13) [90]. The emulsifier SDS seems to interact cooperatively with Cl Na at the CAC, forming micelle-like aggregates within the Cl Na. This implies that the Cloisite platelet surfaces are relatively hydrophilic and, therefore, are not effective for the adsorption of SDS to partition between Cloisite and the aqueous phase. The ionic strength of the Cl Na-containing SDS solution is, however, much higher compared to the native SDS solution. SDS molecules subjected to a hostile environment with a very high electrolyte concentration show a stronger tendency to aggregate cooperatively with one another to form micellar aggregates. As a result, the CMC will be decreased. The CMC values would be expected to have an influence on the concentration of micelles available for particle nucleation.

Emulsion polymerization was observed to be an effective process for synthesizing the intercalated PSt/Cloisite Na particles [90]. This is due to the fact that the interlayers of Cloisite Na are filled with Na^+ ions and, therefore, the hydrophilic Cl Na platelets are highly swollen in water. The effect would be to promote the penetration of the water-borne primary and oligomeric radicals and radicals derived from emulsifier (formed by chain transfer events) and initiator into the multiple-layered Cl Na.

Variation of the conversion with time and SDS for the styrene emulsion polymerizations in the presence of 1 wt% Cl Na are shown in Figure 13.14. Here, the rate of emulsion polymerization was constant over the conversion range (Interval 2) of 20–50% or 60% conversion.

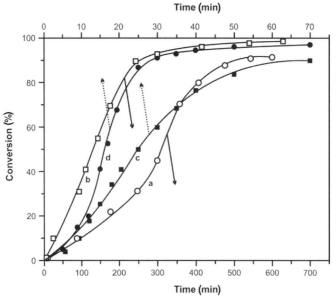

Figure 13.14 Conversion as a function of time for styrene emulsion polymerizations in the presence of 1 wt% Cl Na and various SDS concentrations and the absence of Cl Na. [SDS] = 2 mM (○); [SDS] (mM) = (●) 17, (■) 9, (□) 2 [90].

The maximum rate of polymerization with and without Cloisite varies with [SDS] as follows:

$$Rp \times 10^{-4} (\text{mol dm}^{-3} \text{s}^{-1}) \text{ (with/without Cloisite Na)/[SDS](mM)}:$$
$$(7.5/5.5)/9, (8.7/6.3)/11, (10/7.5)/13, (11.5/8.9)/15,$$
$$(12.3/10.0)/17, (12.4/10.5)/19 \qquad (13.19)$$

and the ratio of rates with and without Cloisite Na $R_{p,Cl}/R_p$ varies with [SDS] as follows:

$$R_{p,Cl}/R_p : 1.36/9, 1.38/11, 1.33/13, 1.29/15, 1.23/17, 1.18/19 \qquad (13.20)$$

The correlations based on the kinetic data are:

$$R_{p,Cl} \sim [SDS]^{0.79} \text{ and } R_p \sim [SDS]^{0.85} \qquad (13.21)$$

for the runs with and without 1 wt% Cl Na, respectively. The results can be discussed in terms of variation of rate of initiation (see above) and the rate of particle nucleation with the ratio [Cl Na]/[SDS]. First, the reaction orders 0.79 and 0.84 indicate that the synergistic effect of [Cl Na] decreases with increasing [SDS] or decreasing the ratio [Cl Na]/[SDS]. Increasing the ratio [Cl Na]/[SDS] leads to a decrease in the interaction between Cl Na and the initiator, which disfavors the formation of radicals. Second, at constant [SDS], R_p is faster for the run containing Cloisite Na compared to the counterpart without Cloisite. The enhanced R_p for polymerizations containing

Cloisite Na was attributed to the larger number of micelles (N_m) available for particle nucleation, provided that [SDS] is above its CMC.

In addition, a significant increase in R_p was achieved when 1 wt% Cl Na was added to the polymerization in the absence of micelles ([SDS] = 2 mM). At [SDS] = 2 mM, micelles do not exist in the polymerization system; however, at [SDS] < CMC the formation of particle nuclei via homogeneous nucleation is very slow in the absence of Cloisite. In contrast, polymerization in the presence of Cloisite does not exhibit the characteristically slow R_p, which implies that Cl Na platelets may serve as additional reaction loci for the increased generation of radicals and their reactions with monomer and they predominate in the early stage of polymerization. Metal cations of Cl Na are supposed to form the redox-initiator system, due to which increases in the rate of radical and particle formation also occur. Furthermore the costabilizing efficiency of the polymer-modified Cloisite platelets can contribute to the formation of additional number of polymer particles.

The particle diameter (d) decreases with increasing [SDS] for both the polymerizations, with and without Cl Na. At each [SDS], the run with Cl Na has a smaller particle size compared to the counterpart without Cl Na:

$$d(\text{nm})/[\text{SDS}]/\text{m}M : 74/9, 70/11, 68/13, 66/15, 66/17, 67/19 \qquad (13.22)$$

$$d_{Cl}(\text{nm})/[\text{SDS}]/\text{m}M : 66/9, 65/11, 64/13, 61/15, 56/17, 55/19 \qquad (13.23)$$

This is attributed to the larger population of particles originating from micelle nucleation as a result of the lowered CMC, along with the participating of Cloisite on the particle stabilization (Pickering stabilization). The formation of Cloisite/PSt nanostructures and their localization on the particle surface increases both the total particle surface and the particle concentration (or decreases the particle size).

Fialova et al. have studied the influence of selected layered silicates, MMT type [Cloisite Na (Cl Na), Cloisite 30B (Cl 30B) and Cloisite 20A (Cl 20A)] with different polarities on the kinetic, colloidal and molecular weight parameters of microemulsion polymerizations of nonpolar styrene or polar butyl acrylate (BA) [131]. Variations in the monomer conversion for several microemulsion polymerizations of St and BA with the reaction time for different types of Cloisite are summarized in Figure 13.15.

The data in Figure 13.15 indicate that the conversion of ca. 80% is reached in the styrene systems in about 15 min. In all cases, the colloidal stable polymer latexes were formed. In the case of styrene, as the difference in Cloisite polarity decreased, the final conversion became higher. The absence of an inhibition period in the microemulsion polymerization of styrene can be discussed in terms of restricted interaction of hydrophobic radicals with the clays. The hydrophobic styrene monomer and polystyrene growing radicals are located within the hydrophobic domain of the monomer-saturated clay interior. In contrast, an inhibition period is observed in the microemulsion polymerization of BA with Cl Na and Cl 20A. The polar BA monomer, as well as its growing radicals, can interact with clay, such that the growth events are restricted. Clays are known to be free-radical scavengers and traps [100]; in fact, clay minerals appear to inhibit the free radical reactions by adsorbing the propagating or initiating radicals onto the Lewis acid surface. The radicals then either

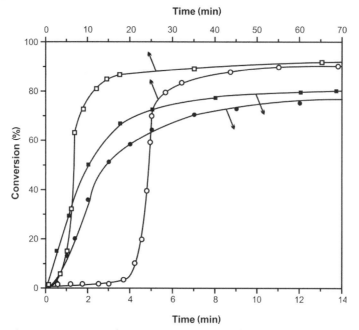

Figure 13.15 Variation of monomer conversion with the reaction time and in the presence and absence of Cloisite Na in the microemulsion polymerization of styrene and BA.
Symbols: □ BuA without Cl Na; ○ BuA with Cl Na; ■ St without Cl Na; ● St with Cl Na. Reaction mixture: 1.74 g styrene or BA; 19.32 g water; 3.86 g SDS; 0.07 g APS; 0 or 0.239 g Cloisite Na [131].

undergo bimolecular termination or form carbocations by electron transfer to the Lewis acid site. Minerals which contain higher levels of aluminosilicates are more effective inhibitors. The polymerization of MMA in the presence of clay (kaolin) increased the molecular weight from 887 K (without clay) to 944 K (kaolin) and 912 K (polyphosphate kaolin); MMT was shown to be less effective than kaolin in this respect. The neutralization and immobilization of charged radicals within the Cloisite interior can also contribute to the deactivation of reaction loci. The conversion of monomer begins to increase at a later time (ca. 20 min), with the final conversion being reached after about 45 min. The highest conversion is reached for the BA systems with Cl Na (in contrast to styrene).

These observations, which involve differences in polarity, can also be related by the swelling capacity with styrene and BA of layered silicates with different hydrophobic character. For example, the swelling capacity of Cl Na with styrene exceeds 100 wt%, whereas by using Cl 20A (which is less polar than Cl Na) the swelling capacity can exceed 300 wt% [67]. By considering the two aspects shown above, it can be appreciated that, as the polarity differences between monomer and the inorganic filler decrease, the polymerization will reach higher conversions.

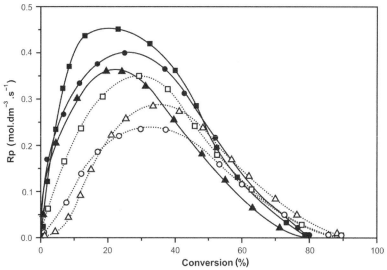

Figure 13.16 Dependence of the polymerization rate (R_p) on conversion in the microemulsion polymerization of styrene and BA in the presence and absence of Cloisite Na.
Symbols: ■ St without Cl Na; ● St with Cl 20A; ▲ St with Cl Na; □ BuA without Cl Na; ○ BuA with Cl 20A; △ BuA with Cl Na.
Reaction mixture: 1.74 g St or BA; 19.32 g water; 3.86 g SDS; 0.07 g APS; 0 or 0.239 g Cloisite [131].

Variations of the polymerization rates of styrene and BA with conversion in the presence of Cloisites are shown graphically in Figure 13.16, which can be described by a curve with two rate intervals. The rate of polymerization (R_p) first increases to a maximum ($R_{p,max}$), and then decreases to the final conversion. The initial increase in R_p can be attributed to the robust particle nucleation and to the monomer saturation condition, with the maximum polymerization rate lying in the conversion range of 15–30%. The values of $R_{p,max}$ vary from 369×10^{-3} mol dm^{-3} s^{-1} (St with Cl Na), over 426×10^{-3} mol dm^{-3} s^{-1} (St with Cl 20A) to 465×10^{-3} mol dm^{-3} s^{-1} (St with Cl 30B).

The kinetic and colloidal parameters of styrene and BA, and their variations with type of Cloisite, are collected in Table 13.1. As can be seen, the diameters of the latex particles (d_p) increase in the presence of layered silicates. However, as the polarity of the inorganic partner increase, the average diameters of the latex particles increase or the colloidal stability of the polymer particles decrease. It is reasonable to expect that the polymer modifier Cloisites would concentrate at the particle surface and, indeed, the surface area of the emulsifier is seen to increase with Cloisite, this being more pronounced with BA. In addition, while the PSt-modified Cloisite can take part in particle stabilization, this is not the case with PBA-modified Cloisite. Such behavior could also be related to the polarity by a higher hydration level. The data in Table 13.1 also show that N_p, which is the number of final polymer particles per unit volume of the continuous phase, decreases with the increasing polarity of Cloisite.

Table 13.1 Variation of kinetic and colloidal parameters with Cloisite type in the microemulsion polymerization of styrene and n-butyl acrylate.

Cloisite	Con.$_{(f)}$ (%)$^{a,\,b}$	$R_{p,max} \times 10^3$ (mol dm^{-3} s^{-1})$^{a,\,b}$	d_p (nm)$^{a,\,b}$	$N_p \times 10^{-19}$ (dm^3)$^{a,\,b}$	$\bar{n} \times 10^3$ (particle)$^{a,\,b}$
–	81, 87	452, 346	26, 33	0.795, 0.418	0.029, 927
Cl 20A	83, 86	426, 238	29, 45	0.585, 0.161	0.0371, 1658
Cl 30B	78, 87	465, 237	34, 55	0.340, 0.089	0.0698, 2955
Cl Na	78, 92	369, 293	38, 94	0.243, 0.019	0.0774, 17324

^{a}Styrene.
^{b}n-butyl acrylate.
Con.$_{(f)}$ = final conversion.

The polymerization rate and number of polymer particles are decreased by the addition of Cloisite. In contrast, the average number of radicals per particle ($\bar{n}$) increases with the decrease in R_p and N_p. The value of $\bar{n}$ is most likely influenced by the decreased monomer concentration with the Cl type at the reaction (this was not taken into consideration). Furthermore, the decreased penetration of monomer into the particle/Cloisite nanoclusters also slows the rate of polymerization. BA, in being polar, is expected more easily to penetrate the interlayer spaces and clusters than would nonpolar styrene. Likewise, the greater the monomer content in the basal spacing and particle/Cl clusters, the slower the polymerization rate.

As the data in Table 13.1 show, $\bar{n}$ increases with the polarity of Cloisite, and it can be seen that the values are relatively high, even for the pure BA. It has been reported that $\bar{n}$ lies well below 0.5 for small PBA particles with diameters $\leq$50 nm [132, 133]. In the present studies, $\bar{n}$ was ~0.9 for the BA/SDS weight ratio of (9.0/20) = 0.45; however, when the ratio was increased to 0.5 and 1.0, $\bar{n}$ was decreased to 0.6–0.3 and <0.1, respectively. A comparison of $\bar{n}$ values as a function of the BA/SDS weight ratio led to the conclusion that $\bar{n}$ is inversely proportional to the BA/SDS ratio. Thus, at high SDS and low BA concentrations, the penetration of BA and/or its radical is increased, which in turn causes the number of radicals to increase.

The data in Table 13.1 also show that the estimated kinetic parameters for polymerization of St are similar to those obtained for the polymerization of BA, although the variations are somewhat less distinct. With increasing polarity of Cloisite, the values of d_p and $\bar{n}$ were increased, but the values of N_p were decreased. The size of the polymer particles was found to increase with polarity, with the increase being more pronounced for poly(butyl acrylate) particles. The formation of larger particles was accompanied by a higher accumulation of radicals in the larger particles, and with the decreased desorption of monomeric radicals from particles.

The size of the polymer particles was increased, and the number decreased, with the addition of Cloisite, though this may indicate poor particle nucleation and stabilization. These findings support the results obtained by Chern et al. (see above) for the miniemulsion polymerization, when they used emulsifier concentrations slightly below and above CMC, in addition to homogenization [90]. Chern's group reported that the synergistic effect of [Cl Na] decreases with either increasing the [SDS] or decreasing the ratio [Cl Na]/[SDS]. An increase in the ratio [Cl Na]/[SDS] caused a decrease in the interaction between Cl Na and initiator, which disfavored the formation of radicals. The high concentration of emulsifier disfavored the interactions between initiator or radicals and Cloisite, and also the formation of polymer-modified Cloisite platelets or hydrophobic clay platelets. The location of exfoliated unmodified Cloisite entities at the particle surface could explain the decreased stability and formation of larger particles.

For polystyrene, $\bar{n}$ is well below 0.5, which indicates an extensive desorption of monomeric radicals from polymer particles. The values of $\bar{n}$ increase slightly with increasing polarity and particle size. The data in Table 13.1 also indicate that the $\bar{n}$ values are approximately three to four orders of magnitude larger for the polar BA compared to the apolar styrene. The accumulation of polar poly BA-derived radicals in the interfacial layer of polymer/Cloisite nanoaggregates is considered to be one of the possible reasons for this unexpected behavior. Furthermore, the different polarities of oligomeric PSt and PBA radicals affect their entry into particles. Polar PBA oligomer radicals are thought to enter more easily into particles than are apolar PSt radicals. The different locations of Cloisite in the styrene and BA microdroplets, and also in the PSt and PBA nanoparticles, may also influence both the entry and exit of radicals.

For both BA and St, the desorption constants (k_{des}) are seen to decrease with the addition of Cloisite (Table 13.2). The specific desorption rate constants k'_{des} (cm^2 s^{-1}) and k_{des} (s^{-1}) were calculated with the three models of Ugelstadt–O'Toole, Nomura and Gilbert [134, 135]. The chain transfer to monomer, and desorption of the monomeric radicals from the polymer particle, caused a reduction in the concentration of radicals in the polymer particles [136]. In the presence of layered silicates, however, the number of radicals in the polymer particles was seen to increase, whilst on the other hand the desorption of monomeric radicals was decreased. Together, these data show that layered silicates may act as a trap or barrier for the desorption of radicals. This situation can be also discussed in terms of polymer chain conformation by the intercalation of polymer chains [137], with the depressed reactivity of intercalated polymer radicals leading to depressed rates of both addition and chain transfer events.

Table 13.2 Variation of desorption rate coefficients with Cloisite type in the emulsion polymerization of styrene.

Cloisite	k_{des} (s^{-1}) Ugelstadt[a]	Nomura[a, b]	Gilbert[a, b]
–	0.685	0.168, 15.54	0.115, 0.97
Cl 20A	0.564	0.135, 8.36	0.092, 0.57
Cl 30B	0.274	0.098, 5.60	0.067, 0.39
Cl Na	0.312	0.078, 1.92	0.054, 0.14

[a]Styrene.
[b]n-butyl acrylate.

The above-observed kinetic behavior could be related to previously observed data for the emulsion polymerization of vinyl monomers in the presence of layered silicates [67, 68, 82]. For emulsion polymerization, there is a certain weight ratio of monomer: nanoclay (1 : 0.125) below which the polymerization rates are increased in the presence of a layered silicate. However, above this ratio the reaction rates are decreased. Bearing in mind that the monomers: nanoclay ratio used for these experiments was 1 : 0.138, the decreasing reaction rates could be explained by the large number of layers that can induce a physical barrier effect to prevent radical propagation processes. This behavior must be observed in relation to the desorption constants (for all calculated models), which suggest a lesser probability of the radicals exiting in the presence of layered silicates.

Normally, inverse or w/o emulsions require the addition of low-HLB surfactants to the reaction mix in order to prevent coalescence of the water droplets. In the absence of any stabilizing agent, no stable emulsion can be obtained. It has been well documented that solid particles with special features will self-assemble at the liquid–liquid interface, such that the interfacial energy between the two immiscible liquids is reduced (see Figure 13.17). Nonionic surfactants such as Span 80 and Tween 85, or anionic surfactants such as AOT (sodium bis-2-ethylhexylsulfosuccinate), are among the most commonly used in inverse emulsion polymerization [107, 138, 139]. Attempts have been made to optimize not only the process of inverse emulsion formation at higher monomer contents and lower emulsifier concentrations, but also the colloidal stability of inverse emulsions before and during polymerization of the monomer [140].

Voorn et al. have described the preparation of polymer/clay composite particles by surfactant-free inverse Pickering (mini)emulsion polymerization initiated by the nonionic water-soluble initiator VA-086 and the oil-soluble 2,2′-azoisobutyronitrile (AIBN) initiator [91]. The inverse emulsion was prepared by mixing a solution of acrylamide (AAm), VA-086 (2,2′-azobis[2-methyl-N-(2-hydroxyethyl)propionamide]), and the crosslinker N,N′-methylenediacrylamide (NDA) in water

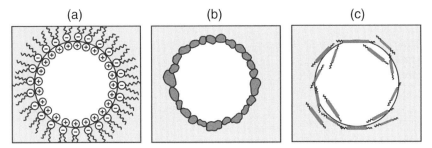

Figure 13.17 Schematic representation of the stabilization mechanism of inverse emulsions. (a) The water microdroplet is stabilized by the surfactant; (b) the solid particles surround the microdroplet; (c) the microdroplet is stabilized by the clay/polymer hybrid platelets [91].

to a dispersion of Cloisite 20A (Cl 20A) in cyclohexane (CH) at room temperature. The average diameter decreased with increasing the concentration of Cl 20A from 0.27 to 0.69 wt%:

$$d(nm)/Cl\,20A\,(wt\%) : 680/0.27,\,600/0.49\text{ and }510/0.69 \qquad (13.24)$$

The final composite nanoparticles appeared to be about 200 nm larger than the initial monomer droplets, while there was no significant change for the polydispersity index (PDI) (ca. 0.3). The values for the latexes obtained using DLS were in good agreement with those obtained with electron microscopy analyses. By increasing the concentration of clay platelets in the initial dispersions, from 2.7 to 8.3 wt% (based on monomer content), a decrease in the particle size from 980 to 700 nm was observed for the crosslinked latexes:

$$\{D_{mon}/D_{polym}\}/Cl\,20A\,(wt\%):$$

$$\{724\,nm/981\,nm\}/2.7,\,\{677/857\}/4.1,\,\{547/802\}/6.4,\,\{508/697\}/8.3 \qquad (13.25)$$

As the polymerization proceeded, the particle size increased continuously, indicating that monomer transportation might be occurring during the polymerization. For example, in the run with 6.4 wt% Cl 20A, the particle size increased from ca. 500 nm (the initial monomer droplet size) to 700 nm at 45 min, and finally to 800 nm after 90 min. Inverse emulsion polymerizations of acrylamide stabilized by the ionic surfactant AOT and the nonionic Span 80 as reference experiments produced latex particles with a final latex particle diameter of about 40 nm. The hydrophobically modified clay platelets alone appeared capable of stabilizing w/o emulsions without the addition of surfactants. The inverse Pickering emulsions formed could be regarded as surfactant-free inverse emulsions. A schematic illustration of the surfactant-free, inverse emulsion polymerization in the presence of hydrophobic MMT is shown in Figure 13.18.

The rugged surface morphology in scanning electron microscopy (SEM) images indicated that the polyacrylamide (PAAm) latex particles were covered by a layer of clay platelets [91]. The clay platelets were also reported to be located at the surface of the latex particle by Bon et al. [64]. Cryo-transmission electron microscopy (TEM) was used to observe the real particle morphology in the liquid state. Although most of these cryo-TEM studies have been conducted with aqueous systems, several investigations using organic solvents have been reported [141]. In one study, a 'fluffy' structure was formed around the particles, whereas on the other hand a partial exfoliation of the clay platelets in the organic medium was observed, indicating that clay platelets are difficult to completely exfoliate in an apolar medium. In some cases, bundles of several stacked platelets were found, although the vast majority of these platelets were shown to be located around the latex particles. The hydrophobic clay platelets appeared to be partially 'stretched' out into the organic medium. The morphology observed using TEM indicated that,

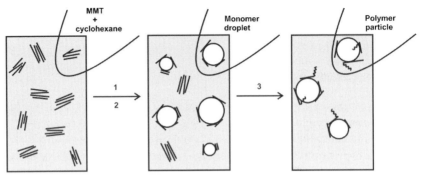

Figure 13.18 Schematic representation of the inverse (mini) emulsion polymerization and stabilizing function of the hydrophobic clay platelets. (1) Water–initiator–monomer mixture; (2) delamination and sonication of the mixture; (3) polymerization of the monomer droplets and stabilization of formed polymer particles by polymer-modified clay platelets [91].

for both initiators VA-086 and AIBN, the clay platelets were mainly located at the surface of the latex particles.

In order to check the influences of different monomers on the formation of composite latex particles by inverse emulsion polymerization, another hydrophilic monomer 2-hydroxyethyl methacrylate (HEMA) was used. The droplet size of the Cl 20A-stabilized HEMA inverse emulsion was in the similar size range to the AAm emulsions. HEMA could be polymerized to form latex particles of about 720 nm; however, the formed latex was less stable than the PAAm latexes. Furthermore, the content of fouling in the reactor formed during the reaction was much higher than with the PAAm-based particles.

PMMA-layered silicate nanocomposites were prepared using *in situ* suspension and emulsion polymerization [43]. The use of inorganic materials such as silicates in suspension polymerization is well known. Organically modified silicates (OLS), as suspension stabilizers, were obtained from Cloisite Na modified with *n*-decyltrimethylammonium chloride (**1**), 2,2′,-azobis(isobutylamidine hydrochloride) (**2**), and [2-(methacryloyloxy)ethyl]- trimethylammonium chloride (**3**). The typical suspension polymerization procedure first involves the dispersion of the layered silicate in water, followed by the addition of an organic modifier. Subsequently, a mixture of MMA and AIBN was added to the aqueous phase with vigorous stirring. A typical volume ratio of water: MMA was 5 : 1. Following addition of the monomer and initiator, delaminated OLS is thought to adsorb to the monomer droplets. In emulsion polymerization, the silicate (natural, unmodified) is added in a post-polymerization step, but because the latex particles have cationic surface charges and the silicate layers anionic charges, electrostatic forces promote an interaction between the silicate and polymer particles. The organic modifiers (**1–3**) used for the suspension polymerization are also employed as surfactants in the emulsion process.

13.4
Collective Properties of Polymer/MMT Nanocomposites

13.4.1
Kinetic and Molecular Weight Parameters

Emulsion polymerization is a relatively new approach to synthesize polymer/clay nanocomposites. Both, PSt/clay and PMMA/clay nanocomposites were synthesized via emulsion polymerization using an anionic surfactant (SDS) as emulsifier [57, 116]. PSt/reactive organoclay intercalated nanocomposites were prepared by emulsion polymerization [142], with the reactive organoclay being synthesized by exchanging the inorganic cations (Na^+) with aminomethylstyrene. As the alkyl chain of aminomethylstyrene is short, it is not very effective in expanding the clay interlayer [142]. PMMA/clay nanocomposites were synthesized by emulsion polymerization with decyltrimethylammonium chloride and [2-methacryloyloxy)ethyl]- trimethylammonium chloride to modify the surface of dispersed clay [43].

Both, Noh and Lee [116] and Kim et al. [143] have synthesized intercalated PSt/MMT nanocomposites by emulsion polymerization in the presence of pristine MMT. In order to improve the compatibility with hydrophobic polymer, the clay was organically modified. Thus, Laus et al. [142] and Qutubuddin et al. [61] prepared PSt/reactive organoclay intercalated nanocomposites by emulsion polymerization. On the other hand, polar monomers have been introduced to copolymerize with styrene to increase the polarity of the polymer matrix. Choi and coworkers synthesized a series of MMT nanocomposites with PSt [144], poly(methyl methacrylate-co-styrene) [87] and poly(styrene-co-acrylonitrile) [88] by soap-free emulsion polymerization using an ionic comonomer 2-acrylamido-2-methyl-1-propane sulfonic acid (AMPS) to stabilize the polymer latex.

Emulsion polymerization can also be used to prepare the exfoliated polymer/silicate nanocomposites, without using modifiers [19, 62]. The storage moduli of the nanocomposites were enhanced over a large loading range of silicates. A possible reason for this was that the water, monomers and reactive surfactants widened the basal space of silicate before polymerizations, after which the silicate lost regularity and was exfoliated during the polymerization. Several groups have attempted to prepare poly(styrene-co-acrylonitrile) (SAN)/silicate nanocomposites [145, 146], but obtained only the intercalated SAN/silicate nanocomposites. The emulsion polymerization method was also applied to the poly(styrene-co-acrylonitrile) copolymer (SAN) [88].

In order to monitor the effect of different types of MMT on the polymer molecular weights of polymer/clay composites, three different polystyrene composites, incorporated with pristine Cloisite Na and organoclays (Cl_{CTAB} and Cl_{AUA}) and modified with CTAB and AUA, respectively, were synthesized via the emulsion polymerization process [59]. The molecular weight varied with Cloisite type ([St] = 1.75 mol dm^{-3}, [APS] = 8 mM, [SDS] = 6.3 mM, 75 °C, 5 wt.% Cl Na) as follows:

$$M_w \times 10^5/\text{clay}: 8.7/\text{without Cl}, 9.0/\text{Cl Na}, 9.5/\text{Cl}_{CTAB}, 9.8/\text{Cl}_{AUA} \quad (13.26)$$

$$M_n \times 10^5/\text{clay}: 34.5/\text{without Cl}, 36.8/\text{Cl Na}, 35.4/\text{Cl}_{\text{CTAB}}, 39.7/\text{Cl}_{\text{AUA}} \quad (13.27)$$

Based on these results, it can be seen that the loading of Cloisite had no significant influence on the molecular weights of the PSt; however, the addition of 5 wt% Cloisite caused a slight increase in the molecular weight. Furthermore, the PSt molecular weights were highly polydisperse (PDI ca. 4.0, with and without Cloisite), which indicated that there had been a variation in monomer concentration at the reaction loci during the emulsion polymerization. The relatively low molecular weights (ca. 1×10^5) of PSt for the emulsion polymerization indicated a strong deactivation of growing radicals via chain transfer or monomolecular termination within the polymer/Cloisite nanoclusters.

The effect of SDS and Cloisite concentrations on the molecular weight parameters was reported by Chern et al. [90]. The CMC of an aqueous solution of SDS is approximately 8 mM [147], and the incorporation of 1 wt.% Cl Na reduces this to 6.4 mM [90]. The incorporation of 5 wt% Na-MMT would be expected to reduce the CMC to below 6.3 mM. It is postulated that reaction systems with [SDS] = 6.3 mM in the presence of 5 wt% Cl Na are so abundant in micelles that the polymerization kinetics is predominated by micelle nucleation. The polymer chains were produced in particles which originated from micelle nucleation, and those produced in the Cl Na platelets. The polymer chains produced in the Cloisite platelets were smaller, probably due to the increased termination events and the lower monomer concentration at the reaction loci. Chern et al. have also reported that the M_w of the PSt with Cloisite Na is much larger than that of the counterpart without Cl Na, and the difference increases with a decreasing SDS concentration [90] (Recipe: 2.6 mM SDS, St = 16 wt% (relative to water), temperature = 70 °C; the reaction mixture was homogenized at ca. 13 500 rpm for 10 min):

$$M_w \times 10^{-6}/[\text{SDS}]/\text{m}M: 1.25/9, 1.3/11, 1.4/13, 1.45/15, 1.55/17, 1.6/19 \quad (13.28)$$

$$M_{w,Cl} \times 10^{-6}/[\text{SDS}]/\text{m}M: 1.6/9, 1.62/11, 1.72/13, 1.74/15, 1.77/17, 1.8/19 \quad (13.29)$$

Furthermore, the M_w of the PSt in Chern's study was much larger than that reported by Li et al. [59]. The M_w was seen to increase with increasing [SDS] for polymerizations both with (1 wt%, $M_{w,Cl}$) and without Cl Na (M_w). The increased M_w is attributed to the enhanced segregation of radicals inside the particles with increasing [SDS] (i.e. Np) (Np $\sim$ [SDS]$^{0.47}$, for the runs without Cl Na). The segregation effect reduces the probability of bimolecular termination via the entry of a radical into the particle already containing one radical. Furthermore, at constant [SDS], the M_w of the polymerization with Cl Na is larger than that of the counterpart without Cl Na. One possible explanation for this observation is the competitive absorption of radicals by Cl Na platelets that further depresses the bimolecular termination reaction in the latex particles. It should be noted that the difference in M_w between the polymerizations with and without Cl Na increases with decreasing [SDS].

Table 13.3 Variations of the molecular weight parameters with Cloisite type in the emulsion polymerization of styrene and n-butyl acrylate.

Cloisite	$[\eta]\ (cm^3\ g^{-1})^{a,b}$	$M_v \times 10^{-5 a,b}$
–	0.021, 0.816	4.115, 7.07
Cl 20A	0.187, 0.449	0.148, 3.53
Cl 30B	0.364, 0.455	0.230, 3.77
Cl Na	0.221, 0.826	1.843, 4.39

[a]Styrene.
[b]n-butyl acrylate.

This shows supporting evidence for the important role of Cl Na platelets in polymerizations stabilized by relatively low levels of surfactant.

The viscosity-average molecular weights were observed to decrease in the presence of Cloisite (Table 13.3) [131]. This behavior is a complex function of the monomer concentration at the reaction loci, the nature of the reaction loci, and the number of polymer particles, and can result from the interaction of growing radicals with the clays. After a three-month period of aging of polymer (PBA and PSt)/clay dispersions at room temperature, the formation of crosslinked polymer/clay nanocomposites was observed. The presence of the Fe^{3+} and Mn^{2+} ions in the Cloisite, and of free unreacted initiator APS, can lead to the formation of a redox system that continuously generates radicals and chain-transfer reactions to the clay and polymer chains. A combination of these radicals can, in turn, result in the formation of a crosslinked polymer net or a percolating network of MMT nanoparticles in the polymer matrix.

Choi et al. [62] have described a simple and convenient way to obtain the exfoliated PMMA/silicate nanoparticles or nanocomposites through an in situ polymerization with a reactive surfactant [148]. Water causes the basal space of silicate layers to widen, without any chemical treatment, while a reactive surfactant, 2-acrylamido-2-methyl-1-propanesulfonic acid (AMPS) [149], has amido and sulfonic acid contents within the molecule. A strong interaction of the amido moiety with pristine silicates may result in a polymeric, end-tethered active material. A conventional anionic surfactant, in this case dodecylbenzenesulfonic acid sodium salt (NaDBS), was compared with AMPS in the interaction with pristine Cloisite Na. Both surfactants were intercalated into the layers of the pristine Cloisite Na, which had been dispersed in water before polymerization. The nanocomposites with AMPS were exfoliated during polymerization because AMPS caused the polymer to be end-tethered on pristine Cl Na. The Cls were exfoliated up to a 10 wt% content of pristine Cl Na relative to the amount of MMA.

The dependence of the molecular weights of PMMA on the Cloisite Na content is described by a curve with a maximum at a certain concentration of Cl Na [62]:

$$M_n \times 10^5 / A_{2.5} M_{20} - T\% : 1.38/0,\ 1.74/1,\ 1.66/5\ \text{and}\ 1.06/10 \qquad (13.30)$$

$$M_n \times 10^5/A_{0.3}M_{20}-T\% : 4.5/0, 4.8/1, 2.6/5 \text{ and } 1.6/10 \tag{13.31}$$

$$M_n \times 10^5/D_{2.5}M_{20}-T\% : 2.0/0, 2.6/1, 1.6/5 \text{ and } 2.7/10 \tag{13.32}$$

The increase in molecular weight was interpreted as follows. Silicates can act as monomer absorbers, as do micelles, so that local concentrations of MMA and surfactants close to will be elevated and the emulsion particles formed more rapidly than when no Cloisite Na is available. PMMA-modified silicate layers may become involved in the nucleation and stabilization of polymer particles (Pickering stabilization), thus favoring particle nucleation, as well as increases in polymerization rate and polymer chain growth at the reaction loci. At a higher Cloisite Na level, the modification of Cloisite is less pronounced, the polymer emulsion is less stable, and the inhibitory activity of Cl Na more pronounced. These effects may in turn lead a decrease in both the rate of polymerization and the polymer molecular weights. Furthermore, the molecular weights of PMMA shows a broad distribution in both systems (the PDI for A ranges from 2.4 to 4.2; and for D from 2.2 to 2.5) which typical for the emulsion polymerization.

In the suspension and emulsion polymerization of MMA, the M_ws of the polymer/ Cl Na nanocomposites were increased in the presence of an organically modified MMT, as follows [43]:

$M_{w,susp} \times 10^{-3}$/modifier : 3.3/none, melt intercalation,

2.64/**1** (*n*-decyltrimethylammonium chloride),

3.67/**2** [2, 2′,-azobis(isobutylamidine hydrochloride)] and

4.75/**3** [2-(methacryloyloxy)ethyl-trimethylammonium chloride]

and

$M_{w,emul} \times 10^{-3}$/(MMT/modifier) : 3.3/melt intercalation, 5.87/(**1**), 5.65/(**3**)

These results indicate that an emulsion polymerization produces higher molecular weights than does a suspension polymerization; such behavior can be attributed to differences in the reaction loci for both systems.

13.4.2
X-Ray Diffraction Studies

13.4.2.1 Homopolymers

In the PSt/Cloisite Na composite prepared by emulsion polymerization, only a slight expansion of the d_{001} spacing (1.70 nm) was observed in the XRD pattern [59], which suggested that intercalation of the polymer to the clay layers was very limited. This, in turn, indicates that the Cloisite Na is too hydrophilic to absorb larger amounts of styrene. In order to improve the affinity to styrene, the clay was cation-exchanged with CTAB, after which the organically modified clay Cl_{CTAB} (d_{001} spacing 1.84 nm) was dispersed in styrene, and the dispersion introduced to the polymerization system. For the PSt/Cl_{CTAB} composite, a clear diffraction peak near 4.4° was observed, and the corresponding d_{001} spacing was seen to be only 2.01 nm:

$$d_{001}(\text{nm})/\text{additive} : 1.7/\text{Cl Na}, 2.01/\text{Cl}_{\text{CTAB}}, \gg \text{no peaks}/\text{Cl}_{\text{AUA}} \quad (13.33)$$

The PSt composites with Cl Na or Cl$_{\text{CTAB}}$ thus obtained were of the intercalated type.

The exfoliation of Cl$_{\text{AUA}}$ takes place in an alkaline aqueous medium, with the long alkyl chain of the amino acid making favorable the clay's compatibility with styrene. As expected, an exfoliated-type PSt/Cloisite nanocomposite was successfully obtained by emulsion polymerization in the presence of the Cl$_{\text{AUA}}$, and the 001 diffraction peak was shown in the XRD pattern to have completely disappeared. The exfoliation of clay layers in the PSt/Cl$_{\text{AUA}}$ nanocomposite was further confirmed by using TEM (Figure 13.19). An examination of this image (where the dark lines correspond to the silicate layers) showed clearly that the clay was well exfoliated, the layers having dispersed evenly in the polymer matrix.

The featureless XRD spectrum of the PSt/Cloisite Na (93.3/6.7, w/w) sample was obtained from the emulsion polymerization of styrene with 9 mM SDS and 1 wt% Cloisite Na [90]. This suggested that the regular structure of Cloisite Na platelets disappeared on completion of the polymerization; hence, an exfoliated-type PSt/Cloisite nanocomposite could be obtained by emulsion polymerization in the presence of 1 wt% Cloisite Na.

X-ray diffraction was also used to study the organization and exfoliation of clay platelets inside a polymer particle matrix obtained by inverse emulsion polymerization [91]. For the crosslinked PAAm nanocomposites, a small peak was observed at about 3.36° which corresponded to a d-spacing of 2.5 nm. As the crosslinking in the PAAm particles prevented the latex particles from film formation during drying, the orientation could only be obtained by stacking neighboring Cloisite 20A sheets on the surfaces of the latex particles. In the uncrosslinked nanocomposites, which were able to form a film upon drying, the d_{001} peak of the clay completely disappeared and only a very small, broad peak was observed at $2\theta = 6.3°$, corresponding to a d-spacing of 1.4 nm, and indicating that the majority of the clays in the film were in an exfoliated state. When the same group attempted to collect XRD patterns in the

Figure 13.19 Transmission electron microscopy image of the PSt/Cl$_{\text{AUA}}$ nanocomposite [59].

2θ range of 0.1–1°, there were some indications of weak peaks at 2θ = 0.5–0.6° (corresponding to d-spacing of 15–20 nm) for the uncrosslinked sample, suggesting that there might be some long-range ordering in the film. By comparison, a smooth curve was observed for the sample containing crosslinked PAAm in the 2θ range of 0.1–1°.

The monomer concentration is proportional to the degree of the basal spacing expansion. If the monomers inside Cloisite particles participate in polymerization under a fixed amount of surfactant (2-acrylamido-2-methyl-1-propane sulfonic acid, AMPS), the degree of expansion produces different structures in the polymer/ Cloisite nanoaggregates and nanocomposites [86]. For polyacrylonitrile (PAN)/ poly(butyl acrylate) PBA/and poly(MMA) (PMMA)/Cloisite nanocomposites – all of which have monomers with high δ_p values that enlarge the basal spacing of Cloisite before polymerization – no peaks were observed within a Cloisite composition range of 20 wt%. Such absence of peaks indicates the presence of exfoliated structures. In addition, styrene has a δ_t value similar to, and a δ_d value higher than, those of other monomers. However, while styrene/Cloisite shows a small basal spacing, PSt/Cloisite nanocomposites show XRD peaks at 5.69° (3 wt%) and 5.96° (5 wt%):

$$d_{001}(\text{nm})/\text{additive}: 1.6/\text{PSt}(3\,\text{wt\%}), 1.5/\text{PSt}(5\,\text{wt\%}) \tag{13.34}$$

These results indicate the presence of intercalation, the degree of which seems to increase with PSt concentration up to a critical value, and then to reach a plateau; alternatively, the intercalation may slightly decrease with increasing concentrations of PSt. The swelling of hydrophilic Cl_{AMPS} organoclay by hydrophobic styrene does not enlarge the basal spacing of Cloisite before polymerization. Rather, the slight decrease in d_{001} with increasing PSt content indicates that the reaction loci transfers from the Cloisite domains to the bulk phase.

In systems with HDTMAB and ODTMAB, however, the corresponding basal spacing of PSt/ODTMAB/Cloisite and PSt/HDTMAB/Cloisite nanocomposites is larger than 3.5 nm [80]. The absence of diffraction peaks indicates that the clay has been completely exfoliated [46]. This further increase in basal spacing is due to the swelling in the interlayer space upon the intercalation of styrene and subsequent polymerization. Accordingly, the penetration process of the monomer is facilitated by the attractive forces between the long alkyl chain in the interlayer space and the monomer molecules. The above results shows that basal spacing $d_{001}(\text{nm})$ increases by the presence of additive as follows:

$$d_{001}(\text{nm})/\text{additive}: 1.25/\text{none}, 1.93/\text{HDTMAB-Cloisite},$$
$$2.23/\text{ODTMAB-Cl} > 3.5/\text{PSt-HDTMAB-Cl or PSt-ODTMAB-Cl}$$

In the emulsion polymerization of styrene, the 001 d-spacings of the composites were found to increase up to 1.55 nm as the PSt content in the composites increased [116]:

$$D_{001}(\text{nm})/\{\text{PSt/Cloisite}\}$$
$$= 0.98/\{0/100\}, 1.24/\{70/30\}, 1.38/\{80/20\}, 1.46/\{90/10\}, 1.55/\{95/05\} \tag{13.35}$$

In comparison with PMMA/clay hybrids produced via the same emulsion method, the increment of 001 d-spacings of the composites in all clay contents were slightly decreased. It is proposed, therefore, that penetration into the clay galleries by the styrene monomer, which has a strong hydrophobicity, is more difficult than with the MMA monomer, which has polar group in its side chain. There remain considerable amounts of unextractable organic materials, even after a five-day period of extraction [116]:

$$PSt(wt\%)/\{PSt/Cloisite\} = 21.7/\{70/30\}, 28.8/\{80/20\}, 33.9/\{90/10\}, 45.9/\{95/05\} \quad (13.36)$$

The residual PSt is regarded as the polymer intercalated between the interlayers of clay. This confinement of a polymer in the interlayer is tentatively ascribed to both the ion-induced dipole force acting between the host (clay) and the guest (PSt), and a contraction of the gallery's height as water is removed from it. The obtained M_w values of the extracted polymer are found to be in the order of 10^5 g mol^{-1}, which is comparable with that of pure PSt, and indicates that the clay presence in the reaction medium does not affect the average molecular mass of the polymer being intercalated. Furthermore, the ratio M_w/M_n was reported as ca. 2.0, which indicates that the polymer being formed has a broad distribution.

In the emulsion polymerization of MMA, the nanocomposites formed as a function of surfactant type were investigated using XRD [62]. The pristine Cloisite Na was shown to have a basal space of 1.14 nm, while the sample $A_{2.5}M_{20}T_{10}$ showed no peak with time during polymerization, and the exfoliation of silicates occurred less than 10 min after the polymerization had been initiated. However, the sample $A_{0.3}M_{20}T_{10}$ showed a slow disappearance of the peak, and a slow gap widening of the silicate layers.

Nonetheless, for $D_{2.5}M_{20}T_{10}$, two peaks were observed at about 6.8 and 3.0° throughout polymerization. Moreover, as the polymerization proceeded, the peak at 6.8° shifted to 6.5°, which is the same position for the extracted nanocomposite of $D_{2.5}M_{20}T_{10}$. Thus, the peak at 6.5° is caused by the intercalation of PMMA into the silicate layers. No peak is observed in PMMA/Cloisite Na for the $A_{0.3-2.5}M_{20}$ [A = AMPS (weight component); M = MMA monomer (weight component); T = wt % Cloisite Na relative to MMA] nanocomposites. When NaDBS (D) is used, the peaks at angle 6.5° (1.36 nm) for $D_{2.5}M_{20}T_{10}$ and at 6.0° (1.47 nm) for $D_{2.5}M_{20}T_5$, each appear. $D_{2.5}M_{20}T_1$ shows two peaks: a strong peak at 3.0° (3 nm), and a weaker/broader peak at 5.4° (basal space 1.75 nm), respectively. The latter peak was considered as the d_{001} spacing of pristine Cloisite Na, while the former peak was seen to be caused by the intercalation of NaDBS into the silicate layers. The two peaks indicate the two types of intercalated states of silicate. The peak shifts to a lower angle; when the amount of pristine Cloisite Na becomes small, this shift indicates the intercalation of PMMA into Cloisite Na.

$A_{0.3}M_{20}T_5$ has the residual molecules of 32 wt% after tetrahydrofuran (THF) extraction. In $D_{2.5}M_{20}T_5$, most polymers are removed by the THF extraction, with only about 4.6% of the residual molecules remaining in the silicate layers. Molecules may be removed easily from silicate in the exfoliated state, if there is no interaction between silicate layer molecules. Following the thermogravimetric analysis (TGA) of $A_{0.3}M_{20}T_5$ and $D_{2.5}M_{20}T_5$, a mechanism of exfoliation of Cloisite Na with AMPS was proposed [62]. As $A_{0.3}M_{20}T_5$ has more residual molecules than $D_{2.5}M_{20}T_5$, this suggests that AMPS might remain as an end-tethered form inside the silicate layers. The sodium dissociated from silicates would then associate with the sulfate ions of AMPS to form sulfonic acid sodium salt, which is a surface-active material.

A wide-angle X-ray diffraction (WAXD) analysis of PMMA nanocomposites (Cloisite/1) prepared by suspension polymerization showed a weak, broad peak at about 3° [43]. This peak was suggested to be caused by nonadsorbed silicate layers that may have aggregated and then reformed a multilayered structure with intercalated PMMA chains. After melt pressing at 180 °C, although the position of the basal reflection did not change significantly (d = 2.85 to 2.62 nm), the intensity increased and a secondary reflection (d002, ca. 6.5°) appeared.

For Cloisite modified (OLS) with initiator cation **2** [2,2′,-azobis(isobutylamidine hydrochloride)] or comonomer cation **3** [2-(methacryloyloxy)ethyl]trimethyl-ammonium chloride], the results of a WAXD analysis indicated that the exfoliated structure was preserved after melt pressing. For nanocomposites prepared using an OLS modified with **2** or **3**, the PMMA chains would become tethered to the surface. Polymer chains tethered to the silicate layers would hinder aggregation of the silicate layers and help to preserve the exfoliated structure.

When the nonfunctional cationic surfactant **1** (n-decyltrimethylammonium chloride) was used in the emulsion polymerization of MMA, an analysis using WAXD suggested that the exfoliated structure was not preserved after processing [43]. However, if the polymerization surfactant **3** [2-(methacryloyloxy)ethyl]trimethylammonium chloride] was used, then exfoliated PMMA nanocomposites could be prepared that remained stable during melt processing.

13.4.2.2 Copolymers

The XRD spectra of the Cloisite/copolymer [(poly(ether-block-amide) copolymer); PEBA] nanocomposites show multiple diffraction peaks, but the higher-order diffraction peaks merge into the peak of the matrix polymer and become difficult to identify [78]. The first-order diffraction peak of hybrid PNa (PEBA/Cloisite Na) is small and is almost combined with the broad peak from the polymer. For hybrids using surfactant with hydroxyl groups (i.e. P30B; PEBA/Cloisite 30B), the first-order peak is barely identifiable, which indicates that the clay is almost completely exfoliated (Figure 13.20). The intensities of the peak for the hybrids modified by single tail surfactants are also negligibly small; in fact, they are smaller than the intensity of the polymer diffraction peak. The low intensities of the peaks essentially indicate that the number of ordered diffraction centers is relatively low, and that there may be a considerable amount of single layer silicate – or, equivalently, a high extent of exfoliation – in these hybrids. In contrast to hybrids of the single tail family, both the first- and second-order diffraction peaks

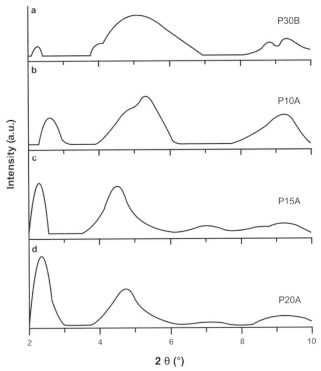

Figure 13.20 The X-ray diffraction patterns of poly(ether-block-amide) copolymer hybrids. (a) P30B; (b) P10A; (c) P15A; (d) P20A [43].

of the hybrids of the swallow-tail family, such as P15A, P20A and P93A, are relatively sharper and more intense.

The diffraction pattern of P20A resembles that of crystal, namely that the first-order peak is sharper and higher than the second-order peak, whereas for P15A the second-order peak – though broader – is of virtually the same height as the first order. A partitioning of the silicate layers may occur with P15A, as the first- and second-order peaks of this hybrid are almost the same height. The partitioning of silicate layers, which is possible in theory, might simply be a coexistence of two phases [150]. The sharp and intense first-order peaks in the swallow-tail hybrid family reveal that, in comparison with the single-tail hybrid family, the number of diffraction centers is higher and the stacks of silicate layers are more ordered. Consequently, it is reasonable to claim that the isolated silicate layers in these hybrids are fewer in number, or the extent of exfoliation is less. The partitioning of silicate layers in P15A could make the extent of exfoliation worse, and consequently the extent of exfoliation for P20A and P15A should follow the sequence of P20A > P15A.

The positions of the first-order peaks were translated into the peak d-spacing values and the following observations were noted:

- The d-spacing of the clay layered structure in the composite using Cloisite sodium is enhanced, which means the polymer molecules are able to penetrate into galleries between the hydrophilic silicate surfaces and reside there.
- There exist stacks of layered silicate within all composites, although the amount of stacks may be small in hybrids of the single-tail family. The d-spacing of the hybrids, or the gallery height between the silicate platelets, ranged from 3.13 to 3.87 nm after hybridization.

Two peaks (corresponding to two types of basal spacing) were observed for these hybrids with a high surfactant loading. Moreover, the intensity of the peak at a smaller 2θ value (larger basal spacing values) relative to that at a larger 2θ value (smaller basal spacing values) increased with the extent of surfactant loading [151–153]. This indicated the presence of heterogeneous interlayers for these hybrids, and with the increase of surfactant loading amount, the interlayers with smaller basal spacings were eventually expanded. The heterogeneous interlayer structures indicate that the adsorbed surfactant aggregated heterogeneously in different interlayers.

13.4.3
Thermal and Mechanical Properties

13.4.3.1 Polystyrene and Poly(methyl methacrylate) Nanocomposites

The exfoliated clay is expected to be valuable in reinforcing the polymer. A major improvement in modulus was achieved in the exfoliated polystyrene/Cloisite nanocomposite, but less efficiently in the intercalated composites [59] (Figure 13.21B):

$$\text{storage modulus at } 40°C(GPa)/\text{clay}: 1.08/\text{without}, 1.26/Cl\,Na, 1.36/Cl_{CTAB}, 1.65/Cl_{AUA} \tag{13.37}$$

Diagrams of tan δ versus temperature for polystyrene (PSt) and PSt/Cloisite composites were used to estimate the glass transition temperature (T_g) (Figure 13.21A). The influence of clay on T_g was of minor importance in the intercalated composites PSt/Cl Na, PSt/Cl$_{CTAB}$ and Cl$_{AUA}$ [59]:

$$T_g(°C)/\text{clay}: 127/\text{without}, 129/Cl\,Na, 130/Cl_{CTAB}, 139/Cl_{AUA} \tag{13.38}$$

By contrast, in the exfoliated nanocomposite PSt/Cl$_{AUA}$, the T_g was 12 °C higher than that of the neat polymer. It is clear that movement of polymer segments was greatly restricted by the clay layers [154].

The thermal stability of the polystyrene/Cloisite composites was studied using TGA. Here, the temperature at which a 20 wt % loss of the polymer occurred was taken as the onset degradation temperature (T_d), which varies with the Cl type as follows (Figure 13.22A):

$$T_d\text{onset}(°C)/\text{clay}: 408/\text{without}, 419/Cl\,Na, 426/Cl_{CTAB}, 454/Cl_{AUA} \tag{13.39}$$

The clay increased the thermal stability of PSt significantly, with the exfoliated silicate being the most effective, raising the T_d value by 11 °C, 18 °C and 46 °C for

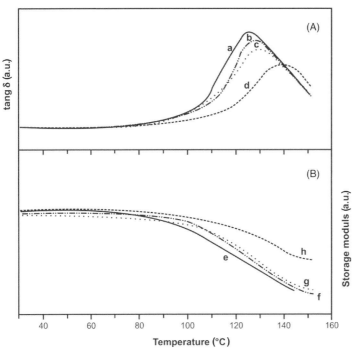

Figure 13.21 Tan δ (A) and storage modulus (B) versus temperature for pure polystyrene and polystyrene/Cloisite composites. (a, e) PSt; (b, f) PSt/Cl Na; (c, g) PSt/Cl$_{CTAB}$; (d, h) PSt/Cl$_{AUA}$ [59].

PSt/Cl Na, PSt/Cl$_{CTAB}$ and PSt/Cl$_{AUA}$, respectively. In this situation, it is likely that the clay layers hindered the out-diffusion of volatile decomposition products [155].

An approximate 34 wt% weight loss was monitored at 700 °C for Cloisite 20A by using TGA, and this was in agreement with the theoretical presence of 95 mEq. of the organic ammonium salt ($M_w = 551.6$ g mol^{-1}) [91]. The pure PAAm latex particles began to decompose at a temperature of about 200 °C, and were completely combusted at approximately 500 °C (Figure 13.22B, curves e–h). Crosslinked nanocomposite PAAm/Cloisite 20A particles with a total clay concentration of 12.4 and 9.5 wt%, respectively, showed a delayed decomposition compared to the polymeric sample, even though the delayed decomposition was much less significant compared to conventional polymer/clay nanocomposites that are prepared by melt intercalation of reactive blending and which have a more homogeneous distribution of platelets inside the polymer matrix. The homogeneous distribution of the clays hinders the diffusion of the volatile decomposition materials, and this in turn contributes to the increased thermal stability [13]. After heating the nanocomposites to 700 °C and leaving them isothermally for 30 min, residues of 8.2 wt% (for C$_{298}$) and 6.3 wt% (for C$_{305}$) of the solid material remained in the chamber. These values were well in accordance with the inorganic content that had been incorporated into the composite particles (8.3 wt% and 6.4 wt% for two samples, respectively,

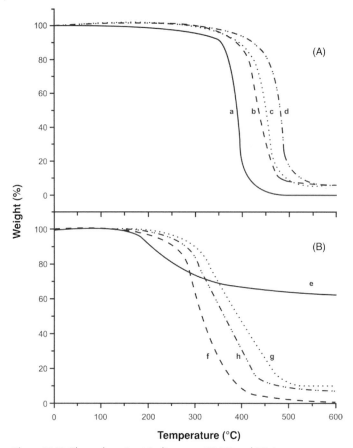

Figure 13.22 Thermal gravimetric diagrams. (A) PSt and PSt/Cloisite composites: (a) PSt; (b) PSt/Cl Na; (c) PSt/Cl$_{CTAB}$; (d) PSt/Cl$_{AUA}$ [59]. (B) (e) Cloisite 20A; (f) polyacrylamide (PAAm) homopolymer latex (C$_{284}$); PAAm clay nanocomposite latexes (g) C$_{298}$ and (h) C$_{305}$ [91].

after deducting the amount of the modifying agent). This means that either no free clay, or only a negligible amount of clay, was present in the composite latex dispersions.

The data in Figure 13.22 show that the addition of clay increases the thermal stability of both the oil-soluble (PSt) and water-soluble (PAAm) polymers; however, the thermal stability of the PSt/clay composite was higher than that of the PAAm/clay composite. It is of interest to note that the PSt/clay particles began to decompose and were completely combusted only at a high temperature, whereas the PAAm/clay particles began to decompose at a low temperature and were combusted at high temperature (comparable with PSt/clay composite).

If the nonfunctional cationic surfactant (**1**, *n*-decyltrimethylammonium chloride) is used in the emulsion polymerization, then a WAXD analysis indicated that the exfoliated structure was not preserved after processing [43]. However, if the polymerization surfactant **3** ([2-(methacryloyloxy)ethyl]trimethylammonium chloride) was used, then exfoliated PMMA nanocomposites could be prepared that remained stable during melt processing; this nanocomposite displayed a 15 °C increase in T_g and a 62 °C increase in the temperature for 20% weight loss. Elsewhere, Bandyopadhyan et al. [156] observed a 6 °C increase in T_g and a 50 °C increase decomposition temperature for a PMMA nanocomposite prepared by an emulsion process when the silicate was present during the polymerization.

The TGA analysis of $A_{0.3}M_{20}$ series demonstrates the thermal stability of the nanocomposites up to 300 °C [62]. An analysis of the $D_{2.5}M_{20}$ series showed the thermal decomposition of PMMA over the range from 350 to 420 °C. The series $A_{0.3}M_{20}$ showed an increase in T_g while the series $D_{2.5}M_{20}$ showed a decrease in T_g according to the contents of the Cloisite Na:

$$T_g(°C)/A_{0.3}M_{20}-T\% : 149/0,\ 154/1,\ 151/5\ \text{and}\ 156/10 \tag{13.40}$$

$$T_g(°C)/D_{2.5}M_{20}-T\% : 154/0,\ 152/1,\ 146/5\ \text{and}\ 142/10 \tag{13.41}$$

Based on the results of the TGA study, both intercalated and exfoliated PMMA nanocomposites showed higher decomposition temperatures when compared to a macrocomposite. The nanocomposite with OLS prepared from Cloisite/2 showed a 50 °C increase in decomposition temperature for a 20% weight loss, while the intercalated nanocomposites exhibited a 15 °C increase for the same weight loss. An increase in T_g of 5–14 °C was observed for the exfoliated PMMA nanocomposite prepared using either **2** [2,2′,-azobis(isobutylamidine hydrochloride)] or **3** [2-(methacryloyloxy)ethyl-trimethyl ammonium chloride] as silicate modifiers. A negligible increase in T_g was observed for both the PMMA macrocomposites and the intercalated nanocomposite prepared using **1** (n-decyltrimethylammonium chloride) as an organic modifier. This increase in decomposition temperature may be associated with barrier properties, although slower heating rates did not affect the presented results to any significant degree [43].

13.4.3.2 Poly(ethylene oxide) Nanocomposites

The intercalation of poly(ethylene oxide) (PEO) with Cloisite exchanged with various cations was first reported by Ruiz-Hitzky and coworkers [157]. These nanocomposites had conductivities ranging from 10^{-9} to 10^{-4} S cm^{-1}. The authors postulated that the PEO chains retain their helical structure in the intercalated state, forming a type of tunnel that increases the mobility of the cations. In all cases, the conductivity of the pristine clay was much lower than for the composite material at temperatures below 600 K. For Li^+-exchanged clays, the conductivity at 550 K was one to two orders of magnitude greater than for the Na^+-exchanged forms. Thin films of clay/polymer nanocomposites have a wide range of potential applications as barriers for gas transport; for example, a Li^+-Laponite-PEO nanocomposite had a conductivity of 3.2×10^{-7} S cm^{-1} at 80 °C, while Cloisite Na/polyethylene (PE) and -polyethylene-block-poly(ethyleneglycol) (PE-PEG) thin film nanocomposites had a maximum conductivity of 2.5×10^{-5} S cm^{-1} [158]. In contrast to the studies of Ruiz-Hitzky on the conformation of intercalated PEO, Wu and Lerner proposed the presence of one or two layers of polymer chains with nonhelical conformations in the galleries of Cloisite Na composites [137], with the number of polymer layers rising with an increasing polymer/clay ratio. It was proposed that the intercalation of PEO would increase the mobility of cations by reducing their interactions with the negatively charged clay surface, and that the observed decrease in conductivity with increasing temperature below the PEO decomposition temperature was due to a change in polymer conformation. Both, Vaia and coworkers [159] and Bujdak and coworkers [160] showed that a layer charge could influence PEO intercalation for polystyrene/mica and PEO/smectite composites.

The intercalation of PEO in the interlayers of natural (Cl_N) and synthetic (Cl_S) Cloisite is verified by the increased d_{001}-spacing for the nanocomposite samples in comparison to the pristine clays [161]. The basal spacings of the pristine Cl_N and Cl_S samples correspond to a single water layer hydration state [162]. The (001) peak of pristine Cl_S is broader than that of Cl_N, which suggests that the restacking of clay platelets during drying on the glass slide is more uniform for Cl_N. Intercalation occurs for Li^+-, Na^+- and K^+-exchanged clay samples, with maximum basal spacings in the range 1.79 to 1.94 nm being achieved with overall clay: PEO ratios of 50:50 (w/w). For the Cl_N composites, the maximum basal spacings are independent of the exchange cation, despite the K^+ form of the pristine clay being ~0.04 nm more expanded. For the Cl_S composites, the maximum basal spacing of the Li form is about 0.14 nm greater than that for the Na form, at a 50:50 clay: polymer ratio. These basal spacings suggest that the polymer is incorporated either as a double layer of chains parallel to the silicate layer, or as a single layer of PEO which takes a helical conformation parallel to the layers [157]. More recent studies have pointed towards a double-layer positioning of the polymer, with no specific orientation of the chains in the mineral layer plane [163].

The nuclear magnetic resonance (NMR) observations suggest that Li^+ remains in the same structural environment, regardless of the amount of intercalated PEO in the interlayer, and that the ion interacts only weakly with PEO [161]. In contrast, ^{23}Na NMR results have shown the presence of Na^+ in multiple hydration states. These environments are distinguished by the inner-sphere or outer-sphere coordination of clay oxygen atoms to the metal cation [164]. PEO intercalation causes the sodium to take on an inner-sphere coordination. In such inner-sphere sites, the ion has a partial shell of water molecules, while the remaining sites are probably occupied by oxygen atoms in the clay basal plane [164], since Cl_N has a permanent negative structural charge of 0.7 per formula unit and almost 40% of this is developed by tetrahedral Al for Si substitution. The reduction in interlayer water content with PEO intercalation in the Cl_N samples depends on the interlayer cation, with the greatest effect for Li^+, and the least for Na^+. For the Li^+-exchanged samples, the reduction from 8.3 to 3.9 water molecules per formula unit does not affect the ^7Li NMR resonance significantly, indicating that the large hydration energy of Li^+ allows it to retain its hydration shell. In contrast, for Na^+ the reduction from 7.0 to 5.3 water molecules per formula unit results in a conversion of all the ions to inner-sphere sites, in turn resulting in an overall decrease in Na^+ hydration. The differences in hydration

13.4 Collective Properties of Polymer/MMT Nanocomposites

behavior between Li^+ and Na^+ correlate well with the observation that the conductivity of Li Cloisite/PEO nanocomposites is up to one order of magnitude higher than the conductivity of Cl Na/PEO nanocomposites [157]. The formation of inner-sphere complexes by Na^+, as supported by NMR results, retards its diffusional motion relative to Li^+, for which the local hydration environment does not change substantially in the presence of intercalated PEO. Indeed, the comparison of pristine Li^+- and Na^+-exchanged clays, and also of some other cation-exchanged clays, showed that their conductivity used to be of the same order of magnitude [165]. Reported conductivities for pristine Li^+ and Na^+ have ranged from 2×10^{-4} to 2×10^{-3} S cm^{-1}, depending on the clay [165], to $(6-9) \times 10^{-9}$ S cm^{-1} [166].

When sufficient PEO exists in the PEO/Cloisite composite, an intercalated structure is formed (with d spacings distributed around 1.7 nm, which corresponds to a PEO bilayer of about 0.8 nm thickness) [163] For composites with extremely small amounts of PEO ('polymer-starved' composites at Cloisite loadings of ($\phi_{Cl} > 90\%$), an intercalated monolayer of PEO can also be observed, with an intercalated d spacing of about 1.37 nm. For the Cloisite loadings of $\phi_{Cl} = 1-10$ wt%, the layered silicates retain their pristine parallel registry, but there is an increase in the d spacing due to the intercalation of PEO in the interlayer gallery (Figure 13.23c) [167]. Successive single layers self-assemble in stacks (tactoids; Figure 13.23a), and in a highly parallel stacking that can give rise to 00l XRD diffraction peaks up to the 11th order [168]. These micrometer-size tactoids are dispersed in the PEO matrix (Figure 13.23b), separated by regions of pure polymer.

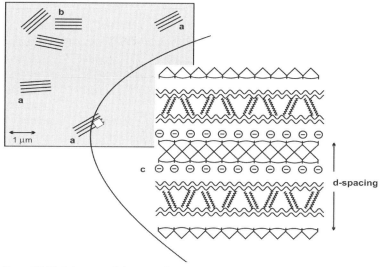

Figure 13.23 Schematic of the PEO/Cloisite Na intercalated nanocomposites. (a) Single tactoid (Cloisite layers assemble); (b) tactoid assembles; (c) within the tactoid, MMT layers are separated by a 0.8 nm film of PEO [163, 167].

The most interesting finding of these studies was most likely that the introduction of Cloisite inorganic fillers slows down the polymer crystal growth in the vicinity of the filler. This contrasts the usual behavior of semicrystalline polymers, where fillers normally result in heterogeneous nucleation, promoting crystal formation in their vicinity. Such crystal nucleating effects are in fact observed also for Cloisite when incorporated into other semicrystalline polymers, such as poly(vinyl alcohol) [169], polypropylene [22] and nylon-6 [170]. The unusual behavior observed herein for PEO originates from the specific manner that Cloisite interacts with PEO.

The addition of small cations, in the form of salts, has been shown to reduce or completely destroy the crystallinity of PEO [171]. This behavior is attributed to the strong coordination of PEO to small cations, such as Na^+ and Li^+, which promote a 'crown-ether'-type of backbone conformations coordinated to the cations [172]. Such crown-ether conformations deviate from the helical PEO conformations typically found in bulk PEO crystals, and therefore amorphize the PEO. Because the Cloisite surfaces bear large numbers of cations (approximately one Na^+ per 70 $Å^2$), PEO chains in their vicinity are highly coordinated to the Na^+, adopting conformations with many crown-ether arrangements, which are highly amorphous. Similar behavior with cations, promoting an amorphous PEO structure, has also been seen in the interlayer gallery between the Cloisite layers in PEO/Cloisite Li composites [17].

The slowing of the crystal growth rate is due to this amorphization of the polymer in the vicinity of the silicate, which forces the spherulite to grow around the dispersed tactoids, resulting in 'broken' lamellar pathways and geometrically anisotropic shapes. Differential scanning calorimetry (DSC) further corroborates these optical microscopy observations, because it is found that the crystallization temperatures (T_c) are shifted to lower values with Cloisite loadings. This reduction in T_c is additional evidence that crystallization is inhibited with the addition of silicate fillers, as larger undercoolings are now needed to begin the crystallization process. If this is the case, then reducing the surface density of Na^+ cations on the Cloisite surfaces should result in enhanced PEO crystallization. Finally, putting all these pieces together, one can trace the crystallization behavior of PEO in the presence of Cl Na fillers: primary nucleation takes place in the bulk away from the Cloisite surfaces, and initially spherulites grow normally until they encounter a filler. At this point, because amorphous PEO structures are promoted in the vicinity of the Cloisite, there is a retardation of the spherulite growth front, resulting in jagged edges and nonspherulitic morphologies. This delay in covering space allows for the nucleation of other spherulites that grow in the same manner until all the volume is filled. These additional nuclei cause the PEO to crystallize faster overall, despite the slower crystal growth rate, and allows for the total volume to crystallize more quickly, albeit with much smaller crystallite sizes than in bulk PEO.

13.5
Polymer–Inorganic Nanocomposites

Generally, organic and inorganic materials are mutually incompatible. In order to prepare fine polymer (synthetic and natural)/inorganic (dispersions) materials, the polymer/inorganic interface must be sufficiently stabilized or modified. This becomes a major requirement, if nanostructured materials with very large interfacial areas are to be prepared. The use of block copolymers makes the construction of such materials straightforward. In order to obtain stable polymer/inorganic dispersions there must be sufficient adhesion between the polymer and inorganic microphase. Adhesion can be provided by polymer-bound functional blocks that bind to the inorganic material through ion binding, acid–base reactions or coordination. For example, the group of ion-binding involves polymers such as poly(styrene sulfonic acid), PMMA, PEO, poly(amino acid), poly(vinylpyridine) and poly(cyclopentadienyl-methyl norbornene). However, the inorganic particles have the problems of poor processability and easy oxidation, combined with a large density. As organic substances have a low density, good processability and chemical stability, many research groups have been guided to integrate one or more inorganic nanoparticles with suitable organic substances (usually polymers), leading to a new special class of nanomaterials, namely, organic–inorganic nanocomposites (or hybrids, OIHs). These materials can take on the unique properties of the trapped nanoparticles, together with the existing quantities of the organic substance, and even show novel properties due to synergistic effects derived from the interaction between them.

The incorporation of inorganic materials into polymeric domains is, at first sight, not a trivial matter. As most polymers are low-surface-energy materials, the adhesion between inorganic compounds and polymers is usually poor. Only through specific interactions such as dipolar interactions, hydrogen bonding, complex formation or covalent bonding can inorganic materials be incorporated into polymers. Those polymers containing polar or ionic groups provide solubility in the desired medium, while simultaneously bonding to the metal colloids [173–177]. Block copolymers are also able to solubilize or adhere to inorganic materials [178], a property which is of special relevance to the controlled synthesis of inorganic colloids or the controlled assembly of hybrid materials.

In order to solubilize inorganic molecules, it is necessary to optimize the organic–inorganic molecules bonding, and in this respect Pearson's hard–soft acid–base (HSAB) principle has proved to be quite valuable. This relates acid–base and donor–acceptor interactions to solid adhesion and surface interactions [179], and represents one of the most important bridges between chemistry and physics of the solid state [180, 181]. On the basis of the HSAB principle, all metals are soft and mostly acidic, all semiconductors are fairly soft and most of them are bases, and most insulators, including polymers, are hard. The HSAB principle states that hard acids prefer to bond to hard bases, and soft acids prefer to bond to soft bases. By incorporation of functional groups of appropriate hardness – for example, polymer–analogue reactions – it is possible to optimize the binding of a polymer to metals or semiconductors. It is also clear that changes in the band structure of semiconductor colloids (quantum–size effects) also change the stability of the polymer–semiconductor interface. The HSAB principle allows one to optimize the bonding block for the stabilization of colloids. Metal colloids are well-stabilized by soft bases such as thiols and phosphines (which serves as standard 'capping agents' for the synthesis of small metal clusters).

The metal salts of hard acids and hard bases, such as hydroxides, sulfates and phosphates, are often water-insoluble substances. In order to stabilize colloids, the appropriate stabilizing polymer should contain hard bases. Until now, poly(vinylalcohol) (PVA), poly(styrene sulfonic acid) and polyphosphates were applied for the preparation of monodisperse metal oxide sols [182]. The HSAB principle also allows for the design of appropriate reaction sequences in the controlled assembly of hybrid materials. The formation of an assembly proceeds through a number of steps that involve the solubilization of inorganic precursors, chemical reactions, and nucleation/growth processes. One can start from a labile hard/soft bonded precursor salt that is easily incorporated into a polymeric domain via the formation of a more stable soft/soft bond. This should not be so stable that it hinders the further transformation (redox reaction, oxide/sulfide formation) of the compound into the desired colloid. Nucleation and growth of the colloid creates a new interface, the structure and stabilization of which again depend on its energy.

The synthesis of nanocomposites based on inorganic layered materials as hosts and polymers as guests has shown a very promising future [2, 183]. These nanocomposites can be considered as reinforcements for polymers, or as the addition of functions for inorganic frames. Several methods have been developed to prepare polymer/layered inorganic compound nanocomposites, such as the intercalation of monomers followed by subsequent polymerization, the direct intercalation of polymer chains, the precipitation of polymers from exfoliated inorganic layers, and the *in situ* synthesis of inorganic materials in polymer solutions [184, 185]. Originally, the MMT-type layered silicates occupied the majority of layered materials (see above), but today the polymer/layered double hydroxide (LDH) nanocomposites represent an emerging class of materials, which may be used as flame retardants, stabilizers, medical materials, and so on [186]. A series of water-soluble polymers have been reported for preparing polymer/LDH nanocomposites, such as poly(acrylate), poly(amino acid), poly(styrene sulfonate), polyaniline and PEO [187, 188]. However, water-insoluble polymers are seldom incorporated into LDH layers. O'Leary et al. [189] has delaminated an organo-modified LDH, $Mg_2Al(OH)_6(C_{12}H_{25}SO_4)$, in poly(2-hydroxyethyl methacrylate) with 10% LDH loading. However, the optical property of the obtained polymer/LDH nanocomposites as opaque solids was sacrificed. In these studies [190], polyethylene-grafted maleic anhydride (PE-g-MAh)/MgAl LDH exfoliated nanocomposites with up to 5% LDH loading were synthesized by refluxing $Mg_3Al(OH)_8(C_{12}H_{25}SO_4)$ in a xylene solution of PE-g-MAh. However, the loadings of LDH were very low in the above-mentioned two types of nanocomposite containing water-insoluble polymers. Therefore, it is a challenge to prepare nanocomposites with good transparency from water-insoluble PMMA with a high loading level of LDHs.

A novel PMMA/MgAl LDH nanocomposite (PMMA-MgAl NCo) with a high transparency has been synthesized *in situ* from the emulsion consisting of a metal ion aqueous solution, SDS as an emulsifier, MMA as a monomer, and benzoyl peroxide (BPO) as thermal initiator [191]. The obtained nanocomposites could be kept transparent at even 50% loading of MgAl LDH, including dodecyl sulfate (DS). The elemental analysis revealed that the contents of Mg, Al and S in PMMA-MgAl

NCo sample were 4.76, 1.88 and 2.19 wt%, respectively, corresponding to the 33.9 wt% loading of $Mg_{2.85}Al(OH)_{7.70}(C_{12}H_{25}SO_4)_{0.98}$. The existence of DS in the PMMA-MgAl NCo sample was also confirmed by the adsorption of the $-OSO_3$ polyhedron vibration at around $1240\,cm^{-1}$ in the Fourier transform infrared spectrum. For comparison, $Mg_3Al(OH)_8(C_{12}H_{25}SO_4)$ (Mg$_3$Al-DS) was prepared using an ion-exchange method. $Mg_3Al(OH)_8(NO_3)$ was obtained by coprecipitation from an aqueous solution of $Mg(NO_3)_2$ and $Al(NO_3)_3$ by adding NaOH aqueous solution. The exchange of NO_3^- with DS was carried out in distilled water to yield a white powder, Mg$_3$Al-DS. A conventional microcomposite (named as PMMA-MgAl MCo) containing 30 wt% Mg$_3$Al-DS and 70 wt% PMMA was also prepared by mixing Mg$_3$Al-DS powder in an acetone solution of PMMA.

Figure 13.24 illustrates the intercalated structure of PMMA-MgAl NCo. The collapse of pillared structure of Mg$_3$Al-DS is due to the dehydration of MgAl hydroxide layers and the partial decomposition of DS alkyl chains at 200 °C [192]. Although, the intercalation of PMMA chains does not obviously expand the basal

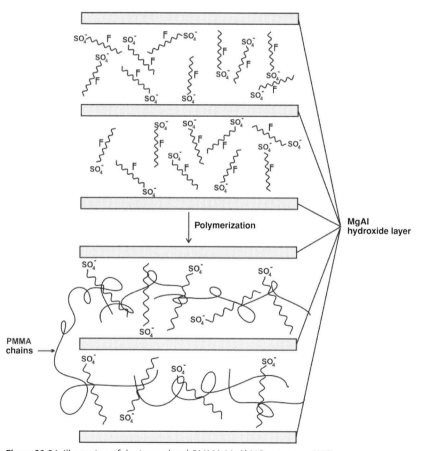

Figure 13.24 Illustration of the intercalated PMMA-MgAl NCo structure [191].

spacing of pillared Mg$_3$Al-DS, it does increase the thermal stability of both DS ions and the intercalated structure due to an interaction between the DS alkyl chains and PMMA chains in the galleries of MgAl LDH.

The obtained image shows that the intercalated MgAl LDH particles are present as hexagonal sheets. The thickness of these intercalated particles was measured as 25–40 nm, which corresponded to 10 to 15 stacked layers. Although the loading of MgAl LDH (including DS) is more than 30 wt%, the good optical property of PMMA matrix is maintained because the intercalated particles are very small and dispersed homogeneously in the matrix at molecular level. This is the unique advantage of *in situ* synthesis techniques.

The thermal decomposition of pure PMMA sample occurs in the weight loss at the range of 200–400 °C, and no residues remain above 400 °C. About 4 wt% weight loss for both PMMA-MgAl MCo and PMMA-MgAl NCo at about 100 °C is due to the evaporation of physically absorbed water in the intercalated layers. The Mg$_3$Al-DS component in PMMA-MgAl MCo sample begins to degrade from ca. 150 °C due to the loss of hydroxide on MgAl hydroxide layers, whereas the hydroxide on MgAl hydroxide layers in PMMA-MgAl NCo sample does not degrade until 200 °C due to the protection of intercalated PMMA chains in the gallery of MgAl hydroxide layers. The residues of PMMA-MgAl NCo samples measured at 230 °C are 6.3% higher than those of the PMMA-MgAl MCo sample. The thermal decomposition of PMMA-MgAl NCo sample took place at the range of 200–450 °C and forms approximately 20% charred residues, while the PMMA-MgAl MCo sample degrades completely at about 400 °C and results in about 10% white powder. The charred layers of PMMA-MgAl NCo sample further degrade at 620 °C to produce about 15% white powder. When the 50% weight loss is selected as a comparison point, the decomposition temperature of PMMA-MgAl NCo sample is about 45 °C higher than that of pure PMMA. These results suggest that the PMMA chains are protected in the dehydrated MgAl hydroxide layers, and therefore the TGA data provide positive evidence for the intercalation of PMMA chains into the galleries of MgAl hydroxide layers.

The stream of colloidal OIHs was pioneered by Kirkland [193] and Iler *et al.* [194], who reported the synthesis of uniform spherical polymer/silica composites with average diameters ranging from 500 to 20 000 nm. Most of the studies of OIHs focus on polymer nanocomposites in the form of films or bulks, for which two main techniques are available for their preparation. In the first technique, the preformed nanoparticles, usually after modification with suitable surfactants, are dispersed in a polymer or polymeric monomer solution. As an example, Mark and colleagues [195] prepared monolithic PMMA/SiO$_2$ nanocomposites by dispersing surface-modified silica nanoparticles in MMA, followed by polymerization of the monomeric continuous phase.

Duan *et al.* have reported [196] on the fabrication of micron-sized three-dimensional (3-D) titanium/polymer structures. Here, the characteristics of two-photon polymerization were evaluated. The size of the polymerized points depends on both the laser power and exposure time. Nonetheless, the resins with a high concentration of titanium(IV) acrylate complexes were difficult to polymerize, even when using UV light irradiation. This fact may be explained by an energy transition between the excited states of initiator and titanium(IV) complexes, which leads to a quenching of the excited initiators. In order to demonstrate the capability of two-photon fabrication with the photopolymerizable resin containing metal ions, the same authors produced a 3-D NCo structure and, to maximize the photonic band gap effect, they selected a covalent bond-type diamond crystal structure which consisted of dielectric atoms and bonded rods [197]. The diameters of the bonded rods and photonic atoms were 500 mm and 580 mm, respectively. The nanoparticles consisted of titanium hydroxide [Ti(OH)$_4$] that had been generated through the reaction between Ti ions and H$_2$O in air. After thermal treatment at 250 °C, the polymer rods showed a slight shrinkage and the particles appeared to become smaller. It could be concluded that the polymers had been degraded slightly and the titanium oxide (TiO$_2$) formatted, and this was subsequently using XRD. This result provided clear evidence that the nanoparticles of TiO$_2$ were generated in the polymer matrix after two-photon polymerization. Therefore, a micro/nanoscale structure of composite materials of polymer and inorganic nanoparticles was successfully fabricated by two-photon polymerization and post-chemical reactions.

Armes et al. [198] reported that this host of conducting polymer nanocomposites had unusual 'raspberry' morphologies that were rich in inorganic components at the surface. Thus, Armes' group synthesized a colloidal dispersion of poly(4-vinylpyridine)/silica nanocomposite particles in high yield by homopolymerizing 4-vinylpyridine in the presence of an ultrafine silica sol using a free-radical initiator in aqueous media at 60 °C. Subsequent TEM and aqueous electrophoresis measurements confirmed that the poly(4-vinylpyridine)/silica nanocomposite particles exhibited 'currant-bun' particle morphologies with the surface being polymer-rich [199].

13.6
General

The Hansen solubility parameters can be used to explain the mixing behavior of solvents, polymers and clays. These consist of three components: the dispersive component (δ_d); the polar component (δ_p); and the hydrogen-bonding component (δ_h). The equation is expressed as [86]:

$$\delta_t^2 = \delta_d^2 + \delta_p^2 + \delta_h^2, \quad \delta_t = (E/V)1/2 \tag{13.42}$$

where δ_t is the total (or Hildebrand) solubility parameter, E is the vaporization energy of a solvent, and V is the molar volume of a solvent.

The basal spacing (d_{001}) from XRD measurements is calculated at peak positions according to Bragg's law:

$$d_{001} = \lambda/(2\sin\theta) \tag{13.43}$$

where d_{001} is the interplanar distance of (001) reflection plane, θ is the diffraction angle and λ is the wavelength.

13.7
Conclusions and Outlook

The exfoliation of pristine Cloisite (MMT) is impeded by an electrostatic attraction between the negatively charged clay layer and the cations in the gallery. Therefore, a pretreatment is usually necessary when preparing exfoliated Cloisite composites. The opportunity to combine – at the nanometric level – clays and a natural or synthetic polymer appears to represent an attractive means of developing new organic–inorganic hybrid materials that possess properties inherent to both types of component. Most natural and synthetic polymer/natural silicate nanocomposites demonstrate an intercalated morphology. However, pristine clay is naturally hydrophilic, and its polymers are often hydrophobic. Because the hydrophilic nature of pristine clay impedes its homogeneous dispersion in polymer matrix, a brilliant suggestion was made to connect silicate layers and polymers covalently, by using modifiers containing reactive sites or by anchoring surface-active additives on the surface of silicates. The modification of clay with alkylammonium cations facilitates its interaction with a polymer, because the alkylammonium makes the hydrophilic clay surface organophilic.

The sodium ions in naturally occurring clays can be exchanged by organic cations to render the platelets dispersible in an organic medium (see Figure 13.25). The long alkyl chains of the quaternary ammonium compounds that are adsorbed on the surface of the hydrophobized clay surface enable the swelling and exfoliation of platelets in an apolar medium. The role of the alkylammonium cation is to improve penetration of the organophilic monomers into the interlayer space, while the role of the monomer is to promote dispersion of the clay particles. The increased basal spacing arises from the expansion of the interlayer space to accommodate the polymer; as a result, the intercalation process of polymers can be distinguished from the difference in basal spacing. The interlayer of natural clays is negatively charged, with the charge being partially neutralized by the presence of mobile cations such as Na^+, K^+ or Ca^{2+}. The attraction here is to speculate on the formulation of the electric 'multi' layer and, indeed, the densely packed layers of natural clay can be discussed in terms of the 'multi' layer concept. The van der Waals attraction within the clay interlayers is changed by the addition of alkylammonium cations; that is, the presence of ammonium cations and bulky alkyls changes both the charge and volume space of the interlayer. The resultant expansion of interlayer spacing can formulate two electric double layers within the interlayer. These double layers might be formed by the modification of clays by a high concentration of a modifier or a modifier with large alkylammonium cations (alkyl = tallow) or charged polymers, due to which the gap between layers is increased and the intercalated structure can shift to its exfoliated counterpart. A higher ionic exchange capacity leads to a larger d spacing, which is reasonable as a greater number of exchanged surfactant molecules will bring about more crowdedness and hence a higher d spacing.

The swelling capacities of clays depends on the solvent or monomer polarity. For Cloisite Na, the swelling capacity was found to be more than 100 wt% for acrylic monomers. In contrast, apolar styrene causes only a slight swelling of the clay.

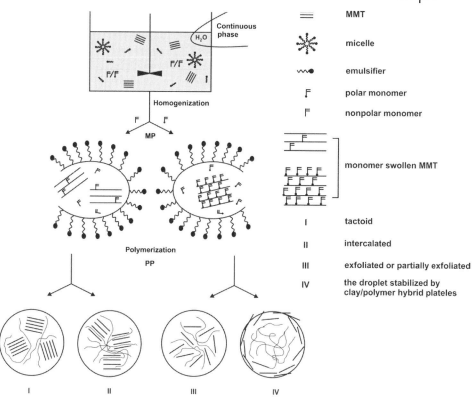

Figure 13.25 The mechanistic routes of polymer/clay nanocomposites formation in the mini-, micro- and emulsion polymerizations of polar and apolar unsaturated monomers.

The solubility parameters depend on dispersion and polar interactions, as well as on hydrogen bonding related to the solvent. The most important parameter, however, is considered to be dispersion, which is determinant for clay exfoliation. Both, the hydrogen components (δ_h) and the polar components (δ_p) of liquids are primary parameters for the dispersion states, and for the basal spacing expansion. The degree of expansion by monomers reflects the expected structure of polymer/Cloisite nanocomposites: a large basal spacing produces exfoliated structures, while a small basal spacing produces interacted structures. Because δ_h and δ_p both involve dipole–dipole interactions, monomer–dipole moments are associated with the basal spacing expansion that is related to the structure of polymer/Cloisite nanocomposites. The liquids with high δ_h (the hydrogen-bonding component) values expand the d_{001} spacing of Cloisite more than do liquids with low δ_h values. The basal spacing expansion depends on polar components (δ_p) and hydrogen-bonding components (δ_h) of organic liquids. Monomers with high dipole moments showed large basal spacings, while those with low dipole moments showed smaller basal spacings. The interlayer space of silicate is widened by water, this being the result of a strong penetration of water into the hydrophilic layers of clay. The addition of an aqueous solution of certain surface active additives led to exfoliation, even with hydrophobic solvents or monomers. These results indicate that a polar solvent has a higher interaction with silicate layers than an apolar one, and the amount of a polar solvent in the interlayer space of silicate is higher than that of its apolar counterpart. Both, monomers and initiators penetrate the interlayer spaces via the aqueous phase of the disperse system.

Monomers with high dipole moments showed large basal spacings before polymerization, and produced exfoliated polymer/Cloisite (MMT) nanocomposites, whereas those with low dipole moments showed smaller basal spacings and produced intercalated polymer/Cloisite nanocomposites. Intercalation implies the insertion of one or several polymer chains into the galleries of the original layer tactoids, which leads to a longitudinal expansion of the galleries. In this case, however, attractive forces are still present between the silicate layers, and these stack the layers within a uniform spacing. Exfoliation (delamination) implies a complete breakage of the original layer stacking order and a homogeneous dispersion of layers in the polymer. Exfoliated nanocomposites are formed when the layer spacing increases to the point where there are no longer sufficient attractions between the silicate layers to maintain a uniform layer spacing.

The presence of buffer (e.g. $NaHCO_3$) slightly increases the d_{001} spacing due to a partial neutralization of the free ions, or their arranging within the Cloisite. The depression of van der Waals forces leads to a partial expansion of the clay matrix. The additional accumulation of SDS micelles is accompanied by a stronger expansion of the d_{001} spacing. This is most likely due to a partial neutralization of cations and the formation of (inverse or direct) emulsifier micelles and mixed clay/emulsifier intercalated aggregates. In an alkaline medium, the terminal anions are accompanied mostly by a strong intercalation, or even by the exfoliation of clays. Derived anionic groups interact repulsively with each other and with negatively charged clay layers, forcing the layers apart and leading to exfoliation of the clay into individual layers. The above-mentioned data indicate that a variation in charge of the interlayer (electric double or triple layer) is directly connected with the d-spacing.

The presence of XRD peak(s) in the direct or inverse emulsions indicates that the clays in the nanocomposite are intercalated and still have the van der Waals force interaction between layers. It is speculated that a multiple electric layer (negative charges are immobilized in clay matrix while cations located between negatively charged groups are mobile) is in existence, while the presence of water strongly favors the expansion of interlayer spacing due to colloidal forces. In some systems, the intensive penetration of water, polar monomers and amphiphiles show no peaks, indicating exfoliation of the clay. The intensive expansion of interlayer spacing of clay could be accompanied by the formulation of electric double layers; indeed, the irreversibility of the intercalation–exfoliation process might support the concept of the formulation of electric double layers.

The basal spacings suggest that the polar polymer (e.g. PEO, polysaccharides) is incorporated as a double layer of chains parallel to the silicate layer, or as a single layer of polymer taking a helical conformation parallel to the layers. A double-layer positioning of the polymer with no specific orientation of the chains appears in the mineral layer plane. The intercalation of polar polymer can increase the mobility of cations by reducing their interactions with the negatively charged clay surface.

'Colloidal' forces, such as double-layer, van der Waals, hydrophobic, depletion and bridging forces may each influence, in complex manner, the interiors of the clay layers in the aqueous dispersion of surfactant, comonomers, polymer and additives. Charged surfactants and additives can initiate the formulation of electrical layers and van der Waals forces. Depletion forces born by the intercalated polymer chains expand the interlayer spacing, while the hydrophobic monomer, together with the hydrophobic chain of a surfactant, may formulate the hydrophobic domains in the clay matrix.

The rate of polymerization of MMA increased monotonically with increasing Cloisite amount. The results were discussed in terms of increased contribution of monomolecular mode of chain termination. Furthermore, the presence of Fe(II) and Fe(III) ions favor the contribution of the redox system BPO/Fe(II), which is known to increase the rate of initiation as well as the rate of polymerization. The metal nanoparticles formed by the reduction of metal ions with primary radicals can take part in the catalysis of the polymerization. In the first step of this reaction, the radicals are captured by metal ions and Lewis acids (the appearance of an inhibitory period), while in the second step the formed metal nanoparticles favor growth events.

Aqueous solutions of emulsifiers are known to present self-aggregation of the emulsifier into micelles at a well-defined CMC. Likewise, polymer micelle-like aggregates begin to form along the polymer chain at the CAC which is, therefore, an analogue of the CMC, albeit in solution with an added polymer compound. Characteristically, the CAC is always lower than the CMC of the corresponding emulsifier, and a lower CAC is particularly pronounced in solutions of polyelectrolytes with an opposite charge to the emulsifier. The emulsifier can interact cooperatively with Cloisite Na at the CAC, forming micelle-like aggregates within the Cloisite. The ionic strength of the Cl Na-containing anionic emulsifier solution, however, is much higher compared to the native emulsifier solution. Emulsifier molecules, when subjected to a hostile environment with a very high electrolyte concentration, show a stronger tendency to aggregate cooperatively with one another to form micellar aggregates, and this can result in a decreased CMC. The interaction of emulsifier molecules with clay can reduce the CMC of an emulsifier. The sonification of a clay/monomer emulsion can disrupt Cloisite, such that exfoliated clay layers can participate in the stabilization of monomer droplets. Furthermore, the clay may exfoliate during polymerization, such that the organically modified clay layers can become involved in the stabilization of polymer particles. The homogenization and delamination of MMT helps in the exfoliation of Cloisite to layers, and their location at the droplet–particle surface.

The emulsion polymerization of unsaturated monomers with different stabilizing systems (anionic, nonionic and protective colloids) in the presence of Cloisite, demonstrated an enhancement of the conversion, the number of polymer particles, and the polymerization rate. An enhanced rate of emulsion polymerization was attributed to a larger number of micelles (N_m) available for particle nucleation and the increased rate of radical generation. Metal ions in the clay are considered to participate in the redox-initiator system, which increases the formation of initiating radicals. A further role of clay is its participation in the stabilization of polymer particles formed by miniemulsion and emulsion polymerization. When clay particles rather than surfactant molecules are used to stabilize the emulsion, the term 'Pickering emulsion' is commonly used. The hydrophobically modified clay platelets alone appeared capable of stabilizing both o/w and w/o emulsions, without the addition of surfactants (see Figure 13.18). The formed Pickering emulsions may then be regarded as surfactant-free emulsions.

In the miniemulsion polymerization of styrene with clay, the silicate layers may locate not only on the surface of the particle but also inside the emulsion–latex particle, depending on the initiator nature. In the suspension polymerization of styrene in the presence of different organophilic MMYs, the silicates are most likely also located inside the particle. In the suspension polymerization of MMA in the presence of organophilic MMT, the clay layers are situated both inside and on the surface of the latex particles. In this case, the kinetics of the monomer polymerization is much influenced by the presence of clay.

In the emulsion and miniemulsion polymerizations of apolar unsaturated monomers, the addition of Cloisite enhanced the rate of polymerization, with the average diameter being decreased with an increasing concentration of clay. Furthermore, the enhanced growth events led to the appearance of larger polymers. In contrast, for the microemulsion emulsion polymerization of apolar and polar monomers, the presence of clay reduced the rate of polymerization and increased the average size of the polymer particles. Furthermore, an inhibition period was observed in the microemulsion polymerization of polar monomers such as butyl acrylate. The absence of any inhibition during the microemulsion polymerization of styrene can be discussed in terms of a restricted interaction of hydrophobic radicals with the clays. The hydrophobic styrene monomer and polystyrene growing radicals are located within the hydrophobic domain of monomer saturated clay interior. The polar unsaturated monomer, as well as its growing radicals, can interact with clay, as a result of which the radical entry and growth events are changed. The clay minerals are known to inhibit free radical reactions by adsorption of the propagating or initiating radicals to the Lewis acid surface. The radicals then either undergo bimolecular termination or form carbocations by electron transfer to the Lewis acid site. The rugged surface morphology suggested that the polymer particles are covered by a layer of clay platelets which were also reported as being located at the surface of the latex particle. A 'fluffy' structure was formed around the particles.

For the preparation of polymer/clay dispersions, the surface character, structure of pristine silicates and compatibility between silicates, solvents and polymers were each considered. With regards to their microstructure, the clay particles are dispersed in the polymer matrix in either intercalated or exfoliated states. In the intercalated composites, the polymer chains are inserted into the interlayer space of the stacking silicate platelets, while the silicate layers remain well ordered, despite the basal space being greatly expanded. In the exfoliated composites, the discrete clay layers are randomly dispersed in the continuous polymer matrix. Exfoliated nanocomposites are formed when the layer spacing increases to the point where there are no longer sufficient attractions between the silicate layers to maintain a uniform layer spacing. In particular, exfoliated polymer/clay hybrids offer improved mechanical and thermal properties due to the homogeneous dispersion of clay in the polymer matrix, as well as a large interfacial area of the clay layers.

The clay also caused a significant increase in the thermal stability of polymers, with the exfoliated silicate being most effective. For example, the T_d of polystyrene was raised 10–50 °C for PSt/Cloisite nanocomposites. Here, it is likely that the clay layers hindered the out-diffusion of the volatile decomposition products. The intercalated and exfoliated PMMA/Cloisite nanocomposites showed higher T_d values compared to a macrocomposite. In fact, the nanocomposite with OLS showed a 50 °C increase in T_d for a 20% weight loss, whereas the intercalated nanocomposites exhibited a 15 °C increase for the same weight loss.

The hysteresis of both Cl_N and Cl_S indicates the presence of antiferro- and ferrimagnetic phases in both, and an associated remnant magnetization, M_r. The remnant magnetization, M_r, of Cl_N is significantly larger than that of Cl_S, which is indicative of a much larger contribution of antiferro and/or ferrimagnetic components. Investigations conducted at different temperatures showed an increase in magnetization from 300 to 5 K as expected (Bloch's law) and, for the majority of the antiferro/ferrimagnetic phase-in studied systems, a T_C in excess of 500 °C. The addition of various reducing agents and stabilizers may lead to the formation of different metal nanoparticles within the interlayer of clay. The shape and/or size of nanoparticles is a function of the type of clay and the concentration of reducing agent and stabilizer.

Acknowledgments

This research is supported by the Slovak Grand Agency (VEGA) through the grant number 2/7013/27 and 2/7083/27, SAV-FM-EHP-2008-01-01 project, and Science and Technology Assistance Agency through the APVT projects (0173-06, 0362-07, 0030-07, 0562-07, 0592-07).

Abbreviations

AAm	acrylamide
AFM	atomic force microscopy
AIBN	2,2′-azoisobutyronitrile
AMPS	2-acrylamido-2-methyl-1-propane sulfonic acid
AN	acrylonitrile
AOT	sodium bis-2-ethylhexylsulfosuccinate
APS	ammonium peroxodisulfates
AUA	amino-undecanoic acid
BA	butyl acrylate
BPO	benzoyl peroxide
CAC	critical aggregation concentration

CEC	cation exchange capacity
CH	cyclohexane
Cl 15A	Cloisite 15A
Cl 30B	Cloisite 30B
Cl Na	Cloisite Na or NaMMT
Cl_{AUA}	Cl (or MMT) modified with AUA
Cl_{CTAB}	Cl (or MMT) modified with CTAB
Cl_N	natural Cloisite (MMT)
Cl_S	synthetic Cloisite (MMT)
Cl_{SDS}	Cloisite (MMT) modified with SDS
CMC	critical micelle concentration
CTAB	cetyltrimethyl ammonium bromide
d_{001}	basal spacing
dBPO	dibenzoylperoxide
DLS	dynamic light scattering
d_p	diameters of the latex particles
DS	dodecyl sulfate
E	vaporization energy of a solvent
ER	electrorheological
f	initiator efficiencys
HD or OD	hexadecyl or octadecyl
HDTMAB	hexadecyltrimethylammonium bromide
HEMA	2-hydroxyethyl methacrylate
HLB	hydrophilic-lipophilic balance
HSAB	hard–soft acid–base
HT	hydrogenated tallow
[I]	concentrations of initiator
IP	induction periods
IPA	isopropyl alcohol
k'_{des} (cm^2 s^{-1})	specific desorption rate constants
k_{des} (s^{-1})	specific desorption rate constants
k_d	initiator decomposition constant
k_p	propagation rate constant
KPS	peroxodisulfate potassium
k_t	termination rate constant.
LDH	layered double hydroxide
[M]	concentration of monomer
M	monomer
M_{mag}	magnetization
MCo	microcomposite
MMA	methyl methacrylate
MMT	montmorillonite
M_r	remnant magnetization
M_v	viscosity-average molecular weight
M_w	weight-average molecular weight

N_A	Avogadro's number
NaDBS (D)	dodecylbenzenesulfonic acid sodium salt
NaMMT	natural montmorillonite
NCo	nanocomposite
NDA	N,N'-methylenediacrylamide
N_p	number of polymer particles
$\bar{n}$	average number of radicals per particle
o/w	oil-in-water
OCl	organically-modified Cl
ODTMAB	octadecyltrimethylammonium bromide
OIHs	organic–inorganic hybrids
OLS	organic modification of layered silicates
P30B	PEBA/Cloisite 30B hybrid
PAAm	polyacrylamide
PAN	polyacrylonitrile
PBA	poly(butyl acrylate)
PDI	polydisperse index
PE	polyethylene
PEBA	poly(ether-block-amide) copolymer
PEBA/Cl Na	copolymer/Cloisite hybrid
PE-g-MAh	polyethylene-grafted maleic anhydride
PEO	poly(ethylene oxide)
PE-PEG	polyethylene-block-poly(ethylene glycol)
PLSN	polymer-layer silicate nanocomposites
PMMA	poly(methyl methacrylate)
PSt	polystyrene
PVA	poly(vinylalcohol)
[R$^\bullet$]	concentration of radical
R_p	rate of polymerization
SAN/silicate	copolymer SAN/silicate hybrid
SAN	poly(styrene-co-acrylonitrile) copolymer
SC VSM	superconducting vibrating sample magnetometer
SDS	sodium dodecyl sulfate
SEM	scanning electron microscopy
SPI	soy protein isolate
St	styrene
T	tallow
tan δ	loss factor
T_C	Curie temperature
TEM	ransmission electron microscopy
T_g	glass transition temperature
TGA	hermogravimetric analysis
THF	etrahydrofuran
V	molar volume of a solvent
VA-086	2,2'-azobis[2-methyl-N-(2-hydroxyethyl)propionamide]

VDAC vinylbenzyl- dimethyldodecylammonium chloride
WAXD wide-angle X-ray diffraction
XRD X-ray diffraction
δ_d dispersive component
δ_h hydrogen-bonding component
δ_p polar components
δ_t total (or Hildebrand) solubility parameter
γ_i surface tension

References

1 Morlat-Therias, S., Mailhot, B., Gonzalez, G. and Gardette, J. (2005) *Chemistry of Materials*, **17**, 1072.
2 Alexandre, M. and Dubois, P. (2000) *Materials Science and Engineering*, **1**, 28.
3 Novak, B.M. (1993) *Advanced Materials*, **5**, 422.
4 LeBaron, P.C., Wang, Z. and Pinnavaia, T.J. (1999) *Applied Clay Science*, **15**, 11.
5 Lagaly, G. (1999) *Applied Clay Science*, **15**, 1–9.
6 Lan, T. and Pinnavaia, T.J. (1994) *Chemistry of Materials*, **6**, 2216.
7 Messersmith, P.B. and Giannelis, E.P. (1994) *Chemistry of Materials*, **6**, 1719.
8 Messersmith, P.B. and Giannelis, E.P. (1995) *Journal of Polymer Science, Part A: Polymer Chemistry*, **33**, 1047.
9 Yano, K. and Usuki, A. (1997) *Journal of Polymer Science, Part A: Polymer Chemistry*, **35**, 2289.
10 Alexandre, M., Beyer, G., Henrist, C., Cloots, R., Rulmont, A., Jerôme, R. and Dubois, P. (2001) *Macromolecular Rapid Communications*, **22**, 643.
11 Gilman, J.W., Jackson, C.L., Morgan, A.B. and Harris, R. (2000) *Chemistry of Materials*, **12**, 1866.
12 Yong, Y., Zhn, Z., Yin, J., Wang, X. and Qi, Z. (1999) *Polymer*, **40**, 4407.
13 Burnside, S.D. and Giannelis, E.P. (1995) *Chemistry of Materials*, **7**, 1597.
14 Kim, B.H., Jung, J.H., Kim, J.W., Choi, H.J. and Joo, J. (2001) *Synthetic Metals*, **117**, 115.
15 Lee, D., Lee, S.H., Char, K. and Kim, J. (2000) *Macromolecular Rapid Communications*, **21**, 1136.
16 Gilman, J.W., Kashiwagi, T. and Lichtenhan, J.D. (1997) *Society for the Advancement and Process Engineering Journal*, **33**, 40.
17 Hackett, E., Manias, E. and Giannelis, E.P. (2000) *Chemistry of Materials*, **12**, 2161.
18 Huang, J.C., Zhu, Z.K., Yin, J., Qian, X.F. and Sun, Y.Y. (2001) *Polymer*, **42**, 873.
19 Choi, Y.S., Wang, K.H., Xu, M. and Chung, I.J. (2002) *Chemistry of Materials*, **14**, 2936.
20 Wang, K.H., Choi, M.H., Koo, C.M., Xu, M., Chung, I.J., Jang, M.C., Choi, S.W. and Song, H.H. (2002) *Journal of Polymer Science, Part B: Polymer Physics*, **40**, 1454.
21 Giannelis, E.P. (1998) *Applied Organometallic Chemistry*, **12**, 675.
22 Manias, E., Touny, A., Wu, L., Strawhecker, K., Lu, B. and Chung, T.C. (2001) *Chemistry of Materials*, **13**, 3516.
23 Kojima, Y., Usuki, A., Kawasumi, M., Okada, A., Kurauchi, T. and Kamigaito, O. (1993) *Journal of Polymer Science, Part A: Polymer Chemistry*, **31**, 1755.
24 Byun, H.Y., Choi, M.H. and Chung, I.J. (2001) *Chemistry of Materials*, **13**, 4221.
25 Templin, M., Franck, A., Chesne, A.D., Leist, H., Zhang, Y., Ulrich, R., Schädler, V. and Wiesner, U. (1997) *Nature*, **278**, 1795.
26 Viville, P., Lazzaroni, R., Pollet, E., Alexandre, M. and Dubois, P. (2004)

Journal of the American Chemical Society, **126**, 9007.

27 Robello, D.R., Yamaguchi, N., Blanton, T. and Barnes, C. (2004) *Journal of the American Chemical Society*, **126**, 8118–8119.

28 Zhang, Z., Zhang, L., Li, Y. and Xu, H. (2005) *Polymer*, **46**, 129.

29 Ray, S.S. and Okamoto, M. (2003) *Progress in Polymer Science*, **28**, 1539.

30 Wang, K., Wang, L., Wu, J., Chen, L. and He, C. (2005) *Langmuir*, **21**, 3613.

31 Friedlander, H.Z. and Grink, C.R. (1964) *Journal of Polymer Science, Polymer Letters*, **2**, 475.

32 Mohan, S., Laura, N., Cerini, S.S. and Ghosh, K.I.M. (1996) *Journal of Polymer Science, Part B: Polymer Physics*, **34**, 1443.

33 Hoffmann, B., Dietrich, C., Thomann, R., Friedrich, C. and Mülhaupt, R. (2000) *Macromolecular Rapid Communications*, **21**, 57.

34 Doh, J.G. and Cho, I. (1998) *Polymer Bulletin*, **41**, 511.

35 Akelah, A. and Moet, A. (1996) *Journal of Materials Science*, **31**, 3589.

36 Fu, X.A. and Qutubuddin, S. (2001) *Polymer*, **42**, 816.

37 Weimer, M.W., Chen, H., Giannelis, E.P. and Sogah, D.Y. (1999) *Journal of the American Chemical Society*, **21**, 1615.

38 Vaia, R.A. and Giannelis, E.P. (1997) *Macromolecules*, **30**, 8000.

39 Hasegawa, N., Okamoto, H., Kawasumi, M. and Usuki, A. (1999) *Journal of Applied Polymer Science*, **74**, 3359.

40 Fu, X. and Qutubuddin, S. (2001) *Polymer*, **42**, 807.

41 Chen, G. and Ma, Z. (2001) *Scripta Materialia*, **44**, 125.

42 Giannelis, E.P. (1996) *Advanced Materials*, **8**, 29.

43 Huang, X. and Brittain, W.J. (2001) *Macromolecules*, **34**, 3255.

44 Zeng, C. and Lee, J.L. (2001) *Macromolecules*, **34**, 4098.

45 Chen, G., Chen, X., Lin, Z., Ye, W. and Yao, K. (1999) *Journal of Materials Science Letters*, **18**, 1761.

46 Okamoto, M., Morita, S., Taguchi, H., Kim, Y.H., Kotaka, T. and Tateyama, H. (2000) *Polymer*, **41**, 3887.

47 Kojima, Y., Usuki, A., Kawasumi, M., Okada, A., Fukushima, Y., Kurauchi, T. and Kamigaito, O. (1993) *Journal of Materials Research*, **8**, 1185.

48 Kim, J.W., Kim, S.G., Choi, H.J., Suh, M.S., Shin, M.J. and Jhon, M.S. (2001) *International Journal of Modern Physics B*, **15**, 657.

49 Kim, J.W., Noh, M.H., Choi, H.J., Lee, D.C. and Jhon, M.S. (2000) *Polymer*, **41**, 1229.

50 Trlica, J., Saha, P., Quadrat, O. and Stejskal, J. (2000) *Physica A*, **283**, 337.

51 Usuki, A., Hasegawa, N. and Kato, M. (2005) *Advances in Polymer Science*, **179**, 138.

52 Tombacz, E. and Szekeres, M. (2004) *Applied Clay Science*, **27**, 75.

53 Dekking, H.G.G. (1967) *Journal of Applied Polymer Science*, **11** (1), 23.

54 Hasegawa, N., Tsukigase, A. and Usuki, A. (2005) *Journal of Applied Polymer Science*, **98**, 1554.

55 Wang, C., Wang, Q. and Chen, X. (2005) *Macromolecular Materials and Engineering*, **290**, 920.

56 Lin, Y., Skaff, H., Emrick, T., Dinsmore, A.D. and Russell, T.P. (2003) *Science*, **299**, 226.

57 Lee, D.C. and Jang, L.W. (1996) *Journal of Applied Polymer Science*, **61**, 1117.

58 Putlitz, B., Landfester, K., Fischer, H. and Antonietti, M. (2001) *Advanced Materials*, **13**, 500.

59 Li, H., Yu, Y. and Yang, Y. (2005) *European Polymer Journal*, **41**, 2016.

60 Sun, Q., Deng, Y. and Wang, Z.L. (2004) *Macromolecular Materials and Engineering*, **289**, 288.

61 Qutubuddin, S., Fu, X. and Tajuddin, Y. (2002) *Polymer Bulletin (Berlin)*, **48**, 143.

62 Choi, Y.S., Choi, M.H., Wang, K.H., Kim, S.O., Kim, Y.K. and Chung, I.J. (2001) *Macromolecules*, **34**, 8978.

63 Pickering, S.U. (1907) *Journal of the Chemical Society*, **91**, 2001.
64 Cauvin, S., Colver, P.J. and Bon, S.A.F. (2005) *Macromolecules*, **38**, 7887.
65 Huber, G., Kealhofer, G. and Plank, J. (2005) U.S. Patent 2005-0113261.
66 Binks, B.P., Clint, J.H. and Whitby, C.P. (2005) *Langmuir*, **21**, 5307.
67 Corobea, M.C., Uricanu, V., Donescu, D., Radovici, C., Serban, S., Garea, S. and Iovu, H. (2007) *Materials Chemistry and Physics*, **103** (1), 118.
68 Corobea, M.C., Uricanu, V., Donescu, D., Deladi, S., Radovici, C., Serban, S. and Constantinescu, E. (2006) *e-Polymers*, **60**, 1.
69 Donescu, D., Radovici, C., Petcu, C., Serban, S., Corobea, M.C. and Ghiurea, M. (2008) *Journal of Dispersion Science and Technology*, **29** (3), 340.
70 Tolbert, S.H., Firouzi, A., Stucky, G.D. and Chmelka, B.F. (1997) *Nature*, **278**, 264.
71 Murray, H.H. (2000) *Applied Clay Science*, **17**, 207.
72 Leu, C.M., Wu, Z.W. and Wei, K.H. (2002) *Chemistry of Materials*, **14**, 3016.
73 Beyer, F.L., Tan, N.C.B., Dasgupta, A. and Galvin, M.E. (2002) *Chemistry of Materials*, **14**, 2983.
74 Ginzburg, V.V. and Balazs, A.C. (2000) *Advanced Materials*, **12**, 1805.
75 Limary, R., Swinnea, S. and Green, P.F. (2000) *Macromolecules*, **33**, 5227.
76 Ren, J., Silva, A.S. and Krishnamoorti, R. (2000) *Macromolecules*, **33**, 3739.
77 Vaia, R.A., Jandt, K.D., Kramer, E.J. and Giannelis, E.P. (1996) *Chemistry of Materials*, **8**, 2628.
78 Yang, I.K. and Tsai, P.H. (2006) *Polymer*, **47**, 5131.
79 Vaia, R.A., Teukolsky, R.K. and Giannelis, E.P. (1994) *Chemistry of Materials*, **6** (7), 1017.
80 Zeng, Q.H., Wang, D.Z., Yu, A.B. and Lu, G.Q. (2002) *Nanotechnology*, **13**, 549.
81 Bongiovanni, R., Chiarle, M. and Pelizzetti, J. (1993) *Journal of Dispersion Science and Technology*, **14**, 255.
82 Donescu, D., Corobea, M.C., Uricanu, V., Radovici, C., Serban, S., Garea, S. and Iovu, H. (2007) *Journal of Dispersion Science and Technology*, **28** (5), 671.
83 Corobea, M.C., Donescu, D., Serban, S., Ducu, C., Malinovschi, V. and Stefanescu, I. (2006) *Revue Roumaine de Chimie*, **51** (1), 39.
84 Kim, S.S., Park, T.S., Shin, B.C. and Kim, Y.B. (2005) *Journal of Applied Polymer Science*, **97** (6), 2340.
85 Ho, D.L. and Glinka, C.J. (2003) *Chemistry of Materials*, **15** (6), 1309.
86 Choi, Y.S., Ham, H.T. and Chung, I.J. (2004) *Chemistry of Materials*, **16**, 2522.
87 Xu, M., Choi, Y.S., Kim, Y.K., Wang, K.H. and Chung, I.J. (2003) *Polymer*, **44**, 6387.
88 Choi, Y.S., Xu, M. and Chung, I.J. (2003) *Polymer*, **44**, 6989–6994.
89 Barton, J. and Capek, I. (1994) in *Radical Polymerization in the Dispersion Systems: A Paramount Communications Company*, (ed. T.J. Kemp), Ellis Horwood Limited, Veda, Bratislava.
90 Chern, C.S., Lin, J.J., Lin, Y.L. and Lai, S.Z. (2006) *European Polymer Journal*, **42**, 1033.
91 Voorn, D.J., Ming, W. and van Herk, A.M. (2006) *Macromolecules*, **39**, 2137.
92 Yudin, V.E., Divoux, G.M., Otaigbe, J.U. and Svetlichnyi, V.M. (2005) *Polymer*, **46**, 10866.
93 Wagener, R. and Reisinger, T.J.G. (2003) *Polymer*, **44**, 7513.
94 Levin, E.M., Hou, S.S., Budko, S.L. and Schmidt-Rohr, K. (2004) *Journal of Applied Physics*, **96**, 5085.
95 Callaway, W.S. and McAtee, J.L. (1985) *American Mineralogist*, **70**, 996.
96 Koerner, H., Hampton, E., Dean, D., Turgut, Z., Drummy, L., Mirau, P. and Vaia, R. (2005) *Chemistry of Materials*, **17**, 1990.
97 Okamoto, M., Morita, S., Kim, Y.H., Kotaka, T. and Tateyama, H. (2001) *Polymer*, **42**, 1201.
98 Uskov, I.A. (1960) *High Molecular Weight Compounds*, **6**, 926.
99 Solomon, D.H. and Rosser, M.J. (1965) *Journal of Applied Polymer Science*, **9**, 1261.

References

100 Solomon, D.H. and Swift, J.D. (1967) *Journal of Applied Polymer Science*, **11**, 2567.

101 Liu, G.D., Zhang, L.C., Qu, X.W., Wang, B.T. and Zhang, Y. (2003) *Journal of Applied Polymer Science*, **90**, 3690.

102 Billmayer, F.W. (1971) *Textbook of Polymer Science*, 2nd ed. Wiley, New York.

103 Liu, G., Zhang, L., Zhao, D. and Qu, X. (2005) *Journal of Applied Polymer Science*, **96**, 1146.

104 Capek, I. (2006) in *Nanocomposite structures and dispersions* (eds D. Mobius and R. Muller), Elsevier, London.

105 Lianos, P., Lang, J. and Zana, R. (1983) *Journal of Colloid and Interface Science*, **91**, 276.

106 Capek, I. (1999) *Advances in Colloid and Interface Sci.*, **82**, 253.

107 Barton, J. (1996) *Progress in Polymer Science*, **21**, 399.

108 Capek, I. (1999) *Advances in Colloid Interface Science*, **80**, 85.

109 Guo, J.S., Sudol, E.D., Vanderhoff, J.W., Yue, H.J. and El-Aasser, M.S. (1992) *Journal of Colloid and Interface Science*, **149**, 184.

110 Ostwald, W. (1901) *Physical Chemistry*, **37**, 385.

111 Capek, I. (2004) *Advances in Colloid Interface Science*, **107**, 125.

112 Kabalnov, A.S. and Shchukin, E.D. (1992) *Advances in Colloid Interface Science*, **38**, 69.

113 Higuchi, W.I. and Misra, J. (1962) *Journal of Pharmaceutical Science*, **51**, 459.

114 Capek, I. and Chern, C.S. (2001) *Advances in Polymer Science*, **155**, 101.

115 Wang, S.T., Schork, F.J., Poehlein, G.W. and Gooch, J.W. (1996) *Journal of Applied Polymer Science*, **60**, 2069.

116 Noh, M.W. and Lee, D.C. (1999) *Polymer Bulletin*, **42**, 619.

117 Harkins, W.D. (1947) *Journal of the American Chemical Society*, **69**, 1428.

118 Smith, W.V. (1948) *Journal of the American Chemical Society*, **70**, 3695.

119 Smith, W.V. and Ewart, R.H. (1948) *Journal of Chemical Physics*, **16**, 592.

120 Smith, W.V. (1949) *Journal of the American Chemical Society*, **71**, 4077.

121 Priest, W.J. (1952) *Journal of Physical Chemistry*, **56**, 1077.

122 Roe, C.P. (1968) *Industrial and Engineering Chemistry*, **60**, 20.

123 Fitch, R.M. and Tsai, C.H. (1971) in *Polymer Colloids* (ed. R.M. Fitch), Plenum, New York, p. 73.

124 Gilbert, R.G. (1995) *Emulsion Polymerization: Mechanistic Approach*, Academic, London.

125 Yoshida, K., Morishima, Y., Dubin, P.L. and Mizusaki, M. (1997) *Macromolecules*, **30**, 6208.

126 Liaw, D.J., Huang, C.C. and Kang, E.T. (1999) *Current Trends in Polymer Science*, **4**, 117.

127 Halay, S.W., McCormick, C.L. and Butler, B.G. (eds) (1991) Water-soluble polymers: Synthesis, solution, properties, and application. ACS Symposium series No. 467, American Chemical Society, Washington, DC.

128 Winnik, F.M. (1991) *Langmuir*, **7**, 905.

129 Xie, W., Hwu, J.M., Jiang, G.J., Buthelezi, T.M. and Pan, W.P. (2003) *Polymer Engineering and Science*, **43** (1), 214.

130 Voorn, D.J., Ming, W. and van Herk, A.M. (2006) *Macromolecules*, **39** (24), 4654.

131 Fialova, L., Capek, I., Ianchiş, R., Corobea, M.C., Donescu, D. and Berek, D. (2008) *Polymer Journal*, **40** (2), 163.

132 Potisk, P. and Capek, I. (1994) *Angewandte Makromolekulare Chemie*, **218**, 53.

133 Capek, I. and Potisk, P. (1995) *Journal of Polymer Science, Part A: Polymer Chemistry*, **33**, 1675.

134 Ugelstad, J., Mork, P.J. and Aasen, J.O. (1967) *Journal of Polymer Science, Part A-1.*, **5**, 2281.

135 O'Toole, J.T. (1965) *Journal of Applied Polymer Science*, **9**, 1291.

136 Nomura, M. and Harada, M. (1981) *Journal of Applied Polymer Science*, **26**, 17.

137 Wu, J. and Lerner, M.M. (1993) *Chemistry of Materials*, **5**, 835.

138 Ge, L. and Texter, J. (2004) *Polymer Bulletin (Berlin)*, **52**, 297.

139 Capek, I. (2005) *Polymer - Plastics Technology and Engineering*, **44**, 539.
140 Candau, F. and Buchert, P. (1990) *Colloids and Surfaces*, **48**, 107.
141 Gu, G., Zhou, Z., Xu, Z. and Masliyah, J.H. (2003) *Colloids and Surfaces A*, **215**, 141.
142 Laus, M., Camerani, M., Lelli, M., Sparnacci, K. and Sandrolini, F. (1998) *Journal of Materials Science*, **33**, 2883.
143 Kim, T.H., Jang, L.W., Lee, D.C., Choi, H.J. and Jhon, M.S. (2002) *Macromolecular Rapid Communications*, **23**, 191.
144 Kim, Y.K., Choi, Y.S., Wang, K.H. and Chung, I.J. (2002) *Chemistry of Materials*, **14**, 4990.
145 Yoon, J.T., Jo, W.H., Lee, M.S. and Ko, M.B. (2001) *Polymer*, **42**, 329.
146 Ko, M.B. (2000) *Polymer Bulletin*, **45**, 183.
147 Jain, A.K. and Singh, R.P.B. (1981) *Journal of Colloid and Interface Science*, **81**, 536.
148 Shouldice, G.T.D., Choi, P.Y., Koene, B.E., Nazar, L.F. and Rudin, A. (1955) *Journal of Polymer Science, Part A: Polymer Chemistry*, **33**, 1409.
149 Aota, H., Akaki, S.I., Morishima, Y. and Kamachi, M. (1997) *Macromolecules*, **30**, 2874.
150 Vaia, R.A., Liu, W.D. and Koerner, H. (2003) *Journal of Polymer Science, Part B: Polymer Physics*, **41** (18), 3214.
151 Zhu, L., Zhu, R., Xu, L. and Ruan, X. (2007) *Colloids and Surfaces A: Physicochemical and Engineering Aspects*, **304**, 41.
152 He, H., Frost, R.L., Xi, Y. and Zhu, J. (2004) *Journal of Raman Spectroscopy*, **35**, 316.
153 Lee, S.Y. and Kim, S.J. (2002) *Journal of Colloid and Interface Science*, **248**, 231.
154 Sinha, R.S., Okamoto, K. and Okamoto, M. (2003) *Macromolecules*, **36**, 2355.
155 Zanetti, M., Camino, G., Thomann, R. and Mulhaupt, R. (2001) *Polymer*, **42**, 4501.
156 Munzer, M. and Trommsdorff, E. (1977) in *Polymerization Process* (eds. C. E. Schildknecht and I. Skeist), John Wiley & Sons, New York.
157 Aranda, P. and Ruiz-Hitzky, E. (1992) *Chemistry of Materials*, **4**, 1395.
158 Liao, B., Song, M., Liang, H. and Pang, Y. (2001) *Polymer*, **42**, 10007.
159 Vaia, R.A., Vasudevan, S., Krawiec, W., Scanlon, L.G. and Giannelis, E.P. (1995) *Advanced Materials*, **7** (12), 154.
160 Bujdak, J., Hackett, E. and Giannelis, E.P. (2000) *Chemistry of Materials*, **12**, 2168.
161 Reinholdt, M.X., Kirkpatrick, R.J. and Pinnavaia, T.J. (2005) *Journal of Physical Chemistry B*, **109**, 16296.
162 Yamada, H., Nakazawa, H., Hashizume, H., Shimomura, S. and Watanabe, T. (1994) *Clays Clay Minerals*, **42** (1), 77.
163 Kuppa, V. and Manias, E. (2002) *Chemistry of Materials*, **14**, 2171.
164 Kim, Y. and Kirkpatrick, R.J. (1997) *Geochimica et Cosmochimica Acta*, **61** (24), 5199.
165 Fan, Y. and Wu, H. (1997) *Solid State Ionics*, **93**, 347.
166 Mandair, A.P., McWhinnie, W.R. and Monsef-Mirzai, P. (1987) *Inorganica Chimica Acta*, **134**, 99.
167 Strawhecker, K.E. and Manias, E. (2003) *Chemistry of Materials*, **15**, 844.
168 Vaia, R.A., Vasudevan, S., Krawiec, W., Scanlon, L.G. and Giannelis, E.P. (1997) *Journal of Polymer Science, Part B: Polymer Physics*, **35**, 59.
169 Strawhecker, K. and Manias, E. (2001) *Macromolecules*, **34**, 8475.
170 Licoln, D.M., Vaia, R.A., Wang, Z.G., Hsiao, B.S. and Krishnamoorti, R. (2001) *Polymer*, **42**, 9975.
171 Edman, L., Ferry, A. and Doeff, M.M. (2000) *Journal of Materials Research*, **15**, 1950.
172 Müller-Plathe, F. and van Gunsteren, W.F. (1995) *Journal of Chemical Physics*, **103**, 4745.
173 Henglein, A., Ershov, B.G. and Malow, M. (1995) *Journal of Physical Chemistry*, **99**, 14, 129.
174 Belloni, J., Amblard, J., Marignier, J.L. and Mostafavi, M. (1994) *Clusters of Atoms and Molecules*, Vol. **2**, Springer, Berlin.

175 Keita, B., Nadjo, L., De Cointet, C., Amblard, J. and Belloni, J. (1996) *Chemical Physics Letters*, **249**, 297.

176 Remita, S., Mostafavi, M. and Delcourt, M.O. (1994) *New Journal of Chemistry*, **18**, 581.

177 Strelow, F., Fojtik, A. and Henglein, A. (1994) *Journal of Physical Chemistry*, **98**, 3032.

178 Forster, S. and Antonietti, M. (1998) *Advanced Materials*, **10**, 195.

179 Lee, L.H. (1990) in *Fundamentals of Adhesion* (ed. L.H. Lee), Plenum, New York.

180 Pearson, R.G. (1987) *Journal of Chemical Education*, **64**, 561.

181 Phillips, J.C. (1970) *Reviews of Modern Physics*, **42**, 317.

182 Matijevic, E. (1996) *Current Opinion in Colloid and Interface Science*, **1**, 176.

183 Liu, Z.H., Yang, X.Y., Makita, Y. and Ooi, K. (2002) *Chemistry of Materials*, **14**, 4800.

184 Oriakhi, C.O. and Lerner, M.M. (1996) *Chemistry of Materials*, **8**, 2016.

185 Sukpirom, N. and Lerner, M.M. (2001) *Chemistry of Materials*, **12**, 2179.

186 Leroux, F. and Besse, J.P. (2001) *Chemistry of Materials*, **13**, 3507.

187 Bubniak, G.A., Schreiner, W.H., Mattoso, N. and Wypych, F. (2002) *Langmuir*, **18**, 5967.

188 Whilton, N.T., Vickers, P.J. and Mann, S. (1997) *Journal of Materials Chemistry*, **7**, 1623.

189 O'Leary, S., O'Hare, D. and Seeley, G. (2002) *Chemical Communications*, 1506.

190 Chen, W. and Qu, B.J. (2003) *Chemistry of Materials*, **15**, 3208–3213.

191 Chen, W., Feng, L. and Qu, B. (2004) *Solid State Communications*, **130**, 259.

192 Leroux, F., Adachi-Pagano, M., Intissar, M., Chauviere, S., Forano, C. and Besse, J.P. (2001) *Journal of Materials Chemistry*, **11**, 105.

193 Kirkland, J.J. U.S. Patent No. 3 782 075 (1974).

194 Iler, R.K. and McQueston, H.J. U.S. Patent No. 4 010 242 (1977).

195 Pu, Z.C., Mark, J.E., Jethmalani, J.M. and Ford, W.T. (1997) *Chemistry of Materials*, **9**, 2442.

196 Duan, X.M., Sun, H.B., Kaneko, K. and Kawata, S. (2004) *Thin Solid Films*, **453–454**, 518.

197 Kaneko, K., Sun, H.B., Duan, X.M. and Kawata, S. (2003) *Applied Physics Letters*, **83**, 1426.

198 Gill, M.P., Armes, S.P., Fairhurst, D., Emmett, S.N., Idzorek, G. and Pigott, T. (1992) *Langmuir*, **8**, 2178.

199 Percy, M.J., Barthet, C., Lobb, J.C., Khan, M.A., Lascelles, S., Vamvakaki, M. and Armes, S.P. (2000) *Langmuir*, **16**, 6913.

Index

a
A-B-A block copolymer   11, 43, 54
– PHS-PEO-PHS   43
ab initio emulsion polymerization   209, 217
absorption spectroscopy   173
2-acrylamido-2-methyl-1-propanesulfonic acid (AMPS)   252, 281, 283, 286
acrylic polymer   211
– butyl acrylate (BA)   211
– clay nanocomposites   226
– methyl methacrylate (MMA)   211
adsorption process   12
aliphatic epoxides   109
– dimethylsulfoxide (DMSO)   109
alkylammonium cation(s)   248, 302
amino-undecanoic acid (AUA)   250, 256
– derivatives   250
ammonium peroxodisulfate (APS) emulsion   268
– polymerization process steps   268–269
amphiphilic dextran(s)   111, 116–118, 123, 125
– derivatives   117, 125
– emulsifying properties   111
amphiphilic lipid molecules   229
– phospholipids   229
amphiphilic polymer   114, 191
– block copolymers   191
– grafted polymers   191
anionic molecular surfactant   128
– sodium dodecyl sulfate (SDS)   128
anionic polymerization   192
anionic surfactant   174, 177, 278, 281
– AOT   278
– SDS   281
anisotropic metal nanoparticles   157, 167–180, 183
– inverse microemulsion   157
– preparation   157
– synthesis   164, 183
armored latex, *see* clay platelets
aromatic epoxide   109
– phenylglycidylether   109
asymmetric straight-through microchannel array   151
– schematic illustration   151
atomic force microscopy (AFM)   59
Avogadro's constant   14
azoisobutyronitrile (AIBN)   108, 278
– 2,2′-azoisobutyronitrile   278

b
Bancroft rule   23, 198, 205
batch miniemulsion copolymerization   215
– formulation   215
B/C network   90, 93
– decoration of nodes   93
B/C system   91, 92
– F_g components   92
benzoyl peroxide (BPO)   260, 262, 298
– concentration   260
blank emulsion latex (BE)   213
blank miniemulsion latex (BM)   215, 223
– small-angle X-ray scattering patterns   223
Bloch's law   259, 307
block copolymer   192–195, 197–199
– biocompatibility   194
– biodegradability   194
– emulsifier(s)   199, 205
– micelles   194
– use   192
Boltzmann constant   10, 59
bovine serum albumin (BSA)   129, 145, 146
Bragg's law   301

breakdown processes 75
– coalescence 75
– flocculation 75
– Ostwald ripening 75
– phase inversion 75
Brownian diffusion 32, 35, 43, 51, 60
Brownian motion 2, 51, 163
bubble pressure technique 48

c

calcein-containing W/O emulsions 235
– fluorescent microphotographs 235
capping agents 177, 297
cation-exchange capacity (CEC) 211, 249
cationic amphiphilic comb-like copolymers 122
cationic exchange process 249
cationic surfactant 145, 172, 292
– cetyl trimethylammonium bromide (CTAB) 145
– tri-n-octylmethylammonium chloride (TOMAC) 145
ceria nanoparticle 179
– self-assembly 179
cerium oxide nanorods 178
– diagrammatic images 178
– selected area electron diffraction (SAED) 178
cetyl trimethylammonium bromide (CTAB) 145, 159, 168, 173, 180, 268, 284
– concentration(s) 169, 179
– derivatives 250
– micelles 172
– monomers 173
– surfactant 164
charge-stabilized emulsions 41
Chern's study 282
clay platelets 246, 258, 306
clay/polymer nanocomposites 260
Cloisite 30B monomer dispersion 220
– small-angle X-ray scattering patterns 220
Cloisite clays 247, 261
Cloisite inorganic fillers 296
Cloisite platelet(s) 271, 273, 277, 282
– polymer-modified 273
cluster insertion energy 84–90
– definition 84
cluster transformation energy 85–90
coefficient of variation (CV) 133, 233
cohesive energy ratio (CER) concept 29–31
– interaction parameters 30
colloidal forces 249, 304
colloidal particles 9

– assessment 201
– stability 201
computational fluid dynamics (CFD) 141
– analysis 141
– calculations 147
– method 153
– results 142
– simulation 141
– studies 154
concentrated emulsions 53
– viscoelastic properties 53
coprecipitation reaction 166
cosmetic emulsions 97
– application 97
– physical chemistry 97
– sensory properties 97
critical aggregation concentration (CAC) 202, 270
critical association concentration, see critical micelle concentration (CMC)
critical coagulation concentration (CCC) 202
– values 202
critical flocculation concentration (CFC) 40
critical flocculation temperature (CFT) 40
critical flotation volume (CFV) 40
critical micelle concentration (CMC) 14, 15, 69, 169, 173, 179, 263, 305
– values 271
critical packing parameter (CPP) 31
cryo-transmission electron microscopy 279
cutting-edge semiconductor microfabrication techniques 152
cyanoacrylate anionic polymerization 127
– soft-chemistry techniques 184

d

Debye attraction forces 7
Debye–Huckel parameter 9
defined nanoparticles preparation 123
– poly(butylcyanoacrylate) 126
– poly(styrene) 123
degree of hydrophobic modification 110
degree of polymerization (DP) 68, 193
depletion flocculation 37
– schematic representation 37
Deryaguin–Landua–Verwey–Overbook (DLVO) theory 10, 71
– calculation 71
– energy-distance curve 10
dextran 108, 110, 126, 127
– amphiphilic derivatives 108, 114, 126
– anionic derivatives 112, 118
– chains 126
– chemical modification 110

- coated poly(alkylcyanoacrylate) nanoparticles 126
- coated poly(butylcyanoacrylate) nanoparticles 129
- coated poly(styrene) nanoparticles 129
- native 108
- T40®sample 109
diblock copolymers 195
- applications 195
differential scanning calorimetry (DSC) 296
diffusion coefficient 22, 38, 116, 122
diffusion/stranding mechanism 193
dimethylsulfoxide (DMSO) 110
- ^{1}H NMR spectrum signals 110
dipole-dipole interactions 252, 303
disproportionation, see Ostwald ripening
DNA strands 184
dodecane-in-water emulsions 113, 117
- droplet size 113
- Ostwald ripening rates 117
dodecylbenzenesulfonic acid sodium salt (NaDBS) 254, 283
donor-acceptor interactions 297
Dougherty–Krieger equation 53
droplet deformation 21
- surfactants role 21–24
droplet generation mechanism 136, 144
droplet generation rate 147
droplets fusion-fission phenomena 161
drybase electrorheological (ER) fluids 245
dry foams 83
dynamic light scattering (DLS) 59, 197, 215, 257

e

edges-to-face agglomeration 220
Einstein limit 51
electrostatic repulsion 9–11
emulsification devices 133
- high-pressure homogenizers 133
- rotor-stator systems 133
emulsification mechanism 16–18, 24
emulsification process 17–19, 112, 133, 135, 138, 139, 144, 146, 150, 192, 195, 203, 205
- aspect ratio 139
- asymmetric straight-through MC array 150
- channel shapes effect 139
- emulsifiers effect 144–146
- experimental set-up 138
- schematic illustration 135
- surfactants effect 144–146
- symmetric straight-through mc arrays 139, 144

- to-be-dispersed phase flux effect 148
- to-be-dispersed phase viscosity effect 146
emulsification set-up 137–139
emulsified fluid foundation 97
- play-time 97
emulsifier-surface interactions 153
emulsifiers selection 25
emulsion aging rate 116, 118
- mechanism 116
- polymerization conditions 118
emulsion breakdown processes 1, 5, 6
- coalescence 3, 6, 43–46
- creaming 2
- flocculation 3, 6
- free energy path 6
- Ostwald ripening 3, 6
- phase inversion 3
- schematic representation 2, 6
- sedimentation 2
- thermodynamics 5–7
emulsion droplet(s) 7, 8, 108, 142
- generation 142
- interaction energies 7–12
- size measurement 108
emulsion films stabilization 67, 68
- correlation 67
- emulsion stability 67
- interaction forces 67
emulsion flocculation mechanism 38–41
- electrostatically stabilized emulsions 38
emulsion formation process 5, 6, 17, 19
- schematic representation 6, 17
- surfactants role 19
- thermodynamics 5–7
emulsion polymerization process 226, 266, 271, 279, 281, 284, 285
emulsion polymerization route 210
- advantage 210
emulsion rheology 53
- droplet deformability influence 53
emulsions 1, 3
- catastrophic inversion 45
- classification system 1
- flocculation 37
- industrial applications 3, 4
- oil-in-oil (O/O) 1
- oil-in-water (O/W) 1
- preparation 58, 108
- sedimentation process 32–37
- sedimentation rates 33, 34
- stability 201, 202
- transitional inversion 45
- types classification 2
- water-in-oil (W/O) 1

emulsions creaming 32–37
– prevention 35–37
– schematic representation 33
emulsion selection 31
– critical packing parameter 31
emulsions rheology 46
– bulk rheology 50
– concentrated emulsions rheology 51
– interfacial dilational elasticity 47, 48
– interfacial dilational viscosity 48
– interfacial rheology 46, 47
– interfacial viscosity measurement 47
– non-newtonian effects 49
emulsion stability investigation 59
– using INUTEC® SP1 59, 60
emulsion systems 4
– physical chemistry 4, 5
entrapment yield determination 232
epoxides 107, 109
– aliphatic 107, 109
– aromatic 107, 109
ethylene glycol (EG) 175
ethylene oxide (EO) units 15
European synchrotron radiation facility (ESRF) 216

f
facial make-up 97
– emulsified fluid foundation 97
Fanning's friction factor 136
film pressure balance technique 69
flocculation kinetics 38, 39
flocculation process 36
flocculation rate 40
Flory–Huggins interaction parameter 11
fluffy structure 306
foam film destabilization 80
food emulsions 3
foundation bulk drying 99
foundation D 102
– drying 99, 100
– viscosity/evaporated mass curve 102
foundation jams 103
fragmentation process 90
free polymer 37
free radical reactions 306

g
γ-gradient 22, 23
gas chromatography (GC) 99
gas constant 20
gel-permeation chromatography (GPC) 212, 218
gel polymer formation 222

gel-sol method 165
giant vesicles (GVs) 229, 234, 236, 238, 240
– average diameters 236
– diameter 241
– entrapment efficiency 229
– mechanical stability 238
– membranes 229
– microphotographs 234
– preparation characteristics 229
– size control 229, 234
– structure 236
– suspension 232
giant vesicles formation 231–232, 239, 240
– characteristics 237
– schematic flowchart 239
giant vesicles preparation process 233, 238
– monodisperse W/O emulsions 233
Gibbs adsorption equation(s) 14, 15, 20
Gibbs adsorption isotherm 13–16
Gibbs approach 13
Gibbs–Deuhem equation 5, 13
Gibbs dividing line 4
Gibbs elasticity 43, 44, 47, 48, 62
Gibbs free energy 13
Gibbs–Marangoni effect 23, 24
– schematic representation 23
Gibbs model 4
gold nanoparticles 168
graft copolymer 76
gram-scale synthesis 183
gravitational field system 90
– evolution 90
gravity force 93

h
Hamaker constant 8, 12, 38, 50, 71
Hansen solubility parameters 301
Harkens spreading coefficient 79
hexadecyltrimethylammonium bromide (HDTMAB) 249
hexane continuous-phase removal process 240
high-internal-phase emulsions, see dry foams
high-molecular-weight polymers 35
high-molecular-weight surfactants 28
high-speed stirrer(s) 16, 18, 58, 76
– colloid mills 18
– high pressure homogenizers 18
– Silverson mixer 16
– ultrasound generators 18
– Ultra-Turrax 16, 18, 58, 76
homogeneous nucleation 269
homogenization/degradation mechanism 266

HPPS instrument   59, 108, 109
hydrating shower cream   77, 79, 80
– compositions   77
hydrogenated tallows (HTs)   248
hydrogen-bonding component(s)   301, 303
hydrophilic block copolymer   200
hydrophilic chains, see PEO chains
hydrophilic-lipophilic balance (HLB) concept   25–27
hydrophile-lipophile balance (HLB) number(s)   23, 26–28, 29, 46, 197, 198
– chemical structure   198
hydrophilic-lipophilic balance (HLB) value(s)   27, 28, 263
hydrophilic molecule entrapment   234
hydrophobically modified dextrans synthesis   109–111
hydrophobically modified inulin (HMI)   43, 57, 58, 65, 81
– INUTEC® SP1   43, 57, 58, 65, 81
hydrophobic clay platelets   220
2-hydroxyethyl methacrylate (HEMA)   280

i

incipient flocculation   40
inner-sphere sites   294
in situ polymerization   283
interdroplet mass transfer   265
interfacial dilational modulus   21
interfacial rheology correlation   49
– emulsion stability   49
– mixed surfactant films   49
– protein films   49
interfacial tension   19, 47, 85, 86
– definition   86
– gradient   21
intermicellar exchange rate   166
intermicellar interactions   183
INUTEC® SP1   57, 60–63, 69, 73, 76, 79–81
– adsorption   79
– application(s)   57, 81
– concentration   69
– conformation   79
inverse emulsion(s)   278, 280
– polymerization   280
– stabilization mechanism   278
in vivo drug carriers   158
ionic stabilizer   180
– CTAB   180
ionic surfactant(s)   9, 14, 15
– sodium dodecyl sulfate (SDS)   14
isotropic metal nanoparticle(s)   157, 166, 167, 179
– inverse microemulsion   157
– preparation   157
– synthesis   164

k

Karl Fischer coulometer   232
Karl Fischer method   99
Keesom attraction forces   7
kinetic exchange process   44, 160
Kolmogorov theory   237

l

laminar flow (LV)   18, 19
– laminar/viscous regimes   19
– turbulent/inertial regimes   19
– turbulent/viscous regimes   19
Laplace's law   237
Laponite composite system   246
latex suspensions   128
– colloidal properties   128
Leica DMLB optical microscope   99
Lewis acids   305
Lifshitz–Slesov–Wagner (LSW) theory   61
light-scattering measurements   221, 265
light-scattering techniques   196
lipid hydration process   232
liquid foundations characterization methods   98, 99
– characterization methods   98, 99
– drying rate determination   99
– flow rheology   99
– foundations drying   99
– selection   98
liquid/liquid interface   46
London dispersion constant   7, 8
London dispersion interactions   7, 8
L/W ratio   237, 238, 240

m

macromolecular surfactants   29
magnetic nanoparticles   182, 184
magnetic resonance imaging (MRI) agents   182
MALDI-ToF mass spectrometry   196
Marangoni effect   22
Mark–Houwink–Sakurada constants   216
massage lotion   76–79
– compositions   77
– formulation   76–79
membrane emulsification technique   134
membrane lysis tension   237
metal nanoparticle(s)   167, 181, 183
metal nanoparticles preparation method   158
– chemical reduction   158
– coprecipitation   158

metal particles formation   162
– mechanism   162
metal salt/precursor   165
– Co(AOT)$_2$ 165
methyl methacrylate (MMA)   218, 252, 254, 260, 270, 274, 287
– dispersion state   254
– monomer   218, 287
– polymerization   260, 274, 287
– suspension polymerization   270
micellar systems   263–281
– radical polymerization   263
micellar template mechanism   168, 169
– parts   170
micelle   263
– aggregates   271, 305
– nucleation   268, 269, 282
– rod-like structures   179
– schematic illustration   263
– template-based primary nanorods   179
– template growth   168
– template mechanism   170
microchannel array (MC)   134–137, 139, 151, 153
– devices   139
– plate(s)   144, 231, 233
microchannel emulsification
– devices   134
– principles   135–137
– process   134–136, 230, 231
– study   146
microchannel module   230
microemulsion polymerization mechanism   264, 265
microemulsion(s)   159, 178, 180, 181, 264, 273, 306
– based approach   179
– droplet dimensions   160
– mediated synthesis process   177, 180
– microwave method   181
– polymerization   273, 306
– production   158
– reduction technique   182
– role   264
– system   167, 169
– technique   179
microfluidic channel devices   134, 229
microinterferometric technique   68
microscope video system   232
miniemulsion aging rate   122
– temperature effect   122
miniemulsion polymerization
   processes   114, 115, 120, 124, 129, 210, 211, 219, 267, 276

– mechanism   114, 267
miniemulsions stability   114
– polymerization duration   114
molecular diffusion process   122
molecular surfactants   118
– sodium dodecyl sulfate   118
molecular weight distribution (MWD)   216
monodisperse emulsion(s)   133, 139, 140, 148, 151
– production developments   133
monodisperse water-in-oil (W/O) emulsions   229, 230, 233
monomer-coemulsifier interactions   264
monomer droplets composition effect   119
monomer polymerization   270
monomer-swollen micelles   269
montmorillonite (MMTs)   219, 225, 243
– clays   209
– platelets   225
multiple emulsion(s)   57, 64
– INUTEC® SP1   63, 64
– optical micrographs   64
– preparation methods   58
– stability   64

n

nanocomposite latex films   215, 219, 223–225
– transmission electron microscopy images   219, 224, 225
– wide-angle X-ray diffraction patterns   224
nanoemulsions   57
– emulsions stabilization   57, 60
– using INUTEC® SP1   60
nanoemulsions preparation methods   58, 191
– poly(caprolactone)-b-poly(ethylene oxide) block copolymers   191
– spontaneous emulsification   191
nanometer-scale magnetic particles   158
nanoparticle formation process   163
– steps   163, 164
nanoparticle growth   172
– kinetic parameters   172
nanoparticle preparation   107, 162–164, 179
– key parameters   179
– miniemulsion polymerization   107
nanoparticle suspensions   129
– colloidal properties   129
nanoprecipitation process   193, 198
nanorods   170, 179
– self-assembly   179
nanoscale magnetic materials   183
– physical properties   183
nanoscale particles   157, 167
– properties   157, 158

natural clay   248, 302
natural Cloisite (Cl$_N$)   258, 294
negatively stained nanocomposite latexes   225
– transmission electron microscopy images   225
N,N'-methylenediacrylamide (NDA)   278
nonadsorbing polymer, *see* free polymer
nonionic surfactant(s)   11, 15, 26, 27, 46, 278
– alcohol ethoxylates   15
– Span 80   278
– Tween 85   278
nonseeding method   170, 183
nuclear magnetic resonance (NMR)   294
nucleation-growth kinetics   172
nucleation/growth mechanism   193, 211

o

oil emulsions   203
– emulsification process   197
– spontaneous emulsification   203
oil-in-water (O/W) dispersions   191
– emulsions   191
– latexes   191
– suspensions of polymer   191
oil-in-water (O/W) emulsion(s)   23, 27, 28, 32, 34, 52, 107, 108, 123, 129, 143, 146, 148, 193, 200
– activation energy   123
– droplet generation   143
– interfacial tension variation   28
– sonication   108
oil-in-water (O/W) interface   42, 44, 49, 51, 65, 68, 78, 141, 143
– time course   143
oil-in-water (O/W) miniemulsions preparation   111
– droplet size control   111
– polymer structure   112
oil-in-water (O/W) system   31, 136, 145–147, 151, 153
– interfacial properties   145
– isoelectric points   146
– preparation conditions   145
organically modified clays (O-MMT clay)   213, 214, 284
– wide-angle X-ray diffraction patterns   213
organically modified silicates (OLS)   247, 280
organic latex particles   246
oscillating bubble technique   48
Ostwald ripening   3, 41–43, 61, 63, 107, 116, 122, 265

– rate   117, 120, 123
– rate constant   62
– schematic representation   42

p

particle aggregation process   102
peanut-like hematite (α-Fe$_2$O$_3$) crystals   165, 176
– synthesis   165
Pearson's hard-soft acid-base (HSAB) principle   297
percolation process   161
personal-care formulations   75, 76
– hydrating shower cream   76
– massage lotion formulation   76
– polymeric surfactants   75
– soft conditioner   76
– sun spray (SPF19)   76
phase inversion process   45, 46
– catastrophic inversion   45
– transitional inversion   45
phase inversion temperature (PIT) concept   27–29
– value   29
photon correlation spectroscopy, *see* dynamic light scattering
pickering emulsion(s)   246, 279, 306
– stabilization   273, 284
plateau angles   84
plateau value   20, 22
play-time   97, 104
– definition   97
Poisson–Boltzmann equation   71
polar polymer   304, 305
– PEO   304
– polysaccharides   304
polyacrylamide (PAAm) latex particles   279
polycaprolactone (PCL)   193, 195
– block   205
– chain   200
– emulsification process   193, 195
– emulsions fabrication   197, 198
poly(caprolactone)-*b*-poly(ethylene oxide) block copolymers (PCL-*b*-PEO)   193, 196, 198, 203, 204
– properties   204
– synthesis   196
polydimethylsiloxane (PDMS) emulsions   53
– η_r–ϕ curves   53
polydisperse emulsion   140, 149
polydispersity index (PDI)   217, 222, 279
poly(ether-block-amide) copolymer hybrids   289
– X-ray diffraction patterns   289

polyethylene-block-poly(ethyleneglycol) (PE-PEG) thin film nanocomposites 293
polyethylene oxide (PEO) 26, 54, 60, 193, 293, 294, 296
– A-B-A block copolymer 54
– block copolymers 195
– block length 200
– block(s) 197, 199, 204, 205
– chain(s) 11, 15, 29, 46, 199, 202, 293
– Cloisite composite 295
– conformations 296
– layer 202
– matrix 295
– nanocomposites 293–296
polyethylene glycol (PEG)-grafted silicones 191
– methyl ether 196
polymer chain conformation 244, 277
polymer chain encapsulation 261
polymer/clay nanocomposites 209, 243, 245, 291
polymer/clay nanocomposites formation 303
– mechanistic routes 303
polymer/Cloisite nanocomposites 252, 303
polymer film formation process 210
polymeric emulsifier(s) 198
– chemical structure 198
polymeric nanoparticles 107, 108
– miniemulsion polymerization 108
polymeric surfactant(s) 20, 21, 57, 61, 63, 68, 75, 80, 107, 115, 122, 123, 126, 191, 192, 195, 205
– biocompatible 107
– mixtures 21
– polyvinyl alcohol 20
polymerization process 108, 114
polymerization system(s) 269, 273
polymer latexes 243
– preparation 243
– stabilized by clays 243
polymer-layer silicate nanocomposites (PLSN) 245
polymer matrix 209–211, 244, 302, 306
polymer/MMT nanocomposites 281–296
– copolymers 288
– kinetic/molecular weight parameters 281
– poly(ethylene oxide) nanocomposites 293–296
– thermal/mechanical properties 290
– X-ray diffraction studies 284
polymer particle formation 268
– micellar model 268
polymer particle(s) 276, 283, 306

– matrix 285
– polymerization rate 276
– stabilization 275
polymer/silicate nanocomposites 243, 244
polymer-to-oil surfactant 111, 112
– droplet diameter 112
poly(methyl methacrylate) (PMMA) 245
– chains 288, 299, 300
– clay nanocomposites 245
– modified silicate layers 284
– nanocomposites 288, 290–293
– wide-angle X-ray diffraction (WAXD) analysis 288
polysaccharide derivatives 115
– amphiphilic polymers 115
polysaccharide surfactants 109
polystyrene(Pst) 290
– clay nanocomposites 244, 245
– Cloisite nanocomposite 290
poly(styrene-co-acrylonitrile) (SAN) 281
– copolymer 281
– silicate nanocomposites 281
polyvinyl alcohol (PVA) 20, 205, 298
– concentrations 205
protein film 49, 50
– creep curve 50
pseudoplastic system, see shear thinning system
pulsed drop method 48

q
quantum-size effects 297
quasi-monodisperse emulsion droplets 149
quasi-plateau 101

r
radical polymerization process 260–281
– reactions 123
– solution/bulk polymerization 260–263
radical polymerization theory 260
rapid coagulation regime 202
redox system 263, 305, 306
reducing flocculation 41
– general rules 41
reverse micelles 170, 179
– synthesis 158, 182
– system 160
– water pool 163
reversible addition-fragmentation chain transfer (RAFT) agents 120
Reynolds number 18
rod-like micelle template mechanism 171

s

scanning electron microscopy (SEM)   279
Scatchard equation   113
seeded semibatch emulsion copolymerization   213, 214, 217
– formulation   214
seeding/autocatalytic growth mechanism   164
selected area electron diffraction (SAED)   177
self-assembly process   177
semi-quantitative theory   19
shear thinning system   35
shirasu porous glass (SPG) membranes   133
size-exclusion chromatography (SEC)   216
skin foundation   102
– disposition   102
– play-time   102
slow coagulation regime   202
small-angle light scattering (SALS)   197
small angle X-ray scattering (SAXS)   209, 216, 220, 221
– experiments   216
– measurements   220
– scattering profile   221
Smith–Ewart theory   268
Smoluchowski rate   39
sodium bis(2-ethylhexyl)sulfosuccinate (NaAOT)   159, 160
– containing microemulsion solution   175
– containing microemulsion system   174, 180
– microemulsions   177
sodium dodecyl sulfate (SDS)   23, 49, 268, 271
– ε-values   23
– micelles   254, 304
sodium lauryl sulfate (SLS)   211
sodium montmorillonite (Na-MMT) clay   209, 212, 224
– chemical structure   212
– WAXD diffraction patterns   224
soft conditioner   77, 80
– compositions   77
soft template effect   177
sol-gel methods   181
sorbitan mono-oleate   27, 230
Soxhlet extraction   216
spontaneous emulsification process   193, 198
stabilization process   75
– enhancement   75
steric repulsion   11, 12
steric stabilization process   38, 57
steric stabilization theory   12
sterically stabilized emulsions flocculation   40, 41
– schematic representation   40
Stern/zeta potential   9
stirred-tank reactor   213
Stokes–Einstein equation   38, 59
Stokes' velocity   34
straight-through microchannel array devices   133, 135, 137–139, 146, 148–150, 152–154
– scaling-up   149, 150
straight-through microchannel plate   138, 149, 150, 153
styrene   115, 244, 245, 270, 275, 285, 286
– colloidal parameters   275
– emulsion polymerization   285, 286
– emulsion retardation factor   121
– emulsion(s)   119, 121
– free-radical miniemulsion polymerization   115
– *in situ* polymerization   244, 245
– microemulsion polymerization   275
– miniemulsion kinetic data   115
– miniemulsion polymerization   270
– vinyloxazoline copolymer   245
submicronic colloidal systems   107
– domains   107
sun spray (SPF19)   77, 81
– compositions   77
supra-aggregate   178–180
– formation   178, 180
– self-assembly   179
surface dilational modulus   22, 42
surface viscometers   47
– schematic representation   47
surfactant-based methods   164–166
surfactant film(s)   44, 163
surfactant molecules   12
– orientation   12
surfactant replacement process   232
surfactants adsorption   12–25
– liquid/liquid interface   12
swallow-tail family   289
– P15A   289
– P20A   289
– P93A   289
symmetric bulk crystal structures   167
synthetic Cloisite (Cl$_S$)   259, 294

t

tallow chains   248
template mechanism   172
thermogravimetric analysis (TGA)   288

three-phase cellular fluids 83
– external force field effect 83
transformation energy(ies) 88, 89
transmission electron microscopy (TEM) 174, 217, 218
– images 174, 218
tri-*n*-octylmethylammonium chloride (TOMAC) 145
tris-HCl buffer(s) 230, 232
Triton X-100 solution 232, 233
turbulent flow (TV) 18, 19
turbulent inertial regime 19

u

Ultra-Turrax homogenizer 58
ultraviolet (UV)-visible spectra 172

v

van der Waals attraction 7–10, 37, 38, 40, 43, 67, 128, 253, 254, 302
– types 7
van der Waals energy-distance curve 8
velocity gradient 19
vibrating sample magnetometer (VSM) 259
– data 259
vinylbenzyl-dimethyldodecylammonium chloride (VDAC) 249
vinyl monomers 278
– emulsion polymerization 278
viscosity evolution 100–102
– drying time 101
viscosity plateau, *see* quasi-plateau

w

waterborne acrylic/clay nanocomposites 209
– synthesis routes 209
waterborne nanocomposites 210, 217, 219
– coagulum-free 210
– emulsion polymerization 213, 217
– latexes 214
– miniemulsion polymerization 214
– MMT nanocomposites synthesis 213, 214
water-borne polymers 267
– clay nanocomposites 210
– preparation 267
water-in-oil emulsion(s) 45, 46, 58, 59, 64, 230, 234, 236, 241
– Arlacel P135 58, 59, 64
– preparation 230
water-in-oil emulsion water droplets 235
– size distributions 235
water-in-oil microemulsion(s) 159, 161, 181
– sol-gel route 181
– use 161
water-insoluble materials 266
– organic pigments 266
– polymers 266
– resins 266
water/oil-soluble polymers 260
water-soluble fluorescent dye 230
– calcein 230
water-swollen micelles 264
wide-angle X-ray diffraction (WAXD) 212
– analysis 216, 288, 292
– diffraction patterns 218, 219
WinROOF image analysis software 232
Winsor R_o concept 29
worm-like micelle template 169

x

X-ray diffraction patterns 250, 253, 284
X-ray diffraction spectra 255
X-ray diffraction studies 284
– homopolymers 284
X-ray diffraction (XRD) analysis 212, 255

z

zeolite nanocrystals 181
zwitterionic polymerizations 126